Room 3C-30, R 6/29/79

Room 3C-30, R 6/29/79

GLC and HPLC Determination of Therapeutic Agents

CHROMATOGRAPHIC SCIENCE

A Series of Monographs

Editor: JACK CAZES
Waters Associates, Inc.
Milford, Massachusetts

Other Volumes in Preparation

GLC and HPLC Determination of Therapeutic Agents (in three parts)

Part 1

Edited by

Kiyoshi Tsuji

*Control Analytical Research
and Development
The Upjohn Company
Kalamazoo, Michigan*

Walter Murozowich

*Pharmacy Research
The Upjohn Company
Kalamazoo, Michigan*

MARCEL DEKKER, INC.   New York and Basel

Library of Congress Cataloging in Publication Data

Main Entry under title:

GLC and HPLC determination of therapeutic agents.

(Chromatographic science ; v. 9-   )
Includes index.
1. Drugs--Analysis. 2. Gas chromatography.
3. Liquid chromatography. I. Tsuji, Kiyoshi,
[Date]   II. Morozowich, Walter, [Date]
RS189.G18   615'.1901   78-2740
ISBN 0-8247-6641-5

MARCEL DEKKER, INC.
270 Madison Avenue, New York, New York 10016

Current printing (last digit):
10 9 8 7 6 5 4 3 2 1

PRINTED IN THE UNITED STATES OF AMERICA

FOREWORD

Within the recent past, an impressive array of new and powerful methods has been introduced to aid scientists in providing an improved definition of materials, processes, and products.

The applicability of chromatographic techniques, most particularly gas chromatography and high performance liquid chromatography, has made a dramatic impact in the analysis of organic compounds in general, and therapeutic agents in particular.

The rapid rise in the application of gas and high performance liquid chromatography is almost unrivaled in modern instrumental analytical methodology. Certainly their increased growth in utility parallels that of the then unprecedented growth in technological advances following the introduction of spectroscopy several decades ago.

While the application of GLC and HPLC has been described in an impressive array of original scientific publications, textbooks, and monographs, no single current summary exists as to their usefulness in defining therapeutic agents analytically. The need for such a compendium is obvious in view of the truly remarkable analytical progress made with these techniques.

The present work successfully fills this gap and provides the first critical, well-documented survey on gas and high performance liquid chromatographic methods applied to therapeutic agents. This text will be very valuable to medical and pharmaceutical researchers.

Both of the editors have made truly novel and original contributions to increase the utility and applicability of gas and liquid chromatographic methods. The editors have been fortunate in securing the generous cooperation of such an extensive group of truly eminent scientists to author the individual chapters. Each author is noted for his insight into and contributions to the ever-expanding utility of gas and liquid chromatographic methods.

This book, then, contains detailed methodology and provides, as well, a timely, critical review of the literature in an important area. This comprehensive work

iii

truly meets the primary test of an analytical text: that of providing assay methods of high reliability and dependability with defined accuracy and precision.

A. J. Taraszka, Ph.D.
Vice President, Control
The Upjohn Company
Kalamazoo, Michigan

PREFACE

In 1885, F.F. Runge noted the colorful patterns obtained upon migration of various
dyes on circular filter paper. The intellectual curiosity in this phenomenon has
led to the current widespread acceptance of chromatographic methods in drug analysis.

> *In nature's infinite book of secrecy,*
> *A little I can read.*
> William Shakespeare

It is fitting to recount some of the major contributors who have taken chromatography
from the realm of academic curiosity to its current practical utility as a major
analytical tool.

In 1903, Michael Semenovich Tswett described the use of glass columns packed
with over 100 adsorbents for the separation of colored chloroplast pigments. A mobile
phase was used to elute compounds adsorbed on the column packing material. He applied
the terms chromatogram and chromatographic method to this process in 1906.

In 1941, A. J. P. Martin and R. L. M. Synge suggested the possibility of using
a gas as the mobile phase in a chromatographic system. However, their suggestion was
regarded only as theoretical and was not put into practice until 1952 when Martin and
James constructed the first gas chromatographic instrument. It used an automated
buret for the titrimetric detection and determination of acids and bases. This first
working gas chromatograph was suitable only for those compounds having these two
functional groups. The full potential of gas chromatography was not realized until
the development of the thermal conductivity detector by Ray in 1954; in about 1955,
the first commercial instrument became available.

Also in 1941, Martin and Synge knew that increased speed in liquid column chromatography could be achieved without loss of resolution by use of very small particles and a high pressure drop across the length of the column, but the technique was
not put into practice until about two decades later. Around 1969, a rapid advance-
ment occurred in the technique of liquid column chromatography. This is attributed

to the discussion arising at a special session on liquid chromatography at the Fifth International Symposium on Advances in Chromatography. Shortly thereafter, liquid chromatographs became commercially available. With the recent introduction of small particle (5-10 μm) supports, HPLC columns with 2,000-7,000 theoretical plates have become commercially available.

The practice of modern HPLC is now highly refined and, along with GLC, these two chromatographic techniques have emerged as the most widely used methods of drug analysis. Chromatographic analysis is important in the determination of the potency of drug formulations as well as in the successful completion of research-oriented projects. Many research scientists, though not primarily analytical chemists, employ HPLC and GLC in a variety of research endeavors.

With the rapid proliferation of publications in HPLC and GLC of therapeutic agents, the need for a compendium solely related to drugs is recognized.

The purpose of this book is to provide an up-to-date literature survey on the GLC and HPLC analysis of therapeutic agents, with a critical review of each chromatographic method, plus a detailed, step-by-step description of the procedure of choice from each author's first-hand experience. Where both GLC and HPLC methods are available, and equally desirable, both methods have been described in detail. Sample preparation, such as derivatization, extraction, and separation of drugs from clinical specimens or drug formulations has also been described in detail. A summary table has been provided at the end of each chapter, where applicable, with full details on derivatization reagents, reaction and chromatographic conditions, column, and references. Therefore, this book will serve as a quick reference as well as a laboratory procedural handbook for chromatographic analysis of drugs.

We gratefully acknowledge our chapter authors for their contributions, advice, understanding, and patience, without which this book could not have materialized. We also wish to thank the many people who graciously contributed information, chromatograms, time, and assistance in the preparation of this book. We are indebted to J. T. Sobota for his advice in the format of the book and to The Upjohn Company for assistance and support in the preparation of this book.

Kiyoshi Tsuji
Walter Morozowich

LIST OF CONTRIBUTORS

*Dale R. Baker,** Instrument Products Division, E. I. DuPont de Nemours & Company, Inc., Wilmington, Delaware

Karl J. Bombaugh, Radian Corporation, Austin, Texas

Moo J. Cho, Pharmacy Research Unit, The Upjohn Company, Kalamazoo, Michigan

Jordan L. Cohen, School of Pharmacy, University of Southern California, Los Angeles, California

J. Carter Cook, Jr., Mass Spectrometry Laboratory, School of Chemical Sciences, University of Illinois, Urbana, Illinois

*Bobby G. Dawkins,*** Department of Chemistry, Cornell University, Ithaca, New York

Charles A. Feit, Regis Chemical, Morton Grove, Illinois

Richard A. Henry,[†] Autolab Division, Spectra-Physics, Santa Clara, California

John F. Holland, Biochemistry Department, Michigan State University, East Lansing, Michigan

Howard Ko, Physical and Analytical Chemistry, The Upjohn Company, Kalamazoo, Michigan

Fred W. McLafferty, Department of Chemistry, Cornell University, Ithaca, New York

Frank E. Martin, Mass Spectrometry Facility Manager, Biochemistry Department, Michigan State University, East Lansing, Michigan

Richard M. Milberg, Mass Spectrometry Laboratory, School of Chemical Sciences, University of Illinois, Urbana, Illinois

---

*Presently at Analytical Division, Hewlett-Packard, Avondale, Pennsylvania

**Presently at E.I. DuPont de Nemours & Company, Aiken, South Carolina

†Presently at Altex Scientific Inc., Berkeley, California

Walter Morozowich, Pharmacy Research Unit, The Upjohn Company, Kalamazoo, Michigan

Carter L. Olson, College of Pharmacy, The Ohio State University, Columbus, Ohio

John A. Perry, Consultant, Chicago, Illinois

Edgar N. Petzold, Physical and Analytical Chemistry, The Upjohn Company, Kalamazoo, Michigan

William H. Pirkle, Department of Chemistry, University of Illinois, Urbana, Illinois

Julius Simon, Laboratory for Chromatography, Flushing, New York

Virginia M. Smith, Autolab Division, Spectra-Physics, Santa Clara, California

David Sohn, Department of Pathology, New York Medical College/Bird S. Coler Hospital, Roosevelt Island, New York, New York

Charles C. Sweeley, Department of Biochemistry, Michigan State University, East Lansing, Michigan

CONTENTS

Cumulative indexes appear at the end of Part 3.

CONTENTS OF PARTS 2 AND 3

Part 2

Part 3

GLC and HPLC Determination of Therapeutic Agents

Part I

THEORY AND PRACTICE

Chapter 1

THEORY AND INSTRUMENTATION

Carter L. Olson

College of Pharmacy
Ohio State University
Columbus, Ohio

I. INTRODUCTION

The analysis of therapeutic agents offers problems that challenge the best analytical
techniques and instrumentation available. The measurement of low-level therapeutic
agents and their metabolic or degradative products in biological fluids is particularly difficult. The compounds are usually found in very low concentrations and
other compounds present may cause analytical interferences. Most measurement techniques are not selective enough to measure individual components in a sample mixture.
In many cases it is necessary to separate the desired component from the others.
Good separation reduces selectivity problems and permits the use of maximum detector
sensitivity with minimum fear of interferences.

In the modern analytical laboratory chromatographic techniques have become
widely used. A chromatographic process is one in which a mobile phase passes over
a stationary phase. When a sample compound is placed into the mobile phase, it will

1

distribute between the mobile and stationary phases. The speed with which the sample moves by the stationary phase depends on its distribution ratio between the two phases. The greater the relative concentration of the sample in the mobile phase, the faster it will move. If a mixture of substances is introduced into the mobile phase, the rate of movement of the components will be different if their relative distribution ratios between the two phases is different. In nearly all cases, mobile and stationary phases can be chosen so that the distribution of the mixture components between the phases is sufficiently different to permit component separation.

The chromatographic techniques most commonly used for quantitative chemical analysis at the present time are gas-liquid chromatography (GLC) and high performance liquid chromatography (HPLC). In GLC, the mobile phase is a gas such as helium or nitrogen and the stationary phase is a low-volatility liquid, coated either on the surface of the column packing particles or on the inner wall of the column. In HPLC, the mobile phase is a liquid and the stationary phase consists of small particles which may be coated with another liquid.

This chapter is intended to give a brief introduction to the theory and instrumentation used in GLC and HPLC. Only the concepts and instrumental components which are most widely used for the analysis of therapeutic agents will be discussed.

II. BASIC CHROMATOGRAPHIC THEORY

The type of chromatography used most frequently for analysis is elution chromatography. In elution chromatography a narrow portion of the sample is introduced into the mobile gas or liquid phase. As the sample travels down the column, the components within the band start to separate from each other into separate bands and the bands spread. Finally, the bands elute off the end of the column and are measured by a detector. The two general areas of chromatographic theory concern the band-retention and band-spreading characteristics of the system.

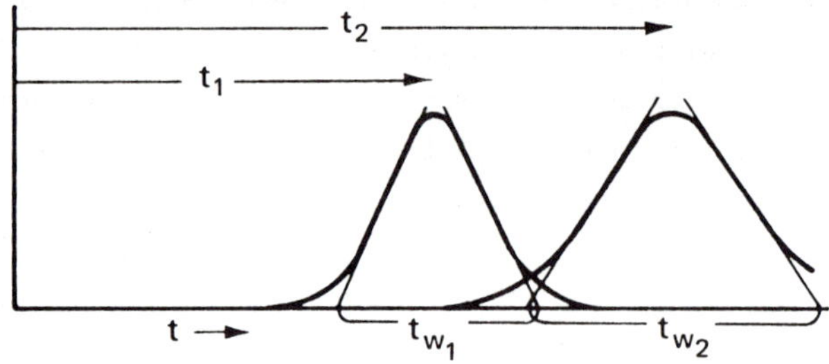

FIG. 1. Elution chromatogram: t = retention time; t_w = base width.

A. Retention Parameters

A representation of an elution chromatogram is shown in Fig. 1. Two constants describing the distribution of a sample component between the mobile and stationary
phases are normally discussed in chromatographic theory: k, the partition coefficient, and k', the partition ratio. The partition or distribution coefficient has
thermodynamic significance and relates the concentration of the sample (X) in the
two phases as follows:

$$k = \frac{\text{Concentration in stationary phase}}{\text{Concentration in mobile phase}} = \frac{(X)_s}{(X)_m} \tag{1}$$

The partition or distribution ratio is a more convenient term because it represents
the ratio of quantities or moles of the sample (M) in the two phases. It is defined
as follows:

$$k' = \frac{\text{Quantity in stationary phase}}{\text{Quantity in mobile phase}} = \frac{M_s}{M_m} \tag{2}$$

The partition ratio can be easily related to the partition coefficient. Since quantity is equal to concentration times volume, k' can be written

$$k' = \frac{(X)_s V_s}{(X)_m V_m}$$

or $\tag{3}$

$$k' = k \frac{V_s}{V_m}$$

The partition ratio is characteristic for a given column since phase volumes are included. The rate at which a sample travels through the column, u_x, is equal to the
mole fraction of the sample in the mobile phase times the velocity of the mobile
phase, u. The quantity u_x can be written as

$$u_x = u \frac{M_m}{M_s + M_m} \tag{4}$$

Since the quantity k' + 1 is equal to $(M_s + M_m)/M_m$, the reciprocal of the mole fraction in the moving phase, u_x, can be written as

$$u_x = u \frac{1}{k' + 1} \tag{5}$$

The retention time t_R is the time required for the sample to pass through the column.
The quantity u_x is equal to the length of the column L divided by the retention time
for the sample.

$$u_x = \frac{L}{t_R} \tag{6}$$

The velocity of the mobile phase is equal to L divided by the time t_0 required for a

$$u = \frac{L}{t_0} \tag{7}$$

Substituting for u and u_x from Eq. (6) and (7) into Eq. (5) yields the relationship between retention time and partition ratio.

$$t_R = t_0 (1 + k') \tag{8}$$

The partition ratio for a given sample can be calculated using Eq. (8) rearranged as follows:

$$k' = \frac{t_R - t_0}{t_0} \tag{9}$$

In some instances it is more convenient to work with retention volumes than with retention times. Retention volume V_R is calculated by multiplying retention time by the mobile phase flow rate F. The column dead volume V_0 is the volume required for a nonretained solute to reach the detector.

$$V_R = t_R F = t_0 F (1 + k') = V_0 (1 + k') \tag{10}$$

Retention volumes for a set of compounds should be adjusted by substracting the column dead volume before comparison. The *adjusted retention volume* V'_R is given by

$$V'_R = V_R - V_0 \tag{11}$$

A pressure gradient is required to force the moving phase through the column. If the mobile phase is compressible, as is the case for GLC, a density gradient exits across the column. Therefore, the volume flow rate varies along the column. The average flow rate inside the column is less than the flow rate measured at the outlet of the column. The flow rate corrected for the pressure drop, called the *corrected retention volume*, V_R^o, is calculated from V_R by the equation

$$V_R^o = j V_R \tag{12}$$

where

$$j = \frac{3}{2} \left[\frac{(P_i/P_o)^2 - 1}{(P_i/P_o)^3 - 1} \right] \tag{13}$$

and P_i and P_o are the inlet and outlet pressures, respectively. The above Eq. (13) applies when the flow rate is measured at the outlet of the column. If flow rate is measured at the column inlet the following equation can be used [1]:

$$V_R = V_R^o \frac{2}{3} \left(1 + \frac{P_o/P_i}{1 + P_i/P_o}\right) \tag{14}$$

In most cases the analyst does not know the exact volumes of the mobile and stationary phases for the columns. It is very difficult to prepare or obtain columns
that are exact duplicates of those reported in the literature. Therefore, absolute
retention values are seldom reported. The use of relative retention values, however,
is of great practical value since the individual column characteristics cancel out
when retention values are divided. The relative retention value α is the ratio of
the retention value for a substance being assayed V_R', to the retention value for a
standard $V_{R,s}$ where both values have been corrected for column dead volume.

$$\alpha = \frac{V_R'}{V_{R,s}} = \frac{t_R - t_o}{t_{Rs} - t_o} \tag{15}$$

Retention values are affected by a variety of factors. It is important that the
sample be entered in a manner that produces as narrow a plug as possible. In addition, the retention values can change with larger sample volumes or concentrations.
Within the limits placed by detector sensitivity and sampling precision, small samples
are preferable for analytical chromatography.

B. Band Spreading

The earliest attempts to describe the shape of chromatograms were based on plate
theory developed to describe distillation-column behavior. Plate theory is also used
to describe counter-current extraction behavior. The separation that occurs along a
length of chromatographic column could be equivalent to that which occurs during an
equilibrium separation stage such as an extraction transfer or distillation stage.
An estimate of the number N of theoretical plates required for a separation can be
calculated. The usual calculation for N uses the following equation:

$$N = 16 \left(\frac{t_R}{w}\right)^2 \tag{16}$$

where t_R equals the retention time and w is the base width of the chromatographic
peak. The column length divided by the calculated value for N gives the length of
column equivalent to a theoretical plate (HETP) or separation stage H.

While the concept of H is used in describing the separation efficiency of a
chromatographic column, the theory describing column efficiency clearly cannot be
reduced to plate theory. The rate theory of chromatographic separation has been
gradually developed and refined.

Several theories have been proposed to describe chromatographic band spreading.
The random walk model considers the behavior of individual sample molecules as they
pass through the column. The nonequilibrium and mass balance models deal with the
spreading characteristics of the sample band as it moves through the column. The
random walk model will be utilized to describe band spreading in this chapter. It

was chosen because space did not permit detailed coverage of all theories, and because many people find it easier to visualize experimental factors contributing to
band spreading using the random walk model. It is not implied, however, that the
random walk model offers the best approach to band spreading theory. A good review
of band spreading theories has been recently presented by Grushka et al. [2]. The
most complete presentation of the random walk theory is that by Giddings [3].

There is a variety of factors that cause spreading of a chromatographic peak.
When a sample is entered into a column, it is usually in the form of a narrow band.
The concentration gradient that exists between the band and the mobile phase in which
it is contained causes the band to spread due to diffusional movement. This process,
called *longitudinal diffusion*, occurs in both the forward and backward directions.
The effect of diffusion is attenuated by the tortuousity of the diffusional path
through the column packing. The tortuosity factor γ_m is dependent on packing geometry
and for a given column is independent of the nature of the gas or liquid used as the
moving phase. The column dead time is the effective time for diffusion of a band to
occur. The band broadening in the mobile phase is given by the expression

$$2\gamma_m D_m t_0 = 2\gamma_m D_m \frac{L}{u} \tag{17}$$

where D_m is the diffusion coefficient of the sample in the mobile phase. The contribution to H is the band spreading divided by column length, which is $2\gamma_m D_m/u$. The
contribution to H is much greater in gas chromatography than in liquid chromatography
because D_m in a gas phase is much larger (0.1-1 cm^2/sec) than it is in a liquid phase
(that is, 1 x $10^{-5} cm^2$/sec).

Longitudinal diffusion can also occur in the stationary phase. The contribution
to H is given by the equation

$$H = \frac{2\gamma_s D_s k'}{u} \tag{18}$$

where

γ_s is the correction factor in the stationary phase and D_s is the diffusion
coefficient of the sample in the stationary phase. The partition or distribution
ratio simply relates the time the sample is in the stationary phase to the time it is
in the mobile phase. In gas chromatography D_s is so small that the contribution of
longitudinal diffusion to H in the stationary phase is negligible compared to the
contribution to H in the moving phase.

The flow patterns which the mobile phase takes as it moves by the stationary
phase provide a second major contribution to band spreading. At normal flow velocities, laminar flow predominates and turbulance that is induced in eddys around sharply
curved surfaces quickly damps out. The velocity profile or speed of a liquid flowing

through a channel has a hyperbolic shape under laminar flow conditions. The velocity
is stagnant at the edge of the channel and becomes faster toward the center. The
sample molecules travel at different rates depending on their radial distance from
the surface of the stationary phase. The average velocity of flow through a narrow
channel will be slower than through a wider channel. A molecule leaving a channel
between particles will encounter channels with differing diameters and lengths leading between downstream particles. The multipath opportunities which the molecules can
take on their passage through the column leads to band spreading and an increase in
H. As sample molecules move through the column, they move abruptly into regions
where the stream velocity changes. If a molecule moves to a region of essentially
stagnant velocity, the molecule must diffuse back into a faster moving stream segment
in order to move on through the column. Modern eddy-diffusion theory, therefore,
includes a coupling of flow and diffusion mechanisms.

If the random walk model proposed by Giddings is used, a molecule travels some
distance at a constant velocity and then suddenly changes velocity by either a flow
or diffusion process. The length of the constant velocity step is equal to the
column length divided by the sum of the velocity changes caused by the flow and diffusion mechanisms. The plate height is proportional to the velocity step length.
The contribution to H can be written as

$$H = \sum \frac{1}{(1/2)\lambda_i\, d_p + D_m/w_i u d_p} \tag{19}$$

$$\underset{\text{contribution}}{\text{Flow}} \qquad \underset{\text{contribution}}{\text{Diffusion}}$$

where d_p is the column particle size, D_m is the diffusion coefficient of the sample
in the mobile phase, and λ_i and w_i are structural parameters related to flow and
diffusion mechanisms.

Five types of mechanisms have been postulated that contribute to band
spreading [4]:

1. Transchannel transfer
2. Transparticle transfer within porous particles
3. Short-range interchannel transfer between neighboring large and small
   channels
4. Long-range transfer
5. Transcolumn transfer due to uneven packing or wall effects

(Magnitudes for these parameters can be found on p. 56 of Ref. 3.)

In order that H be minimal, it is important that the column be packed very uniformly with constant mesh-size particles. Transparticle flow becomes the most significant mechanism contributing to H for porous particles. A small particle diameter
is necessary for high column efficiency.

A third factor that contributes to band spreading is diffusion controlled kinetics. The time required for diffusion in the liquid phase, t_d, is given by the following equation:

$$t_d = \frac{d^2}{2D_s} \tag{20}$$

where d is the thickness of the liquid phase and D_s is the diffusion coefficient of the sample in the stationary liquid phase. The contribution to H by stationary phase diffusion has been shown to be given by the equation

$$H = \frac{qR(1 - R)\ d^2 u}{D_s} \tag{21}$$

where q is a constant dependent on the geometry of the stationary phase R is the fraction of the sample in the mobile phase, and u is the average mobile-phase velocity [5]. Good separation is favored by a thin, low-viscosity stationary liquid phase. The value of D_s becomes larger as the temperature of the liquid phase increases.

Diffusion can also occur between different velocity streamlines in the mobile phase. The contribution of mobile-phase diffusion to H is given by the equation

$$H = \sum \frac{w_i d_p^2 u}{D_m} \tag{22}$$

where w_i are the structural parameters relating to diffusion mechanisms discussed previously under eddy diffusion [6].

In GLC, D_m is large and diffusion controlled band spreading is determined by diffusion in the stationary phase. In HPLC, the thickness of the stationary liquid phase may be much thinner than the particle diameter. Diffusion in the moving phase makes a significant contribution to H in HPLC. When the terms are combined, H can be described as follows:

$$H = \sum_i \underbrace{\frac{1}{(1/2)\gamma_i d_p + D_m/w_i u d_p^2}}_{\text{eddy diffusion}} + \underbrace{\frac{2\gamma_m D_m}{u} + \frac{2\gamma_m D_s K'}{u}}_{\text{horizontal diffusion}}$$

$$+ \underbrace{\frac{qd^2 u R(1 - R)}{D_s}}_{\substack{\text{stationary-phase} \\ \text{diffusion}}} + \underbrace{\sum_i \frac{w_i d_p^2 u}{D_m}}_{\substack{\text{mobile-phase} \\ \text{diffusion}}} \tag{23}$$

It can be seen that horizontal diffusion becomes very important as flow velocity decreases. In GLC, D_m is large and the effect is readily seen. A plot of H versus u shows a minimum for GLC where an optimum flow velocity for best separation exists. The minimum in a plot of H versus u occurs at a velocity much slower than practical flow velocities for HPLC.

Snyder used a simplified equation for HPLC relating efficiency to flow velocity which is written as follows [7]:

$$H = Du^n \tag{24}$$

Here, D and n are characteristic column constants. The value for n is usually about 0.4 but can vary from 0.2 to 0.8. Later investigators developed the following equation [8, 9].

$$H = \lambda \bar{d}_p^{\ B} u^n \tag{25}$$

The symbols B and λ are characteristic column constants and $\bar{d}p$ is the average particle diameter. The value of B is usually about 1.8. Column efficiency is highly dependent on particle diameter and decreases rapidly as diameter increases. Once the specific column constants have been experimentally determined, the effect of flow velocity on column efficiency can be easily predicted. Since chromatograms must be obtained in a reasonable period of time, a compromise between column efficiency and flow velocity is chosen.

The previous discussion has been concerned with band spreading that occurs while the sample moves through the column. Band spreading can also be caused by extra-column effects. A good review of extra-column band spreading has been given by Martin et al. [10]. Three of the factors affecting extra-column band spreading are sample volume, connecting tubing geometry, and detector volume. Since it was shown that band spreading is proportional to the square of the sample volume, the sample volume should be as small as possible. Spreading is also proportional to the fourth power of the diameter of the connecting tubing. Therefore, small diameter connecting tubing is required to prevent excess band spreading. Since spreading is also proportional to the square of the detector volume, low detector holdup volume is desirable. It is clear that for optimum separation careful attention must be given to the design of all the chromatograph components.

C. Resolution of Bands

The major challenge in chromatography is the separation of sample components. An understanding of band retention and band spreading characteristics permits a rational approach to improving resolution.

The separation or resolution R_s can be related to conveniently measured retention parameters as follows:

$$R_s = \frac{(t_{R2} - t_{R1})}{(1/2)(tw_1 + tw_2)} \tag{26}$$

Resolution is the difference in retention times between the center of the two peaks divided by the average peak width. Retention times are related to dead time and

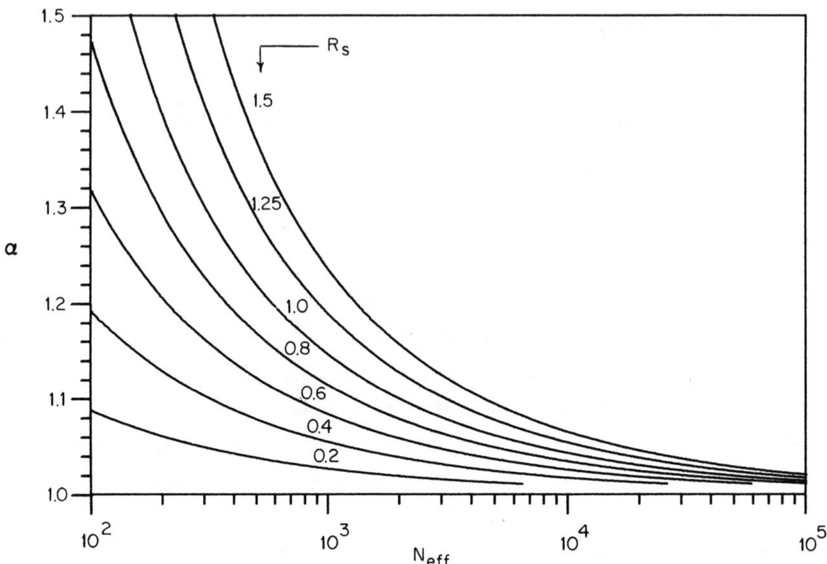

FIG. 2. Plot of relative retention value α versus log N_{eff} for various values of resolution R_s.

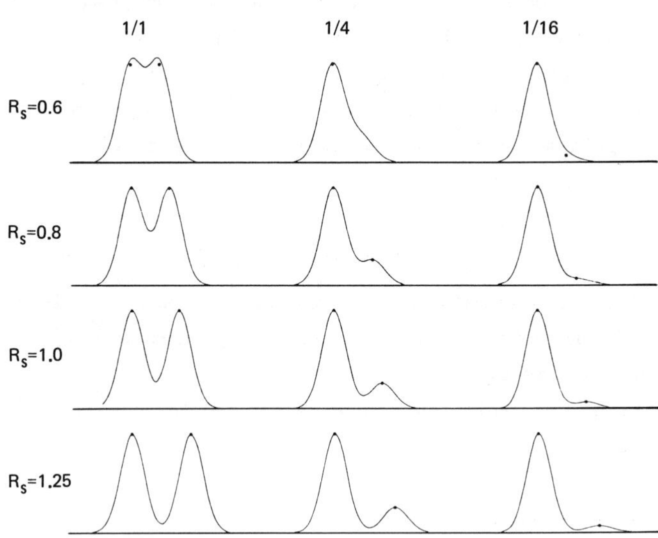

FIG. 3. Band separation versus resolution R_s for different band area ratios [11]. (Reprinted with permission of copyright owner.)

partition ratio in Eq. (8). Peak width is related to deadtime, partition ratio, and theoretical plates in Eq. (16). When the separation factor α is substituted for k_2'/k_1' resolution can be defined as

$$R_s = \frac{1}{4}\left(\frac{\alpha - 1}{\alpha}\right)\sqrt{N}\,\frac{k'}{1 + k'} \tag{27}$$

or

$$R = \frac{1}{4}\left(\frac{\alpha - 1}{\alpha}\right)\sqrt{N_{eff}} \tag{27a}$$

where

$$N_{eff} = N\left(\frac{k'}{1 + k'}\right)^2$$

A plot of α versus $\log N_{eff}$ for several values of R_s is shown in Fig. 2.

A systematic approach for estimating R_s by chromatogram inspection has been given by Snyder [11]. Fig. 3 illustrates the relationship between R_s and band separation. The figures in Snyder's paper are excellent and well worth using for a rapid estimate or R_s.

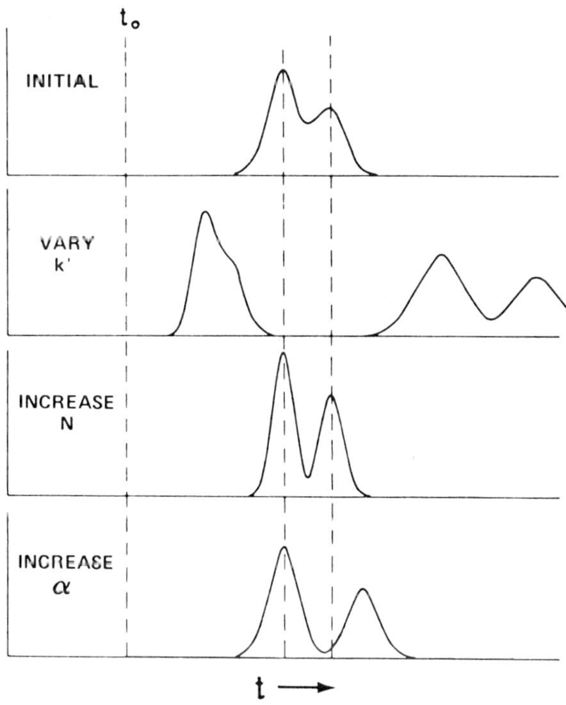

FIG. 4. Band separation as affected by partition ratio k', plate number N, or relative retention value α [12]. (Reprinted with permission of copyright owner.)

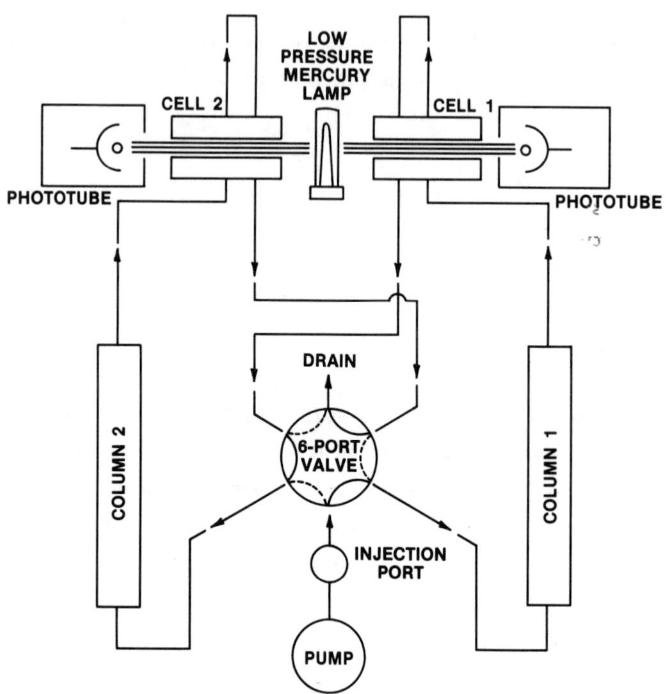

FIG. 5.   Recycle method for HPLC [13].   (Reprinted with permission of copyright owner.)

After an estimate of R_s is obtained, a decision can be made how separation is
best improved [12]. Equation (27) shows that peak resolution depends on the relative
retention values for the sample components α, the partition ratio k', and the number
of theoretical plates. Using experimental data, it is easy to determine N and k'
using Eq. (16) and (9). If R_s and N are known, α can be obtained from Fig. 2.
Figure 4 shows how resolution is affected by altering k', N, or α [12]. The value
for k' should be adjusted to a value between 1 and 10. If k' is too small, resolution suffers. If k' is too large, the chromatogram requires too much time to run,
the peak becomes broad, and analytical sensitivity is diminished. The magnitude of
k' can be altered in GLC by changing column temperature and in HPLC by changing solvent strength or polarity. The change in N_{eff} required to achieve a desired improvement in separation can be calculated or obtained using Fig. 2. A decision can be
made whether changing N, by adjusting flow rate or column length, or changing α is
more practical. Methods for changing α are difficult to predict since they may
involve both mobile- and stationary-phase changes. However, a change in α would
produce the greatest increase in resolution.

A convenient method for increasing N without encountering experimental problems
of undue column length and back pressure is the recycle method for HPLC [13]. One

method for achieving recycle is illustrated in Fig. 5 wherein two columns are alter-
antely employed.

If a sample contains a mixture of components with a wide range of k' values, no
single set of operating parameters is satisfactory. The values of k' can be changed
continuously during the run to achieve better results. In GLC, k' values are changed
by using a temperature-programming technique. At the start of a run the column
temperature is set low enough to yield a suitable k' value for the fastest-moving
sample component. The temperature is elevated during the run such that the slowest-
moving sample component also achieves a suitable k' value. In HPLC, the range of k'
values can be changed during the run by using a gradient elution technique. *Gradient
elution* means the solvent strength or polarity is changed during the run. This is
done by utilizing two solvents of different polarities. The two solvents are continuously mixed to form the moving phase. The volume ratio of the solvents is changed
during the run. Initially a weak solvent is chosen such that the fastest-moving
desired sample component has a suitable k' value. The ratio of stronger solvent is
increased during the run such that the slowest-moving desired sample component also
achieves a suitable k' value.

The chromatographer has a wide variety of variables to control when optimizing
a chromatographic procedure. A number of papers have been published which are concerned with a rational approach to optimization [12, 14-18]. The ACS audio cassette
course on liquid chromatography by Snyder and Kirkland is an excellent introduction
to HPLC theory and practice. It offers a good discussion on various aspects of
optimization.

One of the most intriquing approaches to optimization is the application of the
Simplex technique as described by Morgan and Demming [17]. A wide variety of chromatographs which offer computer or microprocessor control of sampling, instrument
parameters, and data processing are now commercially available. A computer or microprocessor could be programmed to optimize an assay using a Simplex or similar technique. Closed-loop control could be established between data analysis and adjustable
parameters such as temperature, solvent strength, flow rate, and sample size to produce computer optimized assays.

III. INSTRUMENTATION

The basic instrumentation required for GLC and HPLC is relatively simple.
Several components, which are connected in series to form a chromatograph, can
be listed:

   Mobile phase container and delivery system
   Sample inlet system
   Column and column housing

Detectors

Data recording and processing components

A. Mobile Phase Containers and Delivery Systems

In gas chromatography, the mobile phase container is a compressed gas cylinder. The
gases most commonly used are helium, nitrogen, and argon. A compressed gas contains
the energy required to move the sample through a column. The flow rate of the gas
is controlled by means of a pressure regulator valve on the gas cylinder and a needle
valve on the chromatograph.

The volume flow rate of the gas is usually measured with a gas buret, containing a
soap solution, attached to the gas outlet of the chromatograph. The flow rate is
found by timing the rise of a bubble up the buret. Alternatively the flow rate can
be measured with a rotameter which is normally placed on the gas-inlet side of the
chromatograph.

In HPLC, the liquid phase is stored in a container. In some chromatographs the
container is a removable glass jar with a loose-fitting plastic top. The solvent is
pulled from the jar through a tube which has a filter on the inlet end. In other
chromatographs a solvent reservoir, made from stainless steel, Teflon-lined stainless
steel, or glass, is built as an integral part of the chromatograph. The reservoir
may be equipped with a tight top and contain fittings to permit solvent degassing.
In either case, the mobile phase is pumped from the container to the chromatographic
column.

The most simple mobile-phase delivery system is a pressure bottle. The flow of
the liquid phase is controlled by adjusting the pressure from a gas cylinder to the
pressure bottle. The pressure is usually restricted to 1500 lb. The pressure bottle
has the advantage of providing a pulse-free flow of solvent. It is probably the
least expensive solvent delivery system. The pressure of the system can build up
very rapidly. However, changing solvent systems is difficult with a pressure-bottle
delivery system. Furthermore, the flow rate will change if the permeability of the
flow system or the solvent viscosity changes. In some instances the pressure bottle
may take the form of a long coil of tubing that can be filled with the eluting solvent. This type usually has a smaller internal volume, making it easier to refill
or change solvents.

Mobile phase delivery is usually accomplished by a mechanical pumping system.
Three major types of pumping systems are used: pneumatic amplifier, syringe, and
small-displacement reciprocating pumps.

The pneumatic amplifier pump is illustrated in Fig. 6. The moving liquid phase
is pushed with a relatively small piston, which displaces an equivalent volume of
solvent with each piston stroke. At the other end of the piston is a large-diameter

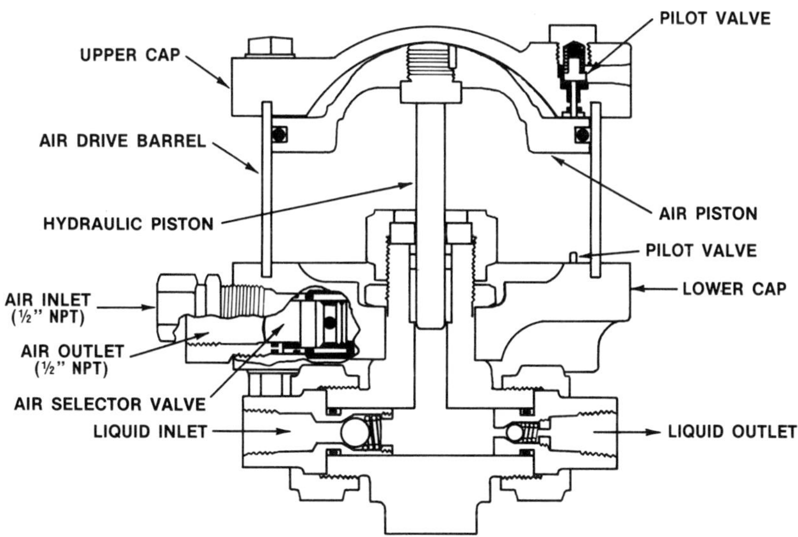

FIG. 6. Pneumatic amplifier pump. (Courtesy of Haskel Engineering and Supply Company.)

piston which is pushed by gas pressure. The pressure applied to the liquid phase is equal to the gas pressure applied to the large piston multiplied by the area ratio of the gas piston to the liquid piston. At the end of each stroke the gas piston is rapidly vented and pushed back. The pneumatic amplifier pump has the advantage of pulse free flow while the piston moves forward. The piston is reset for a new stroke very rapidly so that the break in solvent delivery is very short and not significant for most detector systems. The liquid holdup volume is fairly small, permitting relatively easy change of eluting solvent. The pump can be used with any size solvent reservoir. In addition the gas pressure required for driving the pump is much lower than that required for driving a pressure bottle. The pressure of the liquid is built up very rapidly. The major disadvantage of the pump is that the flow rate will change if the solvent viscosity changes or if the permeability of the flow system changes.

The syringe pump and the small displacement reciprocating pump are both positive-displacement pumps. This type pump provides a flow rate that is not altered by solvent viscosity or moderate changes in flow system permeability, providing that pressure buildup in the system does not exceed pump capacity. The syringe pump has the advantage of providing a smooth flow rate and low mechanical wear on the piston. There are also disadvantages. The syringe pump has limited solvent capacity. It is also difficult to change solvents rapidly. Because the diameter of the syringe is much larger than that of the flow line or column, very tight machining tolerances in the pitch of the drive screw are required to prevent a periodic oscillation of flow

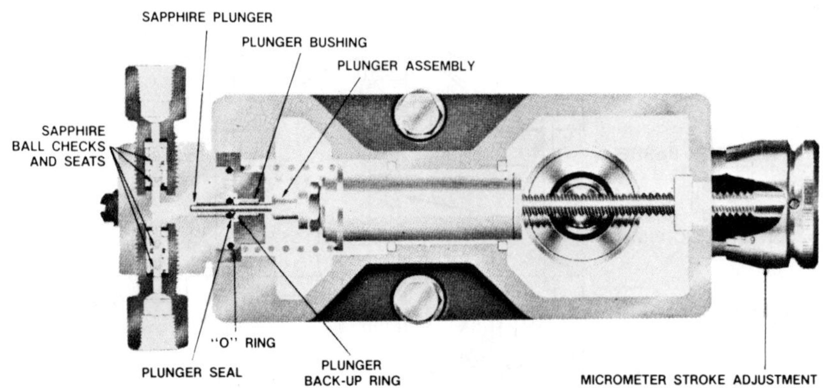

FIG. 7. Small-displacement reciprocating piston pump. (Courtesy of Milton Roy Company.)

rate. The oscillation, if present, has a very low frequency which is very difficult to damp out. If a syringe pump is started in a constant-displacement mode, as much as half an hour is required for the flow rate to stabilize because of liquid compressability [19]. The problem is greatest for a full syringe. It is recommended that the syringe be prepressurized before flow is started or that the pump should be started a sufficient time before the first sample is introduced. Sample introduction should be done with the pump running. The problem can be solved by a variable-speed pump with feedback control from a flow or pressure transducer.

The second category of positive-displacement pumps is the small-displacement reciprocating piston (Fig. 7) and diaphragm pumps. These pumps have several advantages. They have a low internal volume which permits rapid solvent changes. The flow rate is easy to program or control permitting convenient gradient elution studies. Some models are relatively inexpensive. The pump can be used with any size solvent reservoir. The pumps also have several disadvantages. The flow has a pulsating characteristic. Buildup of system pressure is fairly slow. The flow rate is limited by the small piston or diaphragm displacement coupled with moderately slow reciprocating rates. In addition, the pump has the most piston-seal wear. In most instances, the pulsating nature of the flow rate can be adequately damped out. Some pumps use a pair of pistons running in opposite phase to minimize pulsing. Others also couple a rapid response speed control to a flow sensor in a feedback loop to provide a smooth flow rate.

B. Sample Inlet System

Sampling systems can be divided into two categories: syringe injection and sample-loop inlet systems. Syringe injections have the advantage that a narrower band of

sample can be placed on the column than with most sample loops. Greater peak resolution can be obtained with a narrow sample band. The volume of sample delivered by a
syringe can be varied by drawing different volumes of sample into the syringe before
injection. The sample loop volume is fixed. To change the sample volume, different
volume sample loops must be interchanged. The sample loop, however, has two distinct
advantages over the syringe injection system. First, highly reproducible volumes of
sample can be placed on the column permitting high precision quantitative analysis.
Secondly, the sample loop can be switched out of the flow system permitting the sample
to be loaded at atmospheric pressure. A sample loop is shown in Fig. 8.

In the gas chromatography of therapeutic agents, most samples are liquids which
are introduced using syringe injection. Two syringe injection procedures are used
for GLC. The first is to inject the sample into a hot injection port or cavity having
a high temperature and heat capacity, so that the sample evaporates almost instantly.
The rate of volatilization determines the width of the sample band entering the column
and, therefore, affects the separation efficiency of the column. Many compounds of
biological interest are quite labile and will decompose on contact with hot metal
surfaces. The second syringe injection procedure is on-column injection. In this
procedure the syringe needle enters the front end of the column where the sample is
injected and volatilized. If a glass chromatographic column is used, the sample will
not contact a hot metal surface and decompose as much. For this reason, on-column
injection with glass columns is extensively utilized for easily decomposed compounds.

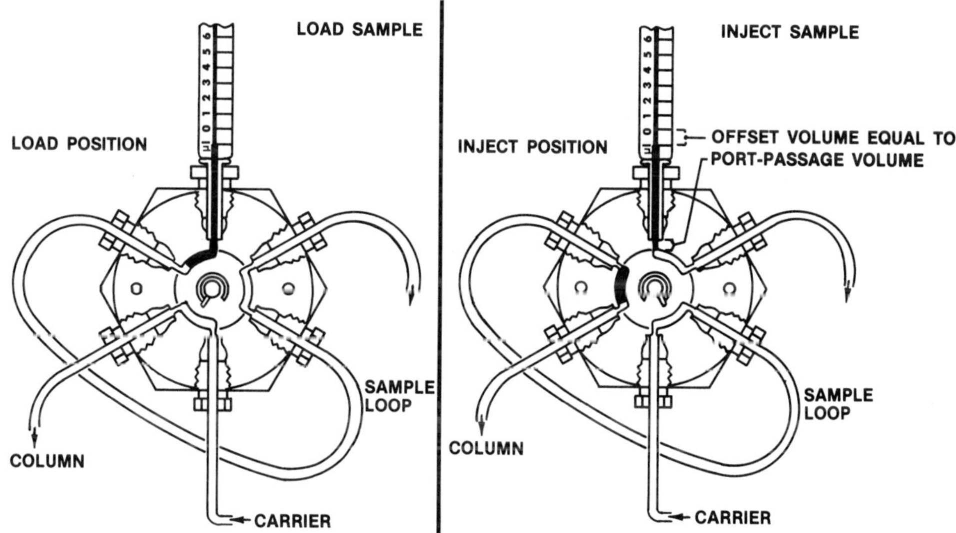

FIG. 8. Sample injection loop. (Courtesy of Valco Instrument Company.)

In HPLC, syringe injection permits the greatest analytical sensitivity since no sample is wasted in the connections necessary with a sample loop. Because of the much higher column pressure, syringe injection through a septum into a flowing stream is more difficult for HPLC than for GLC. The sample loop has become widely used in HPLC because it offers excellent sampling precision and isolation from column pressure. An interesting variation has the sample injected from a syringe into the center of a sample loop that is isolated from column pressure. After injection, the loop is switched into the flow stream. This permits variable volume syringe injection with pumps running and without the back pressure problems encountered with septum-type injection ports.

Capillary columns in GLC have very low sample capacity. Therefore, a splitter system is usually used in front of the column to prevent overloading. An example of a GLC splitter system is shown in Fig. 9.

C. Column and Column Housing

The partitioning behavior of the sample between the moving and stationary phase is very temperature dependent in GLC. Therefore, the GLC column must be placed in an oven where the temperature can be precisely controlled. In HPLC, partitioning behavior is more often adjusted by varying the mobile phase composition. Since the sample is not volatilized the column usually operates at or near room temperature. In many instances the HPLC column is not thermostated in an oven at all.

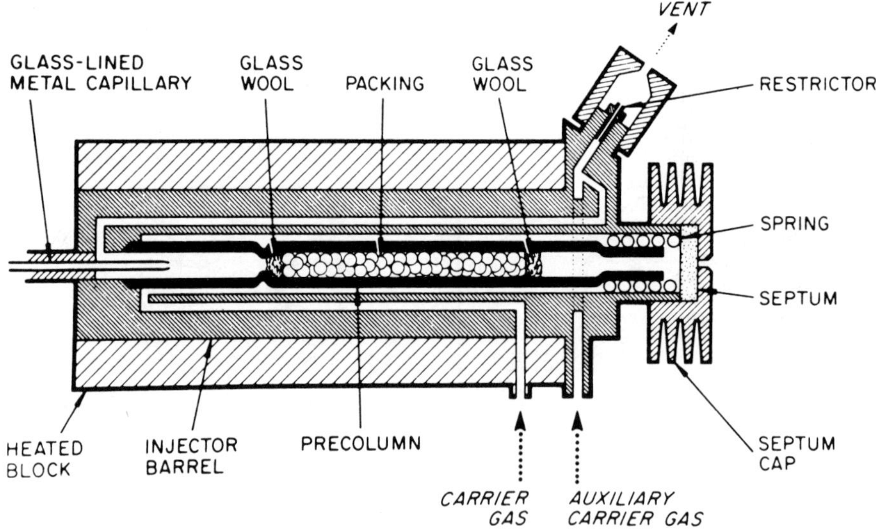

FIG. 9. Capillary column-inlet splitter system. (Courtesy of Perkin Elmer Corporation.)

D. Detectors

1. Characteristics

Chromatographic detectors can be classified into two general categories: those whose response is dependent on the concentration of the sample in the mobile phase such as thermal conductivity or photometric detectors, and those whose response is proportional to mass flow rate or the number of sample molecules entering the detector per unit time such as flame ionization detectors [20]. The response of concentration dependent detectors is independent of flow rate. Since mass is the product of concentration and flow rate, the response of mass dependent detectors increases with increasing flow rate.

Three characteristics used to describe detector behavior are sensitivity, linear dynamic range, and selectivity. Sensitivity can be described as the ratio of the change in detector response to the change in sample concentration or maximum detector sensitivity. Maximum sensitivity (lower detection limit) is usually described as the quantity of sample required to produce a detector signal that is twice the size of the detector noise.

Linear dynamic range is the range of sample concentration over which the ratio of detector response to sample concentration remains constant. Although a long linear range is desirable, some detectors with a relatively short dynamic range have been useful because of another characteristic such as high sensitivity or selectivity. Another detector characteristic is selectivity. Some detectors demonstrate high selectivity by showing response only to compounds containing a particular element such as nitrogen or sulfur. Other detectors are tuneable and can be adjusted to respond only to certain compounds. Tuneable selectivity is provided by wavelength selection in spectrophotometric detectors and voltage change in voltammetric detectors. Separation requirements are often less if detector selectivity permits measurement of desired compounds with little or no interference from other compounds present. In other cases it is desirable that the detector respond to all compounds present in a sample.

2. GLC Detectors

a. Thermal Conductivity Detectors

Thermal conductivity detectors (Catharometers) are extensively used in GLC because of their simple construction design, stable response, and low cost. The detector is non-destructive of the sample, permitting sample recovery. The detector responds to any substance whose thermal conductivity differs from that of the carrier gas. If a very low molecular-weight gas like helium is used as the mobile phase, every known volatile compound of therapeutic interest will produce a detector response.

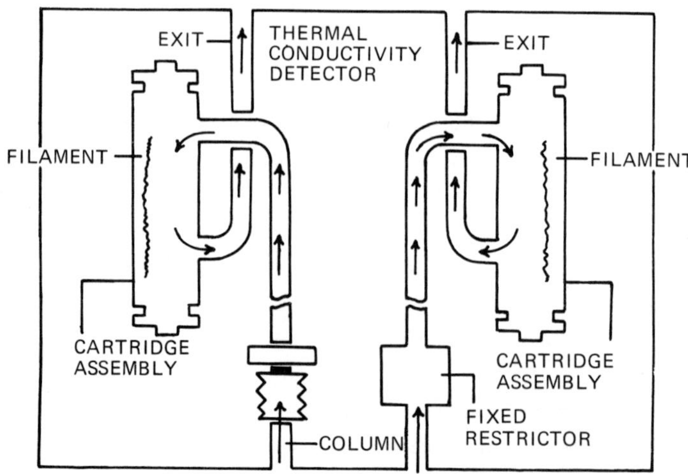

FIG. 10. Thermal conductivity detector. (Courtesy of Hewlett Packard Corporation.)

A typical thermal conductivity GLC detector design is shown in Fig. 10. It consists of a detecting element in a cavity in a metal block. The mobile phase leaving the chromatographic column contacts the detecting element either by flowing through or diffusing into the cavity. Two types of detecting elements are commonly used: a metal wire, such as platinum, tungsten, tungsten-rhenium alloy, or nickel or a thermistor. Metal wires have a positive temperature coefficient of resistance, while thermistors have a negative temperature coefficient of resistance. Both types of detector elements act as resistance thermometers. In most cell designs there are two detector elements: one contacts only pure carrier gas and serves as a reference detector; the other contacts the effluent leaving the column. The two elements are usually mounted in adjacent cavities in a common metal block to maintain a common temperature. The detector elements are usually connected as resistors in a Wheatstone bridge circuit. If the resistance in one of the elements changes, the voltage across the bridge changes and is measured with a high-impedance voltmeter or potentiometric recorder. The bridge circuit is normally powered by a constant voltage source which is adjusted until a desired current is flowing through the detector. The current flowing through the detector causes the temperature of the detector to increase to an equilibrium value. At equilibrium the temperature gradient between the hot element and the detector block ($T_2 - T_1$) causes conduction of heat away from the element which is equal to the heat generated electrically in the element, q. This is described by the following equation:

$$q = \frac{I^2 R}{J} = \lambda G(T_2 - T_1) \tag{28}$$

where R is the detector resistance, J is joule's equivalent, λ is the thermal conductivity of the gas, G is a cell factor which is dependent on cell geometry, T_1 is the cell wall temperature, and T_2 is the detector element temperature.

When the gas contacting the element changes composition, the thermal conductivity of the gas mixture changes, altering the temperature of the detector element. The electrical resistance of the detector element will obey the equation

$$R = R_0(1 + \alpha T) \tag{29}$$

over moderate changes in temperature, where R_0 is the resistance at $0°$ C, and α is the temperature coefficient of resistance.

Normally a light gas such as helium (mol. wt. = M_1) is used as the mobile phase. Most compounds of therapeutic interest have a much larger molecular weight (mol. wt. = M_2) than the carrier gas. It has been proposed that when $M_2 \geq 20M_1$, heat is conducted almost entirely by the carrier gas [21]. Thermal conductivity of the sample mixture decreases because the large sample molecules obstruct thermal conductivity by the carrier gas. The change of thermal conductivity of the gas mixture with mole fraction of the sample can be written as

$$\frac{1}{\lambda_1} \frac{\partial \lambda_{12}}{\partial x_2} = -K\frac{\sigma_{12}^2}{\sigma_1^2} \tag{30}$$

where λ_1 is the thermal conductivity of the carrier gas, λ_{12} is the thermal conductivity of the mixture, x_2 is the mole fraction of the sample, σ_1 and σ_2 are the molecular diameters of the carrier gas and sample, and σ_{12} is the average molecular diameter of the mixture, where $\sigma_{12} = (\frac{\sigma_1 + \sigma_2}{2})$. K is a constant which Littlewood has calculated to be 2.30 for a hard-sphere model [21].

It is much more difficult to predict detector response when a heavier carrier gas such as nitrogen is used. The detector sensitivity is also greater for lower molecular weight carrier gases. For these reasons, helium is the preferred carrier gas for thermal conductivity detectors.

There are other mechanisms by which the detector can lose heat. Heat loss by radiation and normal convection inside the cell is normally very small. End losses through the detector element supports have the effect of lowering the effective power input to the element. The most important heat loss is mass transfer of heat out into the gas stream. A diffusion cell eliminates the problem of mass transfer of heat into the gas stream but does so at the expense of slower cell response time. The flow-through cell has the greatest mass transfer heat loss, but has the fastest response time. Many modern detectors are a flow-through type designed to minimize mass transfer of heat while still affording fast response time.

The thermal conductivity detector's sensitivity is proportional to the temperature gradient between the element and the block. The temperature gradient increases

with increased detector current. The temperature coefficient of resistance is nonlinear and decreases at higher temperatures. Therefore, the gain in sensitivity with
current diminishes at higher currents. The thermistor has a large negative temperature
coefficient of resistance. The sensitivity of a thermistor increases with current for
a while and then decreases because at higher temperatures the resistance becomes so
small that the decreasing rate of power dissipation more than offsets the increase in
thermal conduction. Even though the temperature coefficient of resistance for the
thermistor is much greater than for the resistance wire, the upper temperature at
which a thermistor offers greater sensitivity than a resistance wire is about 100° C.
Increasing the block temperature decreases the temperature difference between the hot
element and the block, lowering sensitivity. The block must be kept hot enough,
however, so that condensation does not occur.

The linear dynamic range is greater for low molecular weight carrier gases. It
has been reported to be about 10^{4} when helium is the carrier gas [22]. The dynamic
range decreases at higher temperatures.

The materials used for thermal conductivity elements are subject to oxidation
especially at elevated temperatures. The elements should receive as little exposure
to air as possible.

Detector sensitivity could be calculated if sample cross sectional areas and
the cell constants were accurately known. However, the usual practice is to determine
the relative molar response for a series of compounds with respect to a reference compound. The reference compound can be used to calibrate the detector or serve as an
internal standard.

b. Flame Ionization Detectors

Ionization detectors are substantially more sensitive than thermal conductivity detectors. The flame ionization detector (FID) is the most popular and widely used ionization detector in GLC due to high sensitivity, very long linear dynamic range, simple
reliable construction, and general utility for a wide variety of organic compounds.

A typical detector design is shown in Fig. 11. The sample passes through a
burner jet into a hydrogen-oxygen flame. Two electrodes are positioned across the
flame. A potential is applied such that any ions formed in the flame will be collected causing a current to flow. Normally the number of ions generated in the flame is
quite small. When a sample containing carbon is introduced into the flame, a large
increase in ion generation occurs, and the current increases.

Extensive studies have been made to determine the mechanism for ion formation in
the flame. The results indicate that neither thermal ionization or thermionic emission
from carbon particles can account for the observed ionization. It is now thought that
chemionization mechanisms are responsible for ion production [23-25]. The primary
chemionization reaction was proposed to be [26]

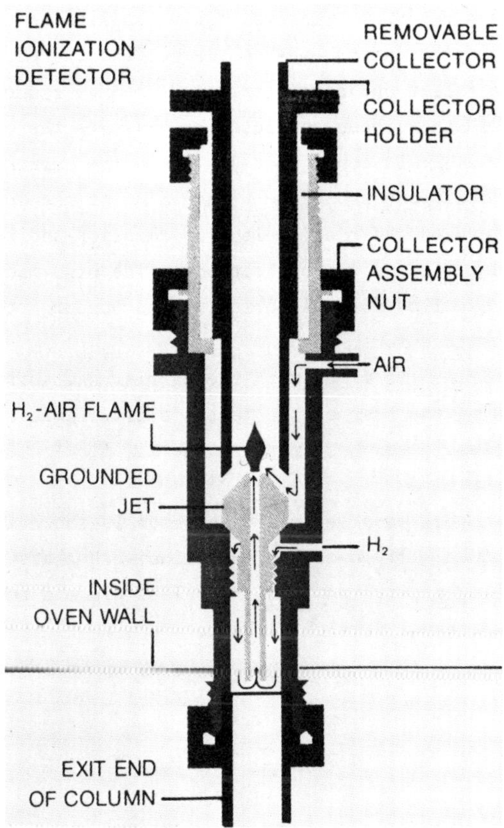

FIG. 11. Flame ionization detector with cylindrical collector electrode.
(Courtesy of Hewlett-Packard Corporation.)

$$CH + O \rightarrow CHO^* \rightarrow CHO^+ + e^- \tag{31}$$

However, the effect of different carrier gases on sensitivity has been difficult to
interpret. Recently a new chemionization mechanism has been proposed:

$$CH + O \rightarrow CHO^*$$

$$CHO^* + X \rightarrow CHO \cdot X^+ + e \tag{32}$$

where the carrier gas X is directly involved in the ionization reaction [27]. The
formation of $CHO \cdot X^+$ was found to predominate over the formation of CHO^+ when xenon or
krypton was used as the carrier gas. This new mechanism may explain the carrier gas
effect of flame ionization detectors.

The prime objective of detector design and operation is to collect all or a constant high proportion of the ions formed. Electrodes are positioned across the flame
with a potential applied across them suitable for complete ion collection. Ion recombination reactions have been shown to be unimportant [28]. As potential is increased
between the electrodes, the current rises to a broad plateau region over a range from

about 10 to several hundred V/cm. The voltage is applied in the plateau region [29-31]. If the potential gradient becomes too large, the charges acquire enough energy to cause an avalanche effect and an electrical discharge occurs.

A number of studies have been conducted to evaluate the effects of electrode position and geometry [32-34]. In configurations where the burner jet serves as one electrode, the collector electrode should be within 5 mm of the jet and have a cylindrical shape. To minimize induced thermionic electron emission and to maximize ion capture, the burner jet should be the positive electrode. The high-mobility electrons can move upstream against the gas flow for collection at the burner jet more easily than the heavy, slow-moving positive ions.

In most FIDs, hydrogen is mixed with the carrier gas just before it enters the burner jet. Surrounding the base of the jet is a flow laminizer through which air flows up around the flame. For best results the air flow rate is usually about 10 times the hydrogen flow rate.

An example of a modern parallel plate detector is shown in Fig. 12. The burner jet is an all quartz tube. If on-column injection with glass columns were used, the system would be an all glass system. This parallel plate detector has a linear dynamic range of greater than 10^7 which is better than that attributed to earlier parallel-plate designs.

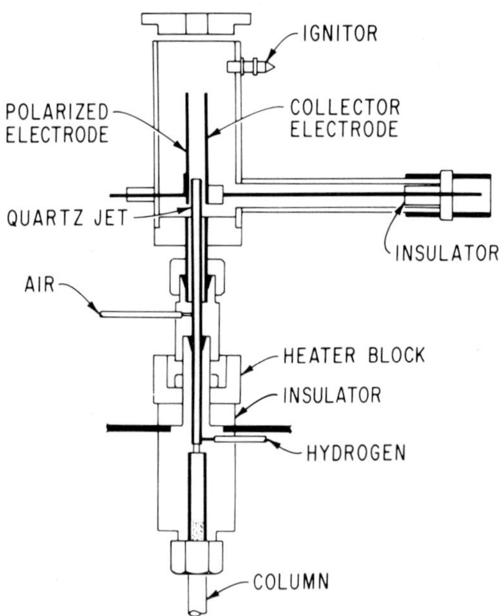

FIG. 12. Flame ionization detector with parallel-plate design. (Courtesy of Tracor Corporation.)

The sensitivity of flame ionization detectors is limited by the inherent noise of the detector which is about 10^{-14} A. If sample ionization is about 10^{-5}, the limit of detection would be about 10^{-13}g/sec of organic sample.

c. Flame Photometric Detectors

Flame photometry is another of the analytical measurement techniques that is done in a flame. The initial development was done by Brody and Chaney [35]. A typical flame photometric GLC detector is shown in Fig. 13. The flame photometric detector shows outstanding selectivity for compounds that contain phosphorus and sulfur, and almost no response from compounds that do not. The sulfur emission peak is at 394 nm and the phosphorous emission peak is at 526 nm. Phosphorus and sulfur may interfere with each other, if both elements are present in the same compound or chromatographic peak. The response ratio of sulfur to phosphorus is high, 1000 to 10,000 when measurements are made in the sulfur mode at 394 nm. In the phosphorous mode at 526 nm the response ratio of phosphorus to sulfur may drop to below 10 [36]. If measurements are made in the phosphorous mode it is possible to have significant interference from sulfur. The presence of sulfur can be easily verified, however, by switching to the 394 nm wavelength.

The quantitative responses for sulfur and phosphorus differ. While detector response is linear with phosphorous concentration, the square root of the detector response is proportional to sulfur concentration. The linear dynamic range varies from about 0.2 to 100 x 10^{-9}g on a log-log scale for sulfur and from 0.2 to 300 x 10^{-9}g on a linear scale for phosphorus. The minimum detection limit is about

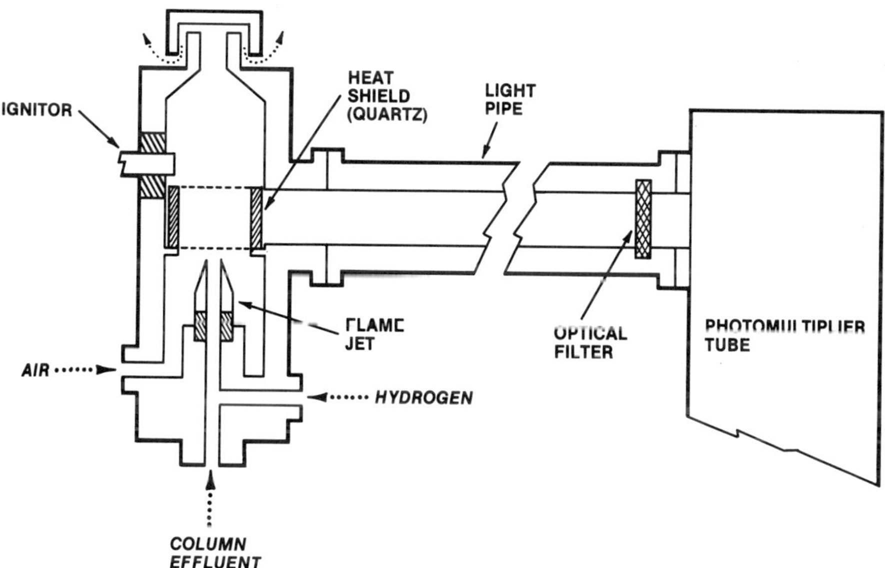

FIG. 13. Flame photometric GLC detector. (Courtesty of Perkin Elmer Corporation.)

200 pg for sulfur and 40 pg for phosphorus [37]. HPO is the species emitting
radiation for phosphorus at 526 nm and S_2 is the species emitting radiation for
sulfur at 394 nm [38].

The flame photometric detector has become widely used and accepted for the analysis of sulfur and phosphorous containing compounds. This is due to the outstanding
selectivity and good sensitivity of the detector and to the fact that it is relatively
dependable and trouble free to use.

d. Thermionic Phosphorus and Nitrogen Detectors

A wide variety of biologically important compounds contain nitrogen or phosphorus.
The thermionic detector offers both high sensitivity and selectivity for nitrogen
and phosphorus-containing compounds. When alkali metal salts are heated, they give
off alkali metal vapors. Decomposition products of nitrogen or phorphorus-containing
compounds can accept electrons from alkali atoms causing ionization and a large current
increase.

The first thermionic phosphorus detector was reported by Karman and Guiffrida
[39]. The first thermionic nitrogen detector was described by Wells [40]. Early
detectors used alkali halide salts which were formed in various geometries. Some
detectors used dual flame jets, one of which heated the alkali salt. The detectors
were difficult to operate and were not very stable.

Recently, better thermionic detectors have been developed. It was shown that
the ionization process in thermionic detectors depended on the pressure of the alkali
metal vapor [41]. The heat required to vaporize alkali atoms is only a small fraction
of the heat liberated by a conventional flame ionization detector. In early designs,
much care was needed in adjusting the flame condition and the position of the alkali
salt to obtain reproducible alkali vapor concentrations. Alkali metal volatilization
can be controlled more precisely by electrically heating the salt. Because alkali
metal salts volatilize, the salt reservoir changes characteristics with time and the
detector becomes contaminated with salt deposits.

A major improvement was made when an electrically heated bead of nonvolatile
rubidium silicate glass was substituted for volatile alkali salts [42]. When a
negative potential is applied to the heated bead, it conducts electrons which can
reduce a rubidium ion to a neutral atom which volatilizes as a thermally excited
rubidium atom. If a species is present that can accept an electron from the rubidium
atom, ionization occurs. The positive rubidium ion formed is attracted back to the
negative bead where it may again be volatilized. An example of this type detector is
shown in Fig. 14. The electrically heated glass bead offers more reproducible results
and a much longer life expectancy than previous detectors using volatile alkali salts.

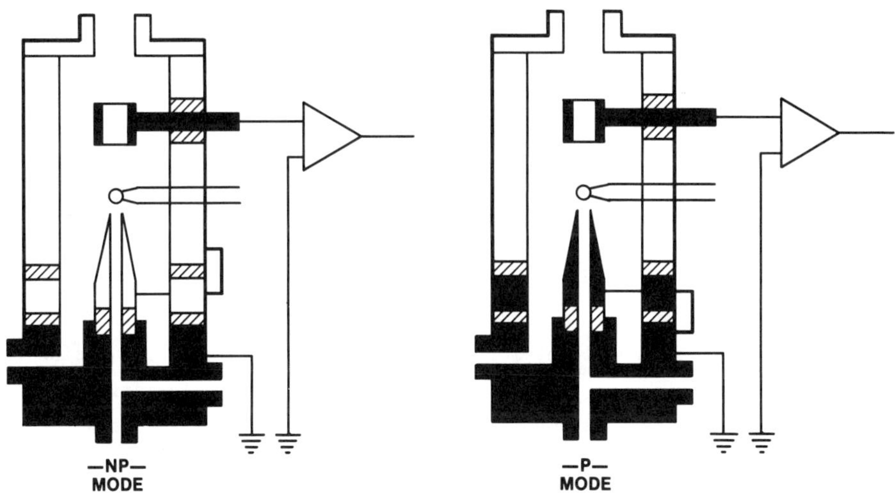

FIG. 14. Thermionic nitrogen-phosphorous detector shown in nitrogen-phosphorous and phosphorous modes. (Courtesy of Perkin Elmer Corporation.)

It has been suggested that background currents may be due to reactions such as a three body collision with hydrogen or by reactions with species such as OH_3^+. The major reaction sequence proposed for nitrogen detection is the following:

$$-\overset{|}{\underset{|}{C}}-N \quad\rightarrow\quad \cdot C\equiv N$$
$$Rb^* + \cdot C\equiv N \quad\rightarrow\quad Rb^+ + [\ C\equiv N\ ]^- \qquad\qquad (33)$$
$$[\ C\equiv N\ ]^- \quad\rightarrow\quad HCN + C^-$$

The formation of the cyano radical is favored by use of a cold diluted flame. Nitrogen-containing compounds with vicinal carbonyls that can prevent the formation of cyano radicals give a much reduced response.

The reaction sequence proposed for phosphorous detection is the following:

$$\cdot P=O + Rb^* \quad\rightarrow\quad [<P=O>]^- + Rb^+$$
$$\text{or} \qquad\qquad\qquad\qquad\qquad\qquad\qquad\qquad\qquad\qquad (34)$$
$$<O=\cdot P=O> + Rb^* \quad\rightarrow\quad [<O=P-O\ ]^- \rightleftharpoons [\ O-P=O>]^- + Rb^+$$

Although optimal operating modes for nitrogen and phosphorus are different, phosphorus can also be detected in the nitrogen mode of operation. For nitrogen, the hydrogen flow rate is greatly reduced. A typical flow rate is 1-3 ml/min H_2 and 100 ml/min air. The flame actually ignites and burns around the hot rubidium silicate bead rather than at the burner jet. The cool dilute flame surrounding the bead favors cyano radical formation. The collector is positive and the jet negative, assuring that the electrons released by the thermionic process reach the collector electrode.

A normal hot flame at the burner jet is used for the optimum phosphorous mode. The flame heats the bead to a red color. The jet is made positive and connected to ground. Any electrons produced in the flame ionization process are collected at the burner jet and separated from the electrons produced by the thermionic process, which are attracted to the collector electrode.

Very recently a new type of thermionic nitrogen-phosphorus detector has been designed which seems to incorporate substantial advantages over earlier designs [43]. The alkali source is an alkali coated ceramic cylinder electrically heated to a dull red. The cylinder is suspended in the center of the collector electrode. The gas mixture, in this case 3.7 ml/min H_2 and 100 ml/min air, passes through the hot cylinder. Instead of igniting a flame, a low-temperature gas plasma forms inside the hot cylinder, which is sufficient for sample decomposition and ion formation. However, the plasma does not produce heat as a flame would and, therefore, does not further heat the ceramic cylinder. The ionization mechanism is not understood but may be substantially different from that occuring at the previously described detector. The collector electrode is set at a negative potential instead of a positive potential. In addition, compounds which contain no HCN bonding such as underivatized barbiturates can be detected with very high sensitivity. The detector shows very high sensitivity to both nitrogen and phosphorus compounds without changing operating mode. The detector is extremely selective showing essentially no response to carbon, sulfur, and chlorine.

Both of the thermionic detectors described have good stability. The linear dynamic range is 10^5 for both nitrogen and phosphorus. The sensitivity is between 10^{-13} and 10^{-14} g/sec for both nitrogen and phosphorus. Recent developments in thermionic detectors have shown dramatic improvements; they are stable, show high selectivity, and offer very high sensitivity.

e. Electron Capture Detector

The electron capture detector was introduced in 1960 by Lovelock and Lipsky [44] The detector shows very high sensitivity for compounds capable of electron capture, such as halogen-containing compounds. Many compounds of biological interest have electron capture capability or can be labeled with electron capturing derivatives.

The electron capture detector contains a radioactive source. The radiation ionizes the carrier gas producing electrons and positive ions. The steady-state baseline current I_s corresponds to the collection of electrons when no sample is present. When a sample capable of capturing electrons enters the detector it reacts with electrons to produce negative ions. The negative ions undergo recombination reactions with the positive ions produced by the radiation. The number of free ions and electrons is, therefore, depleted in the presence of the sample and the current I becomes smaller. The change in current is not linear with sample concentration. If a

steady-state voltage is applied to the collector electrodes, the current is given by
the equation,

$$\frac{I_s - I}{I} = K'a \tag{35}$$

where K' is a constant proportional to the electron capture coefficient of the substance and a is the sample concentration.

The normal mode for detector operation is to apply a series of voltage pulses to
the collection electrodes. When the voltage is continuously applied, the electrons
are collected more rapidly than the slower moving positive ions, which create a space
cloud near the negative electrode. This ion cloud tends to inhibit electron collection and causes nonlinear and unpredictable results.

Electrons can move fast enough to be collected during short voltage pulses only
a few microseconds wide. During the relatively long intervals between pulses, the
electrons are in thermal equilibrium with the gas. This favors electron capture by
the sample and recombination of opposite charged ions. It was shown that the average
current obtained under pulsed conditions was inversely proportional to the sample
concentration and pulse width [45]. The current can be described by the following
equation:

$$I = \frac{bq}{(K_D + aK_1)t_p}\left(1 - e^{-(K_D + aK_1)t_p}\right) \tag{36}$$

Here, q is the charge of the electron, b is the rate of electron production, t_p is the
pulse width, K_1 is the rate constant for the electron capture reaction, K_D is the combined rate constant for all other processes that remove electrons, and a is the sample
concentration. It was shown that if pulse frequency f is substituted for the reciprocal of pulse width, and the current is constant, Eq. (36) can be rewritten as
follows [46]:

$$(K_D + aK_1) = Kf \tag{37}$$

K is a proportionality constant. When a sample is introduced, the pulse frequency
must change to maintain constant current. The following relationship is obtained:

$$f - f_0 = \Delta f - \frac{K_1a}{K} \tag{38}$$

where f_0 is the pulse frequency when no sample is present. It can be seen that the
change in pulse frequency is directly proportional to the sample concentration [46].

A feedback system using voltage-to-frequency and frequency-to-voltage convertors
is used to maintain a constant cell current. This procedure where frequency change
is related to sample concentration has become the standard operation mode for electron
capture detectors. The linear dynamic range can approach 10^5 when the detector is

operated in a variable-frequency pulsed mode. The detector sensitivity is very high and is on the order of 10^{-14} g/sec.

Two isotopes have been widely utilized as radiation sources, tritium and nickel-63. In theory, a detector using tritium should be somewhat more sensitive than one using nickel-63 because it emits softer radiation. The β energy of tritium is 18 keV maximum and the β energy of nickel-63 is 67 keV maximum. To achieve the same amount of ionization, higher tritium activity is required than for nickel-63. Therefore, the current produced by tritium radiation will be less subject to statistical variations.

In practice however, tritium has problems because it is volatile and undergoes exchange reactions with hydrogen gas. The most successful utilization of tritium has been as a rare earth tritide such as scandium tritide [47]. A scandium tritide source can be operated above 300° C. The high temperature is necessary to prevent contamination of the radiation source which would cause erroneous results. The new scandium tritide sources are quite stable although they gradually lose activity and can undergo exchange reactions with hydrogen gas. Nickel-63 is nonvolatile, has a long half-life, and can be operated at high temperatures. At the present time it is considered by many to be the radiation source of choice. The design of an electron capture detector is illustrated in Fig. 15.

Recently Lovelock and co-workers have conducted a systematic investigation of other potential radiation sources [48]. It was found that a detector using iron-55, an auger electron emitter, as the radiation source gave a better signal to noise ratio than either tritium or nickel-63. The auger electron energy is between 5.39 and 5.64 keV which is softer radiation than that of either tritium or nickel-63. It is

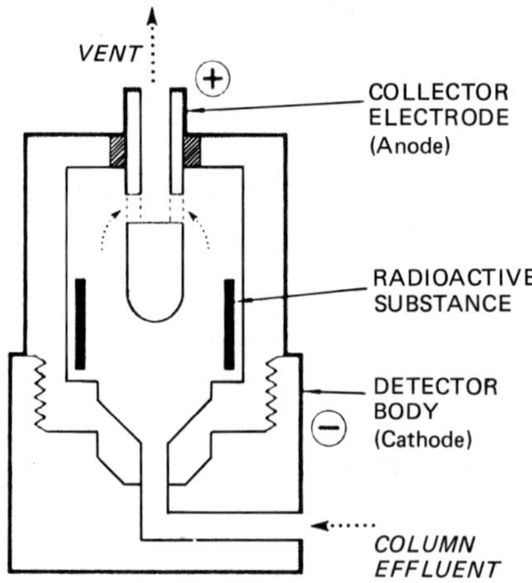

FIG. 15. Electron capture detector. (Courtesy of Perkin Elmer Corporation.)

possible in the future that manufacturers will consider iron-55 as the radiation source of choice for electron capture detectors.

The electron capture coefficient is very temperature dependent. Precise control of detector temperature must be maintained for good quantitative results. It is also possible to optimize sensitivity for a particular compound by adjusting detector temperature [49].

Recent results have shown that increasing the gas pressure inside the electron capture detector increases the response [50]. Future detector designs may utilize higher pressures to maximize sensitivity.

There are two common choices for the carrier gas: nitrogen and an argon-methane mixture. Argon-methane has been the popular choice, but recent detector designs and operating conditions seem to favor nitrogen. The ultimate sample sensitivity is slightly greater with nitrogen. There are significant advantages for simultaneous use of multiple detectors such as a nonspecific flame ionization detector and a selective detector such as a thermionic, flame photometric, or electron capture detector [51]. If argon-methane is used as the carrier gas, a flame ionization detector cannot be used concurrent with an electron capture detector. However, if nitrogen is used, the electron capture detector can be used in combination with other detectors.

f. Photoionization Detectors

One of the first photoionization detectors, utilizing an argon glow discharge as the source of high energy photons, was reported by Lovelock [52]. Other flow-through detectors utilizing either argon or helium glow discharge as the ionization source were developed later. A helium glow discharge detector using a microwave discharge source was reported by Freeman and Wentworth [53].

The most recent photoionization detector designs have a sealed ultraviolet source where the sample is irradiated through a window [54-56]. Figure 16 illustrates the design of a photoionization detector using a sealed ultraviolet source. The recent design of Driscoll and Spaziani is reported to have a linear dynamic range of 10^7 to 10^8 [56]. This is the longest linear dynamic range claimed for any commercially available GLC detector. The sensitivity of the detector was reported to be 2×10^{-12} g. This is about an order of magnitude greater than obtained with a flame ionization detector.

The hydrogen lamp utilized in the photoionization detector emits a high-intensity 10.2-eV Lyman alpha line. Compounds with an ionization potential less than 10.2 eV can undergo the following reaction.

$$S + h\nu \rightarrow S^+ + e^- \tag{39}$$

where S is the sample molecule. When a carrier gas, for example nitrogen, absorbs radiation the following reaction sequence takes place.

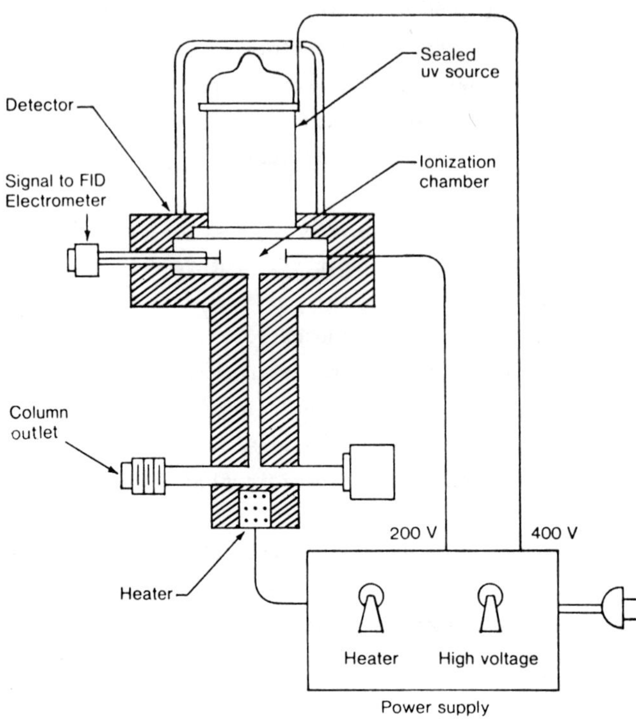

FIG. 16. Photoionization detector. (Courtesy of HNU Instrument Corporation.)

$$N_2 + h\nu \rightarrow N_2^* \tag{40}$$

$$N_2^* + S \rightarrow N_2 + S^+ + e^- \tag{41}$$

Since a helium source emits higher energy radiation than an argon or hydrogen source, it is capable of ionizing a wider variety of sample molecules.

The photoionization detector is nondestructive of the sample. New designs afford both very long linear dynamic range, good sensitivity, and good stability. When compared to the flame ionization detector, the photoionization detector has the advantage of not requiring hydrogen and oxygen supplies and of avoiding complications caused by the flame.

3. Detectors for HPLC

a. Ultraviolet and Visible Light Absorption Detectors

The detector most commonly used with HPLC is based on uv-vis light absorption. The choice of this detector depends on the sample's ability to absorb light in a wavelength region where the solvent is transparent. Absorbance A is directly proportional to sample concentration and obeys Beer's Law, A = abC where a is the sample absorptivity, b is the cell path length, and C is the sample concentration. The cell is designed to provide maximum optical path length with minimum holdup volume. Minor changes in refractive index should not change the measured absorbance readings. Most manufacturers have designed cells which are relatively insensitive to refractive index changes. An example of such a cell is shown in Fig. 17. The tapered geometry allows essentially all the light entering the cell to be transmitted except for that absorbed by the sample [57].

Two types of photometers are used: a fixed-wavelength filter photometer and a variable-wavelength spectrophotometer. Each design has advantages. The main advantages of the fixed-wavelength detector are greater light throughput and wavelength stability. Often fixed-wavelength detectors utilize a mercury vapor lamp as the light source. A low pressure mercury vapor lamp has emission bands at 254, 280, 313,

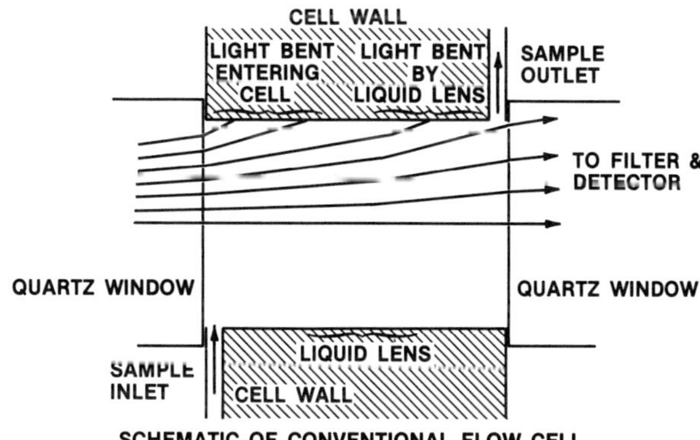

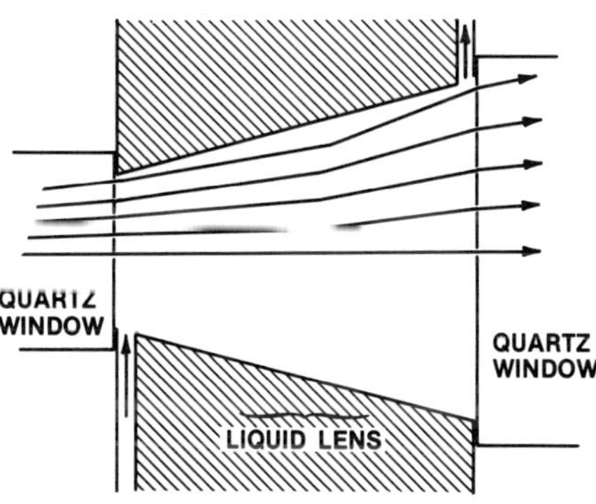

FIG. 17. Comparison of tapered and conventional uv- to visible-absorption cell design [57]. (Reprinted with permission of copyright owner.)

334, and 365 nm, while a medium pressure mercury vapor lamp is normally operated at
254 nm. Detector sensitivities are on the order of 0.005 absorbance units full scale.
Variable-wavelength detectors normally utilize a grating monachromator with a deuter-
ium light source. The advantages of the variable-wavelength detector are that the
wavelength can be adjusted for particular sample components providing optimum sensitivity and selectivity. Most variable-wavelength detectors utilize photomultipliers.
Because of the higher light throughput, most fixed-wavelength detectors are able to
use solid-state sensors. Some of the newest variable-wavelength detectors also use
solid-state detection. If the photometer has a sufficiently large light throughput,
solid-state detectors have more stable dark currents and lower noise characteristics
than photomultipliers.

Although little work has been done, rapid scanning spectrophotometers can be used
with a conventional HPLC flow cell. Rapid scanning permits the entire spectrum to be
continuously monitored. Detector types include vidicons, linear diode arrays, and
rapid scanning grating monachromators. Rapid scanners are useful for HPLC studies
when overlapping chromatographic peaks are a problem. Optimum utilization of rapid
scanners requires the use of computerized data acquisiton.

A Tektronix vidicon system connected to a Perkin Elmer flow cell was demonstrated
at the 1976 Pittsburgh Conference. A system using a Reticon linear diode array has
been described by Dessy et al. [58]. Use of a Harrick rapid-scan grating spectrophotometer as an HPLC detector was reported by Heineman [59]. Figure 18 illustrates
results that can be obtained when overlapping bands with different λ_{max} values are
present in the column effluent.

b. Fluorescence Detectors

Detectors having high sensitivity and selectivity for fluorescent compounds have been
developed. HPLC fluorescence detectors are designed similarly to normal fluorimeters.
Most fluorescence detectors are filter instruments although recently variable-
wavelength grating fluorimeters have been developed specifically for HPLC. The light
source is usually a xenon arc or mercury vapor lamp. At concentration ranges where
the product of sample absorptivity a, cell path length b, and sample concentration C,
is less than 0.01, fluorescence is linear with concentration. Fluorescence obeys the
equation

$$F = I_0 \phi abC \tag{42}$$

where I_0 is the incident light intensity and ϕ is the fluorescence yield. Fluorescence
measurements are succeptible to interferences such as background fluorescence and
quenching.

Flow cells have been designed which can be used for both fluorescence and absorption measurements [60]. An example of such a flow cell is shown in Fig. 19. A

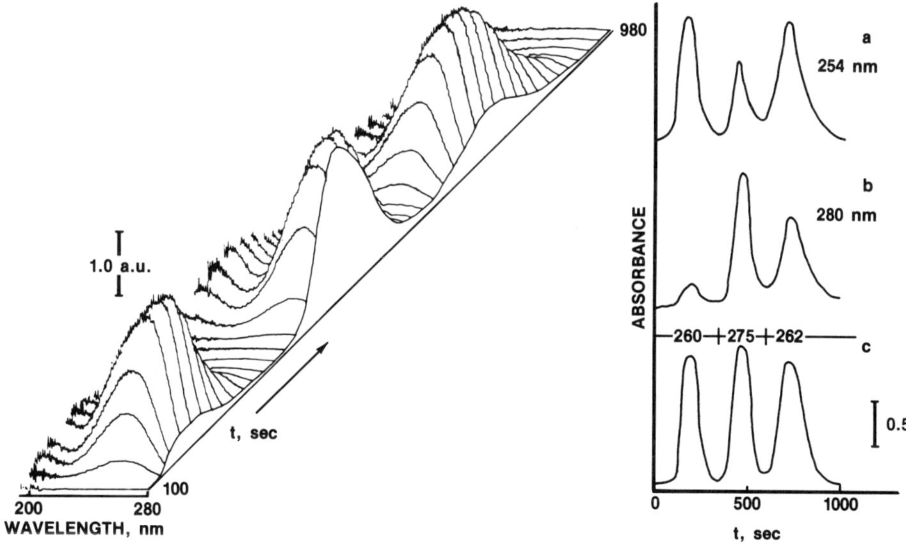

FIG. 18. Chromatograms for separation of uracil, cytosine, and adenine at different monitoring wavelengths [59]. (Reprinted with permission of copyright owner.)

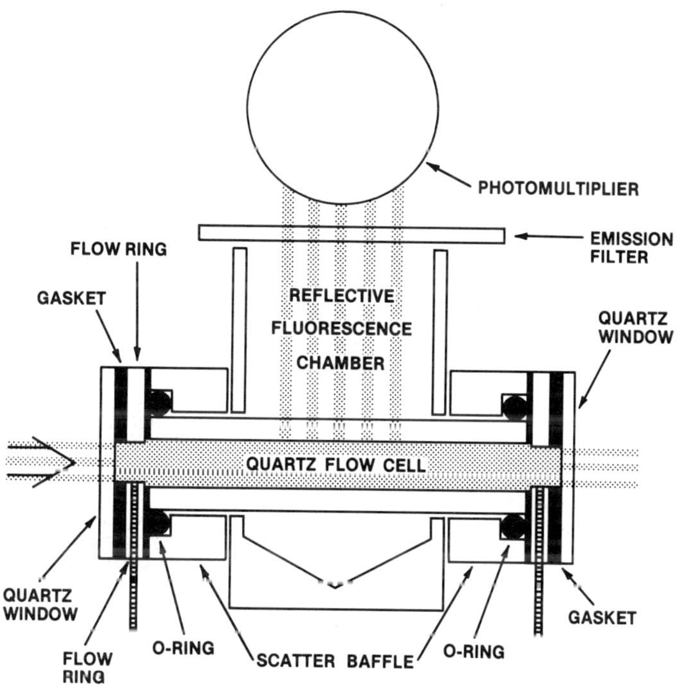

FIG. 19. Fluorescence flow cell. (Courtesy of DuPont Instrument Corporation.)

description of the instrument design permitting concurrent fluorescence and absorption measurements is shown in Fig. 20. Concurrent measurements extend the linear dynamic range for fluorescent samples. At high sample concentrations, where the product of abC is greater than 0.01, fluorescence response becomes nonlinear. At these higher concentrations, however, the light absorbance of the sample is often measureable and linear with concentration. A crossover concentration region exists where both techniques can be utilized.

c. Refractive Index Detectors

The second most often utilized detector for HPLC is the refractive index detector. There are two types of refractive index detectors in common use, the reflection or Fresnel type and the deflection type.

Figure 21 illustrates the reflection-type detector. Fresnel's law of reflection states that the light reflected at an interface is proportional to the angle of incidence of light hitting the interface and the difference in refractive index across the interface. The angle of incident light is adjusted for maximum sensitivity. The optimum angle is slightly smaller than the critical angle for reflection. The detector response measures the difference in light reflected from a reference solution and the sample solution. This difference is proportional to the difference in refractive index of the two solutions. The response can be written

$$R = \Delta RI \times C \tag{43}$$

R is the detector response, ΔRI is the difference in refractive index between the

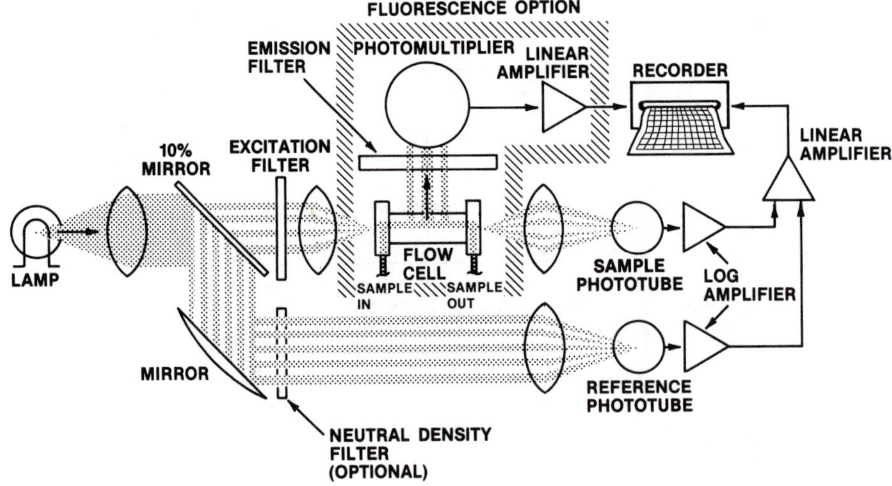

FIG. 20. Instrument for combined absorbance-fluorescence measurements. (Courtesy of DuPont Instrument Corporation.)

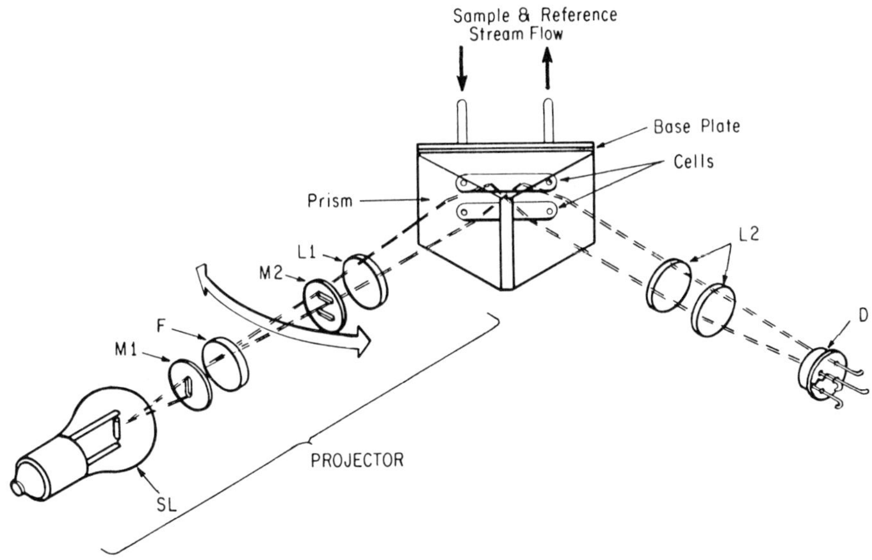

FIG. 21. Fresnel-type differential refractometer. (Courtesy of DuPont Instrument Corporation.)

solvent and sample and C is the concentration of the sample. A low range prism is used to cover a refractive index range from 1.31 to 1.45. A high range prism is used to cover the refractive index range from 1.40 to 1.55.

The operating principle of the deflection-type refractive index detector is illustrated in Fig. 22, where ϕ is the angle of deflection between the incident light and the refracted light. If the difference in refractive index across the interface (Δn) is small, the deflection angle obeys the equation

$$\phi = \tan B\frac{\Delta n}{n} \tag{44}$$

where B is the angle of incident light. Since the light strikes a mirror and passes back through the detector a second time the measured angle of deflection with respect to the incident light is 2ϕ. The positional deflection Δx on a plane located Q distance from the interface is given by

$$\Delta x = 2 \tan B\frac{\Delta n}{n}Q \tag{45}$$

where Q is the optical lever length. If the interface separating the sample and reference solutions is set at a 45° angle with respect to the incident light, tan B = 1 and

$$\Delta x = 2Q\frac{\Delta n}{n} \tag{46}$$

The deflection is directly proportional to the change in refractive index of the

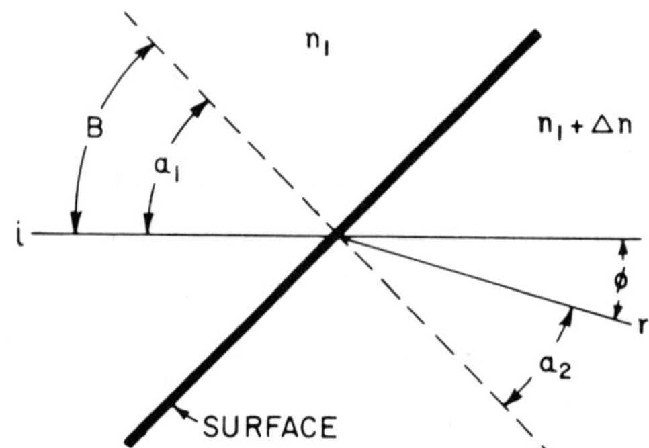

FIG. 22. Principles of deflection refractometry. (Courtesy of Waters Instrument Corporation.)

sample which is proportional to the sample concentration. A deflection-type detector is shown in Fig. 23.

Both types of detectors are capable of detecting refractive index changes of about 5×10^{-8} to 10^{-7} units. The deflection-type detector has the advantage of a smaller cell volume, typically 3 to 5 µl. depending on the thickness of the gasket clamped between the prism and the backing plate. The deflection cell is not as sensitive to contaminant buildup on the cell surfaces. Also, the entire refractive index range can be covered without the need to change prisms. A typical deflection-type cell has a volume of about 10 µl.

The refractive index of a substance changes if its density changes. Since density is dependent on temperature and pressure, both must be maintained as constant as possible.

The linear dynamic range of a refractive index detector is about 10^4. Since large concentrations can be conveniently measured, the detector is quite suitable for preparative as well as analytical chromatography.

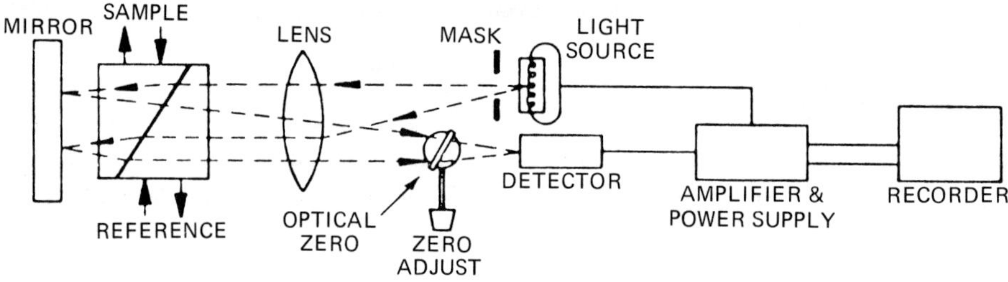

FIG. 23. Deflection-type refractive index detector. (Courtesy of Waters Instrument Corporation.)

Refractive index detectors are very useful for samples that do not absorb light in the ultraviolet or visible range. They also can be used with solvents which would absorb light and interfere with light absorbance measurements, however they cannot be used in gradient elution systems.

d. Electrochemical Detectors

A large variety of biologically active compounds are electroactive and can be oxidized or reduced at electrode surfaces. The Faradaic approaches that seem most successful are constant potential amperometry and coulometry. The coulometric detector offers the largest detector response for a given sample concentration. If the current efficiency for the sample electrolysis is essentially 100 percent, no calibration procedure or reference standard is required. Faraday's law can be used to calculate the results.

In amperometry the potential is applied in the diffusion-limited current region where the current is directly proportional to sample concentration. Only a small fraction of the sample is electrolyzed in the amperometric measurement. Amperometry can be considered a nondestructive method since the bulk solution concentration changes only slightly. The average flux density to the electrode surface is higher for an amperometric measurement than for a coulometric measurement, where the concentration is depleted as the sample moves by the electrode. Consequently the signal-to-noise ratio and the analytical sensitivity should be higher for the amperometric measurement than for the coulometric measurement.

If the surface area of the electrode is constant, charging current rapidly diminishes after the constant potential is applied. The background current becomes small and constant permitting the measurement of very small samples.

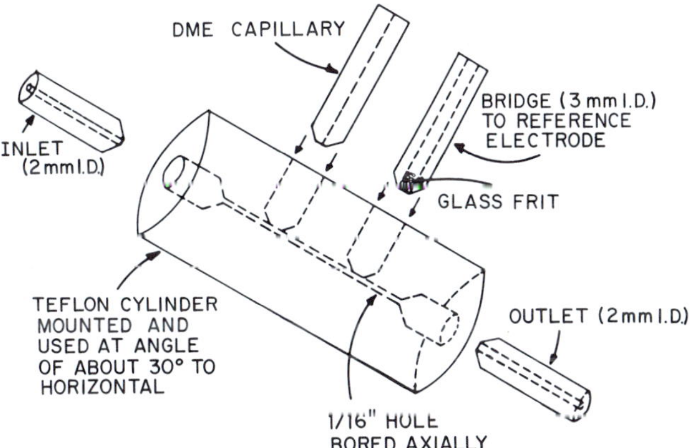

FIG. 24. Dropping-mercury electrode for flowing solution [73]. (Reprinted with permission of copyright owner.)

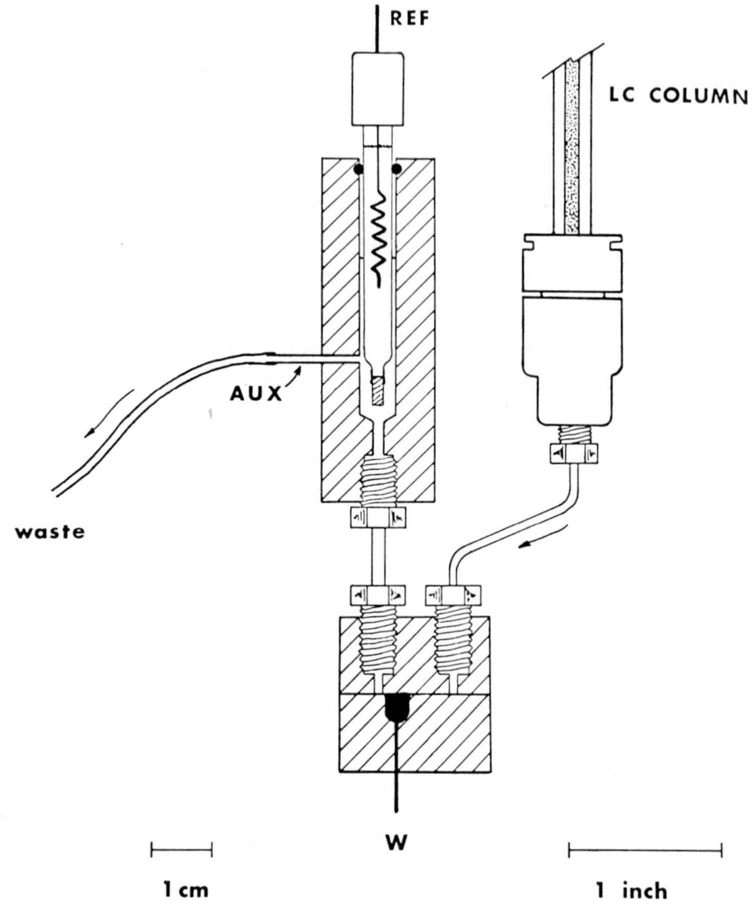

FIG. 25.   Carbon-paste electrode assembly for HPLC.   (Courtesy of Bioanalytical Systems Inc.)

One example of a coulometric detector is that described by Takato and Muto [61]. The working electrode was made from carbon cloth, platinum gauze, or silver wire netting.   Sensitivity as low as 5×10^{-10} mole was reported.   Another type of coulometric detector was reported by Johnson and Larochelle [62].   This detector consisted of a platinum tube packed with small platinum chips.

Amperometric detectors have been made utilizing both dropping-mercury and solid-working electrodes.   An example of a dropping-mercury detector is illustrated in Fig. 24.   The sensitivity of the dropping-mercury electrode is limited by charging current.   Buchta and Papa have reported the use of a mercury-coated platinum wire for amperometric reductions [63].   They also compared amperometric and coulometric techniques.   A wide variety of electrode shapes have been used as amperometric flow stream detectors. In addition to this electrode which was a thin wire intersecting the stream, electrode shapes include short tubes [64], a flat disk referred to by Fleet as a wall-jet electrode [65], and carbon-paste electrodes. The most sensitive electrochemical detector reported to date is the carbon-paste electrode [66]. An example of a carbon-paste chromatographic detector is illustrated in Fig. 25. The detector is very easy and inexpensive to build. Multiple electrodes can be positioned in series. A recent example of a dual detector was reported by Blank [67]. By applying different potentials to the two electrodes, compounds with different redox potentials can be simultaneously measured, even though the chromatographic peaks are not separated. A linear dynamic range of 10^5 and a sensitivity of 6×10^{-13} mole for catecholamine-type compounds was reported.

In general, the following equation holds for a convective diffusion limited current:

$$i_d = KnD^{2/3}C \tag{47}$$

K is an aggregate of constants including geometry and flow velocity terms, n is the number of electrons transferred per molecule in the electrolysis, D is the diffusion coefficient, and C is the bulk solution concentration. Since the current is very sensitive to flow rate, a stable pumping system must be used. High-frequency noise or pump pulsations can be damped out, but low-frequency oscillation of flow rate is difficult to handle. Most investigators have used highly damped small-displacement reciprocating piston pumps for electrochemical measurements since they provide steady positive-displacement flow.

One of the significant advantages of electrochemical detectors is the freedom to choose different electrode geometries. There is no need to worry about cell windows and cell path length. As a result amperometric detectors can be constructed with a cell holdup volume of 1 μl or less. Such small cell volumes permit optimum chromatographic resolution to be achieved. In contrast to spectrophotometric measurements where molar absorptivities vary greatly, amperometric response for most compounds are quite similar. The use of electrochemical detectors for many compounds of therapeutic interest is rapidly increasing

c. Solute Transport Detectors

The solute transport detector provides a mechanism where the sample can be transported from the liquid moving phase to a high-sensitivity GLC-type flame ionization detector. The sample must be volatile enough to enter into the detector, yet not be so volatile as to be lost when the liquid solvent is stripped off. The most common detector of this type is the wire transport detector illustrated in Fig. 26. A flame ionization detector is used to measure the sample. The sample is placed on a freshly cleaned wire which passes through an oven where the sample is oxidized to CO_2. The CO_2 is

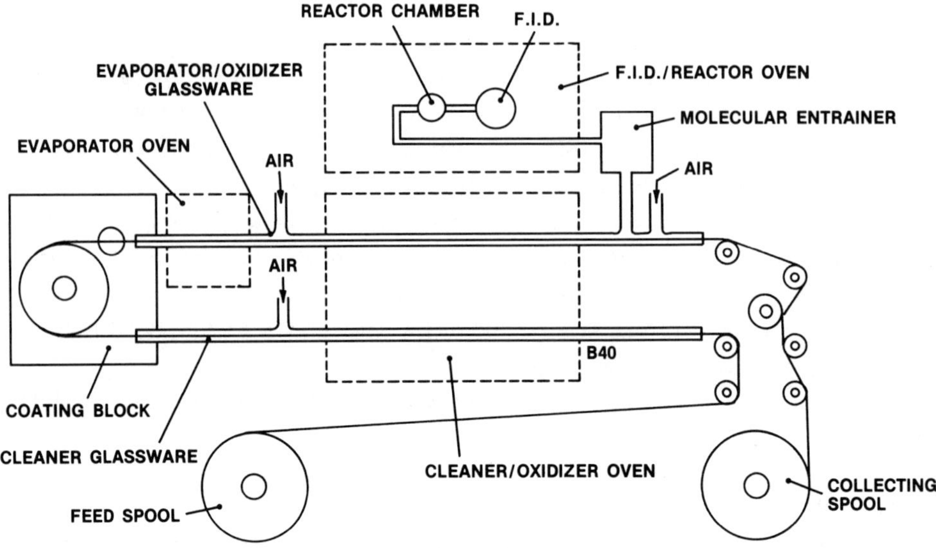

FIG. 26. Moving-wire transport detector for HPLC. (Courtesy of Pye-Unicam.)

then reduced to CH_4 which is measured in the flame ionization detector. A more recent flame ionization system uses a rotating porous alumina disk as the sample transport system [68]. Other ionization detectors such as the electron capture detector have also been used for HPLC [69].

A variety of other types of detectors can be used for chromatography. One of the new detectors that may become utilized in the future is the spray impact detector developed by Mowery and Juvet [70].

IV. QUANTITATIVE ANALYSIS

The objective of most assays is to determine the quantitative amount of the various components present in a sample. The chromatograph detector must put out a signal that is linearly related to the mass or concentration of the sample passing through the detector. Hopefully, the detector has a fast enough response time that the output signal is an accurate indication of the elutant composition at any instant in time. If this is true the integral of the detector output with time, peak area, represents the quantity of sample that passed through the detector. If the detector response is proportional to sample mass, the peak area will be proportional to the total mass or weight of the sample. If the detector response is proportional to concentration, the peak area will represent the concentration time integral but the volume flow rate must be known before the sample mass can be calculated. Alternatively, if the flow rate is reproducible, the results can be compared with results obtained for a reference sample run under the same conditions.

Since it is often difficult to precisely duplicate experimental conditions, it is convenient to add a reference compound or internal standard to the sample so that the sample component quantities can be calculated by direct comparison with the internal standard. Since the ratio of detector response to sample weight varies from compound to compound, a proportionality constant or response factor must be calculated for each component in the mixture if quantitative results are to be calculated by comparing the peak areas. The weight of sample component is calculated by multiplying the ratio of sample to reference peak areas by the reciprocal of their response ratios and by the weight of the reference sample. Concentrations can also be calculated by using a calibration plot prepared using known weights of pure sample. The calibration in some instances is made using a standard addition method, where a known amount of pure compound is added to the sample and the ratio of the original response to the added response is used to calculate the results.

In many instances the fractional or percent composition of a mixture is desired. The fractional amount of a sample component is found by dividing the peak area response factor ratio for the desired component by the sum of the peak area response factor ratios for all the peaks in the mixture. The weight response ratios or response ratios relative to a chosen peak in the sample must be known for all the peaks in the mixture

When sample recovery is uncertain, an internal standard method is usually used. The compound chosen as the internal standard should behave as similarly as possible to the sample. A series of samples containing varying weights of the desired sample and a constant known weight of the internal standard are prepared. The ratio of sample peak area to internal standard peak area is plotted versus sample quantity to prepare a calibration curve. Accurate quantitative results can be obtained as long as the sample and internal standard behave the same.

All the measurements previously discussed have referred to peak area measurements. Accurate and convenient peak area measurements can be made using integrating recorders, digital integrators, or computerized data acquisition equipment. Most modern laboratories have access to this equipment. Peak height measurements are much less reliable and are warranted mainly for very small trace approximations or when peaks are overlapping and deconvolution of their areas is subject to uncertainties. Triangulation procedures offer a simple form of area approximation but are not reliable substitutes for integrated area measurements. An excellent review has been written by Novak for those who are interested in a detailed discussion of quantitative methods [71].

REFERENCES

1. A. B. Littlewood, *Gas Chromatography*, 2nd Ed., Academic Press, New York, 1970, p. 29.

2. E. Grushka, L. R. Snyder, and J. H. Knox, *J. Chromatogr. Sci., 13,* 25(1975).

3. J. C. Giddings, *Dynamics of Chromatography*, Marcel Dekker, Inc., New York, 1965.

4. J. C. Giddings, *J. Chromatogr., 5,* 46(1961).

5. J. C. Giddings, *Anal. Chem, 35,* 439(1963).

6. J. C. Giddings, *Anal. Chem, 35,* 1186(1963).

7. L. R. Snyder, *J. Chromatogr. Sci., 7,* 352(1969).

8. I. Halasz and M. Naefe, *Anal. Chem., 44,* 76(1972).

9. R. E. Majors, *J. Chromatogr. Sci., 11,* 88(1973).

10. M. Martin, C. Eon, and G. Guiochon, *J. Chromatogr., 108,* 229(1975).

11. L. R. Snyder, *J. Chromatogr. Sci., 10,* 200(1972).

12. L. R. Snyder, *J. Chromatogr. Sci., 10,* 369(1972).

13. R. A. Henry, S. H. Byrne, and D. R. Hudson, *J. Chromatogr. Sci., 12,* 197(1974).

14. D. C. Collins and G. F. Freeguard, *Chromatographia, 6,* 404(1973).

15. M. Martin, C. Eon, and G. Guiochon, *J. Chromatogr., 108,* 229(1975).

16. G. Blu, C. Eon, and G. Guiochon, *J. Chromatogr. Sci., 12,* 438(1974).

17. S. Morgan and S. N. Demming, *J. Chromatogr., 112,* 267(1975).

18. J. C. Sternberg, in *Advances in Chromatography* (J. C. Giddings and R. A. Keller, eds.), Vol. 2, Marcel Dekker, New York, 1966, pp. 205-269.

19. M. Martin, G. Blu, C. Eon, and G. Guiochon, *J. Chromatogr., 112,* 399(1975).

20. I. Halasz, *Anal. Chem., 36,* 1428(1964).

21. A. B. Littlewood, *Nature, 184,* 1631(1959).

22. A. E. Lauron, Jr. and J. M. Miller, *J. Gas Chromatogr., 4,* 273(1966).

23. J. E. Lovelock, *Anal. Chem., 33,* 161(1961).

24. H. F. Calcote, *Combustion and Flame, 1,* 385(1957).

25. J. C. Sternberg, W. Galloway, and D. T. L. Jones, preprint of *1961 International Gas Chromatography Symposium*, Lansing, Michigan, pp. 159-184.

26. J. A. Green and T. M. Sugden, *IX Symposium (International) on Combustion*, Academic Press, New York, 1963, p. 607.

27. A. T. Blades, *J. Chromatogr. Sci., 14,* 45(1976).

28. D. A. Thombs, *Chromatographia, 6,* 111(1973).

29. R. C. Condon, P. R. Scholly, and W. Averill, in *Gas Chromatography 1960, Proceedings of the Third International Symposium* (R. P. W. Scott, ed.), Butterworths, Washington, D. C. *1960 p. 30.*

30. L. Onkiehong, in *Gas Chromatography 1960, Proceedings of the Third International Symposium* (R. P. W. Scott, ed.), Butterworths, Washington, D. C., 1960, p. 7.

31. H. Bruderreck, W. Schneider, and I. Halasz, *Anal. Chem., 36,* 461(1964).

32. J. M. Gill and C. H. Hartmann, *J. Gas Chromatogr., 5,* 605(1967).

33. I. G. McWilliam, *J. Chromatogr., 6,* 110(1961).

34. R. A. Dewar, *J. Chromatogr., 6*, 312(1961).

35. S. S. Brody and J. E. Chaney, *J. Gas Chromatogr., 4*, 42(1966).

36. H. W. Grice, M. L. Yates, and D. J. David, *J. Chromatogr. Sci., 8*, 90(1970).

37. M. L. Selucky, *Chromatographia, 4*, 425(1971).

38. R. W. B. Pearse and A. G. Gaydon, *The Identification of Molecular Spectra*, 3rd. Ed., John Wiley and Sons, New York, 1963, pp. 263 and 359.

39. A. Karman and L. Guiffrida, *Nature, 201*, 1204(1964).

40. C. Wells, *USFDA Pesticide Workshop*, Kansas City, Mo., May 1966.

41. M. Scolnick, *J. Chromatogr. Sci., 8*, 462(1970).

42. B. Kolb and J. Bischoff, *J. Chromatogr. Sci., 12*, 625(1974).

43. C. A. Burgett, D. H. Smith, H. B. Bente, J. C. Wirfel, and S. E. Goodhart, Hewlett-Packard application note ANGC2-76 (1976).

44. J. E. Lovelock and S. R. Lipsky, *J. Amer. Chem. Soc., 82*, 431(1960).

45. W. E. Wentworth, E. Chen, and J. E. Lovelock, *J. Phys. Chem., 70*, 445(1966).

46. R. J. Maggs, J. L. Joynes, A. J. Davis, and J. E. Lovelock, *Anal. Chem., 43*, 1966(1971).

47. C. Arnold Hartman, *Anal. Chem., 45*, 733(1973).

48. D. J. Dwight, E. A. Lorch, and J. E. Lovelock, *J. Chromatogr., 116*, 156(1976).

49. C. F. Poole, *J. Chromatogr., 118*, 280(1976).

50. S. Kapila and W. Aue, *J. Chromatogr., 118*, 233(1976).

51. R. Pigliucci, W. Averill, J. E. Purcell, and L. S. Ettre, *Chromatographia, 8*, 165(1975).

52. J. E. Lovelock, *Nature, 188*, 401(1960).

53. R. R. Freeman and W. E. Wentworth, *Anal. Chem., 43*, 1987(1971).

54. J. Sevcik and S. Krysl, *Chromatographia, 6*, 375(1973).

55. N. Ostojic and Z. Sternberg, *Chromatographia, 7*, 3(1974).

56. J. N. Driscoll and F. F. Spaziani, *Res. and Dev.*, May, 70(1976).

57. J. N. Little and G. J. Fallich, *J. Chromatogr., 112*, 389(1975).

58. R. E. Dessy, W. G. Nunn, and C. A. Titus, *J. Chromatogr. Sci., 14*, 195(1976).

59. M. S. Denton, T. P. De Angelis, A. M. Yacynych, W. R. Heineman, and T. W. Gilbert, *Anal. Chem., 48*, 20(1976).

60. J. C. Steichen, *J. Chromatogr., 104*, 39(1975).

61. Y. Takato and G. Muto, *Anal. Chem., 45*, 1864(1973).

62. D. C. Johnson and J. Larochelle, *Talanta, 20*, 959(1973).

63. R. C. Buchta and L. J. Papa, *J. Chromatogr. Sci., 14*, 16(1976).

64. W. D. Mason and C. L. Olson, *Anal Chem., 42*, 548(1970).

65. B. Fleet and C. J. Little, *J. Chromatogr. Sci., 12*, 747(1974).

66. P. T. Kissinger, C. Refshauge, R. Dreiling, and R. N. Adams, *Anal. Lett., 6*, 465(1973).

67. C. Leroy Blank, *J. Chromatogr., 117*, 35(1976).

68. J. J. Szakasits and R. E. Robinson, *Anal. Chem., 46*, 1648(1974).

69. F. W. Willmott and R. J. Dolphin, *J. Chromatogr. Sci., 12*, 695(1974).

70. R. A. Mowery, Jr. and R. S. Juvet, Jr., *J. Chromatogr. Sci.*, *12*, 687(1974).

71. J. Novak, in *Advances in Chromatography* (J. C. Giddings and R. A. Keller, eds.), Vol. 2, Marcel Dekker, New York, 1974, pp. 1-71.

72. L. R. Snyder and J. J. Kirkland, *Introduction to Modern Liquid Chromatography*, John Wiley and Sons, New York, 1974, p. 37.

73. W. J. Blaedel and J. Strohl, *Anal. Chem.*, *36*, 445(1964).

Chapter 2

COLUMN SELECTION IN GLC

Jordan L. Cohen

School of Pharmacy
University of Southern California
Los Angeles, California

I. INTRODUCTION

The success of any chromatographic separation depends in large part upon the chromatographer's selection of an appropriate stationary phase which will resolve the individuals components in a mixture. How critical column selection becomes depends primarily on the nature of the mixture and the effectiveness of any preliminary

separation and isolation procedures prior to chromatography. Of the two general types
of problems facing analysts of therapeutic agents, namely drug dosage forms and analysis of drugs and/or metabolites in biological mixtures, the latter generally poses the
greater chromatographic challenge in terms of column selection. In addition to increased sensitivity requirements, the presence of structurally related metabolites and
a variety of unknown extraneous components requires a careful and sophisticated approach to column selection.

While the primary objective in column selection is to effect symmetrical, reproducible peaks and baseline separation of all components of a mixture, there are a
variety of other considerations that the chromatographer must take into account in
individual circumstances. Among these are accuracy and sensitivity requirements,
heat stability of individual components, turn-around-time requirements for the entire
procedure and relative convenience and cost of the procedure with regard to its application. For example, different chromatographic conditions are often established for
a drug analysis problem depending upon whether it is to become a routine clinical, or
a research procedure. Another consideration may be the availability and sophistication of instrumentation. The presence of an on-line integrator, for example, may
allow a less-than-baseline separation to be utilized for quantitative analysis and
obviate the need for a column providing greater resolution of the two compoments.

Generally most column selection decisions for drug-related problems are done on
an empirical or semiquantitative approach based in part on individual experience and
preference, as well as on available literature chromatographic data. The basic reason
for this confusion and lack of a standardized approach to column selection, is the
large number of stationary phases which are commercially available, and currently in
use. There are over 700 liquid phases cataloged in the recent issue of *Gas
Chromatographic Data Compilation DS, 25 A*, many of which are nearly identical in
terms of actual structure or relative retention data [1]. Although the number of
substrates proliferated in the past 20 years, in part due to a need for a much wider
temperature range, there seems to be little doubt that all chromatographic separations
could be accomplished with a greatly reduced number of "preferred" liquid phases
representative of a wide range of polarities and temperature limits. There is a
significant movement among chromatographers to establish such a preferred list as is
evidenced by recent literature [2-4]. Standardization for column selection would
result if all compounds had retention data reported only on one or more of these
preferred phases.

Presently, absolute retention times, relative retention times, and retention
indexes are all reported and cross-referencing or comparisons are difficult. For
example, it is possible to find compilations of retention times for frequently analyzed drugs such as anticonvulsants [5] or steroids [6] on several different columns,

however, the data are not useful for predicting relative behavior on other specific columns or for predicting separability of other interfering drug substances.

Because the problem of column selection can be extremely time consuming and costly if approached on an empirical basis, it is important for an analyst to utilize some basic concepts of separation theory, column technology, and an understanding of the relationship between molecular structure and relative retention on columns of varying polarity. The overall objectives of this chapter are to outline these concepts, develop and define systems for determining and classifying retention data for stationary phases and solutes; and finally, discuss a practical approach for the selection of columns for analytical problems involving gas chromatographic methods development for therapeutic agents.

II. SEPARATION THEORY

While it is well beyond the scope of this chapter to discuss rigorous theoretical details of the chromatographic process, a brief discussion of terminology and principles of separation is essential to establish a foundation for variables and problems involved in column selection for complex mixtures. For a more detailed theoretical treatment, the reader is referred to a number of basic gas chromatography sources [7, 8].

A solute's retention time in gas chromatography is a complex function of the chemical properties of the solute, the stationary phase (solvent, substrate), and several operating variables of the column. Littlewood [7], separates these conveniently into two classes:
where

    1. Column variables
        (a) Dead volume
        (b) Pressure drop across the column
        (c) Temperature of the carrier gas as it affects flow rate
        (d) Weight of the stationary phase
    2. Thermodynamic variables
        (a) Chemical nature of the solute (vapor)
        (b) Chemical nature of the stationary phase
        (c) Temperature of the column as it affects equilibrium of the vapor
        (d) Chemical nature of the carrier gas (minor)

The column variables affect retention time as a function of how the column is operated and made. In a given chromatographic run, the column variables are essentially the same for all components of the mixture. Provided that a reasonable concentration of stationary phase is used and the column is properly packed and conditioned, separation will depend solely on the thermodynamic variables for each component. If

these are significantly different, a separation or preferential migration of one
solute relative to another will occur; if they are not, adequate separation will not
be achieved and another stationary phase may be required.

The thermodynamic variables for a given compound on a specific stationary phase,
can be characterized as a partition coefficient which, in practice is essentially
independent of column variables.

This partition coefficient β is defined by

$$\beta = \frac{\text{Weight of solute per g of stationary phase}}{\text{Weight of solute per ml of gas at column temperature}}$$

Since solutes migrate in a chromatographic system (only while they are in the vapor
phase), the relationship between vapor pressure of a solute to the concentration of
the solute in a stationary phase (solvent) is defined by Henry's law:

$$p = x\gamma p^{o}$$

where p is the vapor pressure of solute over the solution, p^{o} is the vapor pressure
of the pure solute, x is the mole fraction of solute in solution and γ is the activity
coefficient for real solutes. Since at very low mole fractions, the activity coefficient is constant, this rearranges to

$$\frac{x}{p} = \frac{1}{\gamma p^{o}} = \text{Constant}$$

The quantity x/p is a partition coefficient, although not as convenient as β. The
important point is that this partition coefficient is a function not only of vapor
pressure but also of the activity coefficient of the solute in the stationary phase.
This relationship is the basis of the separation of closely related compounds by the
selection of appropriate stationary phases. Compounds with identical vapor pressure
can be separated if there are differences in their activity coefficients in the
stationary phase.

The retention volume $V_{R}{}^{o}$ for a solute in a chromatographic system is the total
volume of gas required to move the solute completely through the column and is
defined by

$$V_{R}{}^{o} = 1(a + m\beta)$$

where

    1 = column length

    a = volume of gas per unit length of column

    m = mass of stationary phase per unit length of column

The product of a·1 is the total gas holdup or dead volume of the column and m
is the total weight of the stationary phase. If these are symbolized V_{M} and W respectively, the equation for retention volume becomes

$$V_R^{\,o} = V_M^{\,o} + W\beta$$

Since $\beta > 0$ for substances which are retained by the column, $V_R^{\,o}$ is greater than $V_M^{\,o}$ and generally ranges from 3 to 200 times the dead volume with packed columns.

The observed retention time (t_R) can be related to the retention volume by

$$V_R^{\,o} = \dot{V} t_R$$

where $\dot{V}$ is the volume flow rate of carrier gas in appropriate units (such as ml/min). Similarly the mobile phase volume $V_M^{\,o}$ is related to t_M, the retention time for a solute that is not retained at all by $V_M^{\,o} = \dot{V} t_M$.

In gas chromatography, the flow of the gas through the column is impeded by the packing material and because of the viscosity of the gas, a pressure gradient develops along the column. Since gasses are compressible, a velocity gradient also develops and the gas flow varies along the column. Ordinarily the flow measured at the end of the column is greater than at any other point, so the calculated retention volume is greater than that which actually occurs, and must be corrected by a compressibility factor j given by

$$j = \frac{3}{2} \frac{(P_i/P_o)^2 - 1}{(P_i/P_o)^3 - 1}$$

where P_i and P_o are the inlet and outlet pressures, respectively. Applying this correction,

$$t_R \dot{V} = \frac{o(a \mathrel{\text{\tiny I}} m\beta)}{j}$$

or

$$t_R \dot{V} = \frac{V_R^{o}}{j}$$

therefore,

$$V_R = \frac{V_R^{o}}{j}$$

where V_R is the *observed* or *experimentally determined retention volume* and V_R^{o} becomes the *corrected retention volume*. Column variables can be separated from the expression of retention volume by subtracting V_M from V_R with this difference denoted as V_R', the *adjusted retention volume*. If the compressibility correction is applied to V_R' rather than V_R, the product of $j \times V_R'$ equals V_N, the *net retention volume*. If the net retention volume is divided by the total weight of packing material W, and corrected for differences between measurement (T_f) and column (T_o) temperatures, a derived, column independent unit V_g the *specific retention volume* can be defined as [9]

$$V_g = \frac{T_f V_N}{T_c W}$$

The following example is used to illustrate the experimental determination of V_g. The measured retention time for a compound was 14.0 min while the retention time for an air peak under the same conditions was 1.1 min. The column contained 1.6 g of stationary phase and was maintained at 170°. Carrier-gas flow was 30.2 ml/min using a flowmeter calibrated at 0° C and j, the compressibility correction factor was 0.7.

$$V_R^\cdot = 30.2(14.0) = 422.8 \text{ ml}$$

$$V_R^\prime = (14.0 - 1.1)30.2 = 389.6 \text{ ml}$$

$$V_N = 0.7(389.6) = 233.7 \text{ ml}$$

$$V_g = \frac{273(233.7)}{443\ (1.6)} = 90.01 \text{ ml/g}$$

Substituting the expression for V_N and calibrating the flowmeter at 0°C allows V_g to be redefined as

$$V_g = \frac{273.16}{T_c}\beta$$

which allows a simple calculation of either retention volume, independent of column parameters, or β in a given chromatographic system, if one or the other is known.

The process of distribution of solutes in gas chromatography between the mobile and stationary phases can be described by Henry's law under usual conditions where *ideal* behavior is operable. Namely,

$$p_2 = \gamma x_2 p_2^{\ o}$$

where p_2 is the equilibrium vapor pressure of the solute, x_2 is the mole fraction of solute, $p_2^{\ o}$ is the vapor pressure of the pure liquid solute, and γ is an activity coefficient describing deviation from Raoult's law for *perfect solutions*. By combining this expression, the general form of the gas law and the definition for β, the partition coefficient, can be redefined as [9]

$$\beta = \frac{RT}{\gamma M p_2^{\ o}}$$

where M is the molecular weight of the stationary phase; V_g then becomes

$$V_g = \frac{273R}{M \gamma p_2^{\ o}}$$

which expresses the specific retention volume as a function of three variables independent of chromatography. This will hold as long as γ is constant and can be approximated by the activity coefficient at infinite dilution. Molecular weights of most commonly used stationary phases are in the 100-1000 range and vapor pressures range from a few millimeters to several thousand millimeters so that values of V_g generally lie in the range of 10 to 1000 ml/g.

Since retention data is easier to obtain than vapor pressures and activity coefficients particularly for new compounds, V_g is generally calculated from chromatographic rather than thermodynamic data. β can then be calculated for specific compounds from the specific retention data. Since β can be determined independent of chromatography, the specific retention volume (defined per gram of column packing) is independent of column operation.

In principle, this volume should characterize the retention time of a given compound under specific chromatographic conditions, and it should allow for the conversion to any other set of chromatographic conditions provided the same column packing is utilized. It should then be possible to optimize a separation of two closely related compounds from literature data alone, provided chromatographic data has been reported. Unfortunately, much of the literature data is presented in volumes other than the specific retention volume since chromatographic conditions are not specified precisely enough or appropriate corrections for compressibility and variation in gas flow have not been made. This severly limits the usefulness of most absolute literature retention data and requires more careful definition and reporting of absolute and relative data.

III. SOLUTE STRUCTURE AND RETENTION

A. General

In principle, if all of the physical-chemical parameters describing the interaction of a solute with the column were known, the choice of a suitable substrate and set of chromatographic conditions to effect a separation of closely related molecules would be straightforward. Based upon this premise the relationship between specific retention volume, partition coefficient, vapor pressure of the solute and molecular weight of the solvent developed by Littlewood [7] was outlined in the previous section. Since it can be shown [10] that V_g is relatively independent of M_1, the molecular weight of the stationary liquid, the specific retention volume is primarily a function of the vapor pressure of the solute and the activity coefficient describing the net interaction between solute and solvent molecules. Generally the latter is unknown and is an extremely complex function to rigorously calculate. Even the vapor pressures of many important drugs which have been chromatographed are not known or available from the literature. In practice then, calculation of the specific retention volume for only simple solutes is possible. Instead an attempt is made to predict retention

behavior on the basis of the strengths of the relative interactions which contribute
to the partition coefficient in a gas chromatographic system. In some cases this can
be related to solute structure and becomes useful for prediction or precalculation of
retention behavior on a specific stationary liquid phase.

The retention of a solute depends upon the vapor pressure (boiling point) of the
solute and the partition coefficient. Under identical conditions, the separation of
closely related compounds will require differential solvent and solute interactions
as migration occurs down the column. Since the types of molecular interactions in
gas chromatography are typical of all other cohesive and adhesive forces, namely
electronic dispersion, dipole-dipole and induced dipole interactions, and hydrogen
bonding, the chemical nature of the solute and solvent play a major role in retention
prediction. Intuitively one would expect that the greater the intermolecular attrac-
tive forces between solute and solvent the greater the retention. This is illustrated
by the retention data for n-hexane on a variety of stationary liquids at 80° C listed
in Table 1. Retention volumes vary by more than 30-fold and are extremely small on
highly polar stationary phases [7, 11].

As in normal solution processes these interactions are additive and the total
intermolecular energy as well as the molar free energy of solution are also additive
functions of the individual functional groups. The significant result of this additivity is the apparent linear free-energy relationship between the logarithm of the
retention and the chemical nature of the solute for a given stationary liquid, namely

$$\log Z = a'n + b$$

where Z is any relative retention parameter. This relationship was first demonstrated
by James and Martin [12], who demonstrated a linear relationship between the logarithm
of V_g of fatty acids and their carbon number. Subsequently similar correlations were
observed for nearly all other homologous series of organic molecules [13]. Figure 1
shows an example for esters of fatty acids [14]. The most notable among these types
of correlations are the n-alkanes which form the basis for Kovat's retention index
system outlined in the following discussion.

It is generally accepted that this relationship holds for any homologous or
pseudo-homologous series. In addition, it has also been shown for fairly large repeating structural units including phenyl rings [13]. In terms of more complex drug
molecules some excellent work by Lipsky and Landowne [6] demonstrated a similar relationship for C_{19}, C_{21}, and C_{27} steroids using several different stationary phases.
Generally, only lower members of a series deviate markedly and these types of correlations are widely used in predicting retention values and in qualitative identification
of substances using two or three stationary phases.

TABLE 1. Specific Retention Volume for *n*-Hexane in Various Stationary Liquids [7]

Stationary liquid[a]	V_g (ml/g)[b]
Squalane	63.9
Hexadecanol	46.5
Dioctylsebacate	48.3
Dodecanol	44.0
Dinonyl phthalate	36.9
Benzyl diphenyl	18.7
Tricresyl phosphate	16.9
Polyethylene glycol 400	5.1
1,2,3-Tris(cyanethoxy)propane	1.6

[a]In increasing order of polarity.

[b]At 80°.

B. Relative Retention Values

The characterization of a compound's retention relative to a reference substance under the same chromatographic conditions, is generally more useful since this minimizes the dependence of the retention parameter upon column variables as well as minor changes

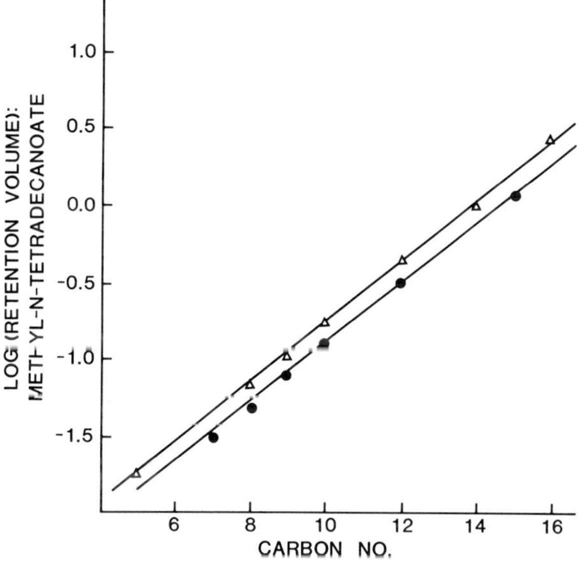

FIG. 1. Relationship between the log of the retention volume, relative to methyl *n*-tetradecanoate for methyl esters of fatty acids, and the number of carbon atoms. (Δ) straight-chain acids and (o) branched-chain acids [14]. (Reprinted with permission of copyright owner.)

in column temperature and gas flow. The relative retention r for two solutes A and B
is defined as [9]

$$r_{AB} = \frac{V_{N,A}}{V_N'{}_B} = \frac{V_{R,A}'}{V_{R,B}'} = \frac{V_{g,A}}{V_{gB}} = \frac{t_{R,A} - t_M}{t_{R,B} - t_M} \tag{1}$$

and still requires a measurement of gas holdup volume, V_M or t_M. When V_A and V_B are
both much greater than V_M, this simplifies to the more commonly used expression

$$r_{A,B} = \frac{t_{R,A}}{t_{R,B}} \tag{2}$$

The major difficulty with utilization of relative retention values is the wide variety
of reference substances used as standards in the literature [15] which makes comparisons difficult since conversion factors are generally not known. It has been recommended to use a single reference standard for comparative purposes. In those instances
of low or high temperatures, a secondary reference could then be used provided that
the retention relation between the secondary and primary reference was known. This
was hoped to allow relative retention data to be presented in standardized fashion.
Evans and co-workers [16, 17] and Kaiser [18] have suggested *n*-nonane to be suitable
as a primary reference standard.

C. Retention Indexes

A more sophisticated and systematic approach to the problem of reporting relative
retention data has been developed by Kovats [19]. This system is based upon the
linear relationship between the logarithm of the partition coefficient of an aliphatic hydrocarbon and carbon number for a homologous series. The *retention index* I is
defined for a given substance as follows:

$$I_A = 100\ (Y - X)\ \frac{\log\ (t_{rA}/t_{rX})}{\log\ (t_{ry}/t_{rx})} + 100X \tag{3}$$

where

$\quad t_{rA}$ = retention time of A

$\quad t_{rX}$ = retention time of X (*n*-alkane eluted before A)

$\quad t_{rY}$ = retention time of Y (*n*-alkane eluted after A)

Since the logarithm of the retention time for an alkane is a linear function of
carbon number, retention indexes for alkanes can be defined as 100Z, where Z is the
carbon number, regardless of the stationary phase. The alkanes then form an equally
spaced group of indexes 100 units apart to which all other compounds can be readily
compared.

Generally $t_{rX} < t_{rA} < t_{rY}$ and most retention indices are calculated by interpolation. To illustrate the use of retention indexes, consider the following example:

$$t_{rX} = 1.83 \text{ min} \qquad \text{(retention time for } n\text{-hexane)}$$

$$t_{rY} = 3.34 \text{ min} \qquad \text{(retention time for } n\text{-heptane)}$$

and

$$t_{rA} = 2.12 \text{ min} \qquad \text{(retention time for an unknown compound)}$$

$$I_A = 100(1) \frac{\log (2.12/1.83)}{\log (3.34/1.83)} + 100 \ (6) \tag{4}$$

$$I_A = 100 \left[\frac{0.06}{0.26} + 600 \right] = 623 \tag{5}$$

Although it is more accurate to determine retention indexes by interpolation as in this example, experimental data for polar compounds, such as drugs, on relatively polar substrates generally require extrapolation. While the same expression can be used, reproducibility and accuracy of these determinations requires selection of reference compounds with retention times as close as possible to the substance of interest.

By definition, the retention indexes of the n-alkanes are 100Z and are independent of substrate. In general, the retention indexes of nonpolar substances are less affected by changing from nonpolar to polar stationary phases than are polar substances. Alcohols, for example show a 370 unit increase in retention index when chromatographed on polyethyleneglycol instead of Apiezon L, while napthalene analogs show only an increase of 20 units. The variation of retention index for polar compounds has led to the compilation of retention indexes for drugs and pesticides on two or three standard stationary phases [20]. Unknown substances in biological samples are then chromatographed on the standard phases and retention indexes form the basis for qualitative identification in toxicology. Phenobarbital for example has characteristic retention indexes of 1750, 2030 and 2385 on OV-1, OV-17, and QF-1 stationary phases, respectively.

Retention indexes also have other general properties which are useful to keep in mind when making column decisions.

(a) On nonpolar stationary phases, substances of similar structure (but not necessarily members of a homologous series) with boiling point difference of X° C will differ in their retention indexes by about 5X units.

(b) Retention indexes are much less affected by column and instrument variables, particularly temperature. Typical variation of retention indices is from 2-5 units per 10° change in temperature. The n-alkanes are invariant with temperature.

(c) Retention values may be determined with an accuracy of about 0.5% if they are obtained by interpolation. As mentioned earlier, the accuracy of extrapolated values depends upon the selection of closely eluting reference compounds.

TABLE 2.   Retention Indexes on a Nonpolar Stationary Liquid Phase (Apiezon L) [18]

Substance	Retention index		
	70°	130°	190°
Pentane	500	500	
Heptane	700	700	
Nonane		900	
2-Methylpentane	570	571	
2,3-Dimethylpentane	570	572	
2,2,4-Trimethylpentane	690	694	
Acetone	447	450	
2-Butanone	547	551	
2-Heptanone		846	
2-Nonanone		1047	1055
6-Undecanone		1231	1237
Acetates of			
Methanol	475	468	
Propanol-1	654	648	
Pentanol-1		855	856
Cyclohexane	676	700	
Cyclohexanol		880	898
Benzene		691	
Naphthalene		1263	1292
Toluene		798	816
o-Dichlorobenzene		1076	1117
m-Dichlorobenzene		1058	1097
p-Dichlorobenzene		1060	1096
Acetophenone		1067	1095
Benzyl alcohol		1010	1044

The gas chromatography literature is replete with retention index data for a multitude of organic compounds on a variety of stationary phases. Kaiser [18] has a notably extensive listing of these and a few are illustrated in Tables 2 and 3 to illustrate some useful properties of this type of retention data.

Although similar data on drugs is still scarce, a recent compilation by Moffat [21] details retention indexes for more than 480 therapeutic agents using SE-30 as the standard stationary phase. This data was collected from several laboratories where instruments, temperatures, flows, and chromatographic supports differed markedly. The data from 11 laboratories for amphetamine, diphenhydramine, and

TABLE 3. Retention Indexes on Polar Stationary Liquid Phase (Carbowax 1500) [18]

Substance	Retention index	
	50°	100°
Pentane	500	500
Heptane	700	700
Nonane	900	900
2-Methylpentane	457	463
Ethylene	160	172
Propylene		393
1-Hexene	642	642
Acetone		865
2-Butanone		942
2-Pentanone		1012
2-Nonanone		1189
Ethyl acetate		910
n-Butyl acetate	990	
Diethyl ether		642
Di-*n*-butyl ether		969

dipipanone are reproduced in Table 4 to illustrate the remarkable consistency of this method of reporting retention data.

D. Retention Index and Solute Structure

For relatively simple nonpolar organic molecules it is possible to relate their structure to the corresponding *n*-alkane and make reasonable predictions of retention indexes. Although this type of analogy breaks down for polar molecules and large complex molecules typical of drug structures, a brief discussion of the approach will provide additional insight into the retention index system, which is becoming the accepted standard for reporting all chromatographic retention data. For a more complete treatment, the reader is referred to Littlewood [22].

In this system a molecule is transformed by one or more operations into a "pseudo-alkane." Generally a methyl or methylene group is simply replaced by a different group. For example, diethyl ether can be thought of as being derived from *n*-pentane with the central methylene being replaced by oxygen. If the change causes a characteristic shift in the retention index from the expected 100Z, the difference can be added or subtracted to estimate the retention index for the compound under study. This approach is reasonably useful if one or two operations are required to derive the molecule from an alkane. It becomes unworkable when three or more such operations

TABLE 4. Retention Indexes on SE-30 [21]

Laboratory	Amphetamine	Diphenhydramine	Dipipanone
1	1110	1855	
2			2467
3	1135	1880	2500
4	1170	1870	2480
5	1165	1895	2515
6	1148	1864	2496
7	1140	1870	2492
8	1168	1897	2492
9	1153	1897	2477
10	1170	1892	2502
11	1140	1850	2507
Mean	1150	1877	2492
Standard deviation	19.4	17.8	14.6
Coefficient of variation	1.7	0.95	0.58

are required since the chemical nature of the molecule becomes too altered to permit predictable behavior based upon the linear free energy relationship.

1. Branched-Chain Hydrocarbons (Isomers)

In all series of isomers studied, the straight-chain alkane has the largest partition coefficient regardless of stationary liquid. The retention index, therefore, of a branched alkane is always less than 100Z. These differences are quite constant and rank in the same order as differences in boiling point between branched and straight chain compounds.

2. Polar Substituents

For *n*-alkanes substituted with a single polar functional group, the retention is greatest when the polar group is in the 1 position. In addition, this difference is most marked in polar stationary phases. This behavior is illustrated in Table 5 by a series of alcohols, esters, and ethers. In each series there were much more marked differences in retention index between the 1-substituted and centrally substituted polar groups on the carbowax and ethylene glycol stationary phases which are quite polar.

TABLE 5. Retention Indexes for Polar Substituted Alkanes [23]

Substance	I^a	I^b	I^c	I^d
1-Pentanol	728	1354	878	1293
2-Pentanol	666	1211	793	1176
3-Pentanol	671	1194	798	1164
Hexane	600	600	600	600
n-Hexyl formate	884	1275	997	1339
n-Pentyl acetate	856	1231	964	1260
n-Butyl proprionate	855	1198	957	1227
Octane	800	800	800	800
n-Propylmethyl ether	497	686	549	708
Diethyl ether	473	661	525	680
Pentane	500	500	500	500

[a]Apiezon L (100°).

[b]Carbowax 400 (100°).

[c]Diisodecylphosphate (120°).

[d]Ethylene glycol adipate (120°).

3. Multiple Substituents

As indicated earlier, when more than two changes are made in an alkane structure, the alteration of the resultant retention index becomes difficult to predict and the general rules often break down. Littlewood [22] has abstracted several examples which are illustrated in Tables 6 and 7. If the principle of additivity were to hold, the differences from unit operations should always be nearly the same. In some instances this holds well; for example, for ether substitution on dioctyl sebacate and Apiezon L. In several other instances changes are relatively unpredictable. Examples of the latter include nearly all changes on Carbowax 400. This uncertainty warrants caution in precalculations of retention data for complex molecules and is the main reason that several stationary phase systems are generally run when trying to verify structures of unknown substances using gas chromatography.

E. Retention and Physical-Chemical Properties

Among the large variety of other correlations between retention parameters and physical or chemical properties of solutes, linear relationships have been found with dipole moments for isomeric compounds, densities, refractive indexes, and molar

TABLE 6. Selected Retention Indexes for Multiple-Substituted Alkanes[a]

Substance	Retention index		
	I[b]	I[c]	I[d]
1. 1-Pentene	483	402	553
2. *n*-Propanol	513	662	1132
3. Allyl alcohol	504	672	1203
4. Butylmethyl ether	594	640	782
5. Ethylene glycol dimethyl ether	607	683	999
6. *n*-Butanol	620	759	1243
7. 1,4-Butanediol	900	1170	2100
8. Isopentane	475	475	475
9. 2-Butanone	552	648	891
10. 3-Methyl-2-butanone	619	708	1008
11. Methylal	476	545	788
12. Methoxymethylal	654	745	1094

[a]Data derived from McReynolds [23].

[b]Apiezon L.

[c]Dioctyl sebacate.

[d]Carbowax 400.

refractions of hydrocarbons [13]. The most important nonstructurally specific relationship for precalculation of retention behavior is that of boiling point,

$$\log Z = a' + bT_b$$

This relationship predicts that the order of elution of members of a homologous series is determined by their boiling points on nonpolar stationary liquids. This has been demonstrated to apply to a number of relatively complex series including the phenothiazines [24], and trimethylethers [25] as well as to simpler classes of organic molecules such as the alkanes [26]. Figure 2 depicts a typical relationship. By combining the derived expression for specific retention volume with Troutons rule, Littlewood et al. [28] derived the following relationship:

$$\log V_g \simeq - 0.5 - \log M_1 - \log \gamma + 4.8 \frac{T_B}{T}$$

where all the terms have been previously defined. This approximate relationship holds for many homologous series since γ remains relatively constant in the same stationary liquid. A plot of $\log V_g$ versus boiling points then will give a series of straight lines of approximately similar slopes for homologous series. While linearity is best

TABLE 7. ΔI Differences for Unit Operations[a]

Structural difference[a]	ΔI[b]	ΔI[c]	ΔI[d]
1. 1-Unsaturation			
$1-n-C_5$	-17	2	53
3-2	- 9	10	71
2. -O- for $-CH_2-$			
$4-n-C_6$	- 6	40	182
5-4	13	14	217
3. OH group added			
$6-n-C_4$	220	359	843
7-6	280	411	860
4. -CO- for CH_2			
$9-n-C_4$	152	248	491
10-8	144	233	533
5. $-OCH_3$ added			
$4-n-C_4$	194	240	382
12-11	178	200	206

[a]Numbers refer to substance in Table 6.

[b]Apiezon L.

[c]Dioctyl sebacate.

[d]Carbowax 400.

for nonpolar stationary liquid phases, the same pattern of nearly parallel straight lines for various homologous series has been observed even for relatively polar solvents. These types of relationships have become useful in identification of unknown substances after preliminary characterization has been performed.

IV. RETENTION PROPERTIES OF STATIONARY PHASES

Because the chemical nature of the solute is often fixed[†] in a given mixture, the significant variable under the chromatographer's control becomes the selection of an

---

[†]Although modification of solutes is possible in many instances by derivatization which alters chemical and chromatographic properties, it will be assumed here that the chemical nature of the solutes is fixed in selecting columns.

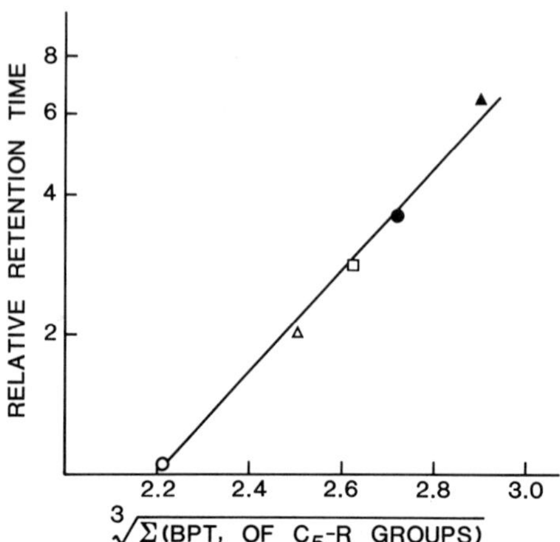

FIG. 2.   Relationship between the log of the relative retention time for
dimethylated barbituric acids and the cube root of the summed boiling point
increments of the 5-substituents: (o)-Barbital, (Δ)-Butabarbital, (□)-Amobarbital,
(●)-Secobarbital, and (▲)-Phenobarbital [27].   (Reprinted with permission of copyright owner.)

appropriate column to effect the desired separation. Throughout this discussion constant reference has been made to the polarity of the stationary phase as an important
consideration for theoretical and observed retention behavior. While the polarity of
available stationary phases can be classified in a number of ways the final selection
is complicated by (1) the variety of phases available, (2) the variability of gas
chromatographic conditions reported in the literature and the lack of good chromatographic reference data for drugs, and (3) an inability to define microscopic differences in partitioning between closely related solutes. The problem of too many
phases has been alluded to several times earlier and will become more obvious throughout this section. The retention index system of Kovats is gradually gaining acceptance as the standard method for reporting chromatographic retention data, so that an
increasing number of drugs are being classified accordingly. The last problem has
always relegated the selection of columns to more of an "art" than a science although
the retention index system has also resulted in more specific classification of stationary phase behavior with respect to standard reference solutes.

A.   Chemical Classification

The polarity of stationary liquid phases commonly employed in gas chromatography can
be generally classified according to two systems. The first is the classical listing
of phases in terms of their chemical nature and their apparent polarity. The second
is a more recent attempt to quantify observed retention behavior of a stationary phase

in terms of standard reference solutes. This latter system is generally associated
with either the Rohrschneider or McReynolds systems. Of practical importance is that
these stationary liquids must be nonvolatile over the intended range of operating
temperatures. Therefore, most liquid phases are high molecular weight polymers which
are not as well characterized as low molecular weight liquids. In addition, theoretical considerations of partitioning behavior of solutes are more complex.

1. Chemical Classification of Stationary Phases

This classification is based upon the relative polarity of the repeating polymeric
units to allow assessment of relative interaction capability with various solutes.
For simplicity these interactions can be divided into (a) nonpolar, (b) intermediate,
(c) polar, and (d) specific. Although it is not possible to place all stationary
liquids precisely in one of these four categories, it does serve as a useful classification of the major structural types of liquids which are in use.

a. Nonpolar

These are generally paraffinic in nature and do not contain polar or polarizable
groups. Included are all paraffins, such as squalane, the Apiezon greases, and the
alkyl silicones. Interaction occurs through dispersion forces only. Nonpolar molecules are generally retained as discussed with respect to the alkanes in the previous
section with separation mainly on the basis of molecular size, although surface area
contact also appears to be important to maximize interaction forces.

Polar solutes separate solely on the basis of size since specific interactions
are not possible. Polar molecules generally elute much more rapidly than nonpolar
molecules with similar boiling points. From the Rohrschneider classification, it will
be seen that there is a large degree of specificity attainable from various nonpolar phases.

b. Intermediate

This represents a large group of solvents having both polar and nonpolar properties
so that retention of most types of solutes is possible. These include many types of
organic esters, ethers, alcohols and aromatic substituted silicones. Nonpolar molecules separate primarily as a function of molecular size with retentions shorter than
in nonpolar phases. Polar molecules separate in part on the basis of size and also
on the basis of the strength of the solute-solvent interactions. There is some selectivity for hydrogen bond donors or acceptors with solvents containing hydrogen bond
donor groups such as hydroxyl and hydrogen bond acceptor groups such as ethers.

c. Polar

These stationary phases are made up primarily of polar groups which interact very
strongly with each other as well with polar solutes. Examples include the polyethylene

glycols and many polyesters. The nature of the methylene portion of the monomer of the latter solvents, imparts considerable variation in polarity and provides a basis for selectivity. Nonpolar solutes are generally excluded and not retained at all on polar solvents, which are not of much value for this type of application. Polar solutes are, of course, retained considerably and polar solvents find their greatest utility for mixtures of polar compounds.

d. Specific Solvents or Phases

Certain solvents are utilized which have very specific interaction ability with a particular group of solutes. These interactions are generally much stronger than the types of interaction forces discussed previously and result in very prolonged retentions. Examples include the silver complex with olefins and the separation of amines on stationary phases containing metal salts of fatty acids [7]. Table 8 illustrates the chemical properties of commonly employed stationary phases.

B. Retention Index Classification

In this system the retention of a standard reference series of compounds on a given stationary phase is compared to the retention behavior on squalane, the least polar of the commonly used liquid phases. Differences in retention can then be expressed in a quantitative manner and specific solvents can be ranked in terms of their relative polarity for several different types of solute. The classification of stationary phases in this way has sharply focused the problem of too many stationary phases since many have been shown to be nearly identical in terms of their retention behavior.

1. Rohrschneider Classification

This type of classification first developed by Rohrschneider [30] defines the difference in retention index for a given solute between a polar and nonpolar phase as

$$\Delta I = I_{polar} - I_{nonpolar} = ax$$

where a represents the polarity characteristics of the solute and x represents the polarity of the polar stationary phase. The same nonpolar stationary phase must be used as a reference; usually squalane. Rohrschneider developed a system based upon 5 solutes: benzene, ethanol, methyl ethyl ketone, nitromethane, and pyridine. The overall ΔI term is thus defined as

$$\Delta I = ax + by + cz + du + es$$

where x becomes the polarity of the column with respect to benzene and is defined as $\Delta I/100$ for benzene. Similarly y, z, u, and s are defined for the other four liquids. Table 9 illustrates the Rohrschneider constants for a number of commonly used stationary phases.

TABLE 8. Examples of Typical Stationary Phases [a]

Stationary phase	Chemical formula	Maximum recommended temperature (^{o}C)
a. Nonpolar liquids		
Paraffins, *n*-alkanes	$CH_3-(CH_2)_nCH_3$	$30^{o}-100^{o}$
Squalane	Branched C_{30} saturated hydrocarbon	150°
Alkyl silicones	$\begin{array}{cc} CH_3 & CH_3 \\ | & | \\ -O-Si-O-Si-O- \\ | & | \\ CH_3 & CH_3 \end{array}$	300°
Apiezon greases		300°
b. Intermediate liquids		
Alkyl, aryl silicones	$\begin{array}{cc} C_6H_5 & CH_3 \\ | & | \\ -O-Si-O-Si-O- \\ | & | \\ CH_3 & CH_3 \end{array}$	300°
Dibutyl phthalate	(benzene ring with $COOC_4H_9$ and $COOC_4H_9$)	100°
Dioctylsebacate	$C_8H_{17}OCOC_8H_{16}COOC_8H_{17}$	100°
Benzyl cyanide	(benzene ring with $CH_2C\equiv N$)	35°
Tricresyl phosphate	(cresyl ring with $O-$, CH_3, and $P=O$, subscript 3)	125°
Tetrahydroxyethylethylene diamine	$(HO-CH_2CH_2)_2-N-CH_2CH_2-N-(CH_2CH_2OH)_2$	150°
c. Polar liquids		
Dethyleneglycol succinate	$[-CH_2-CH_2-O-CH_2CH_2-O-\overset{\overset{\textstyle O}{\|}}{C}-CH_2-CH_2-\overset{\overset{\textstyle O}{\|}}{C}-O-]_n$	190°
β,β'-Oxydipropionitrile	$\begin{array}{c} NC-CH_2CH_2 \\ \searrow \\ O \\ \nearrow \\ NC-CH_2CH_2 \end{array}$	100°

TABLE 8. (Continued)

Stationary phase	Chemical formula	Maximum recommended temperature (°C)
Polyethyleneglycols	$HO-CH_2-CH_2-[O-CH_2CH_2]_n-OH$	$100-200°$
Phenyldiethanolamine	$-[CH_2-CH_2-N-CH_2CH_2-O-\overset{O}{\overset{\|}{C}}-CH_2CH-\overset{O}{\overset{\|}{C}}-O]_n-$	$225°$
d. Specific liquids		
Polypropyleneglycol		
Silver nitrate		$75°$
Perfluorocarbons		$100°$
Metal salts of fatty acids		

[a]For a more detailed listing, see Refs. 18 and 29.

While the sum of all of the ΔI values provides an indication of overall relative polarity, the individual constants often give an indication of specific retention capacity for a particular class of compounds. The high value for QF-1, for example, is indicative of selectivity for ketones, and Supina [31] pointed out its usefulness in the gas chromatography of steroids.

After the Rohrschneider constants have been determined for a large number of stationary phases it becomes possible to characterize the a, b, c, d, and e constants for a given molecule by chromatographing on at least five of the stationary phases. It is then possible to predict the ΔI value on a given column for a specific solute and compare to the experimentally determined difference in retention index relative to squalane. In principle then it would be possible to determine an exact set of conditions for closely related solutes knowing how large a difference in ΔI is required. It has been possible to accurately predict ΔI values for simple organic molecules in this way. The amount of experimental data which would have to be

TABLE 9. Rohrschneider Constants for Common Stationary Liquid Phases[a,b]

Stationary Liquid	X	Y	Z	U	S
OV-1	0.16	0.20	0.50	0.85	0.48
Apiezon L	0.32	0.39	0.25	0.48	0.55
QF-1	1.41	2.13	3.55	4.73	3.04
OV-17	1.30	1.66	1.79	2.83	2.47
XE-60	2.08	3.85	3.62	5.43	3.45
DEGS	4.93	7.58	6.14	9.50	8.37

[a]Abstracted from Supina [31].

[b]Data relative to 20% Squalane on Chromosorb W at $100°$ C.

collected for a complex drug molecule for which chromatographic data is not available, however, makes this an impractical approach since the retention data could be determined directly for the solute under investigation. As more chromatographic data are reported in Kovats and Rohrschneider terms, it may be possible to accurately use literature data for specific column selection.

2. McReynolds Classification

McReynolds [32] has modified the Rohrschneider classification by replacing three of the standard solutes with higher homologs for ease of handling under experimental conditions. In the McReynolds system, butanol, 2-pentanone, and nitropropane were substituted for ethanol, methyl ethyl ketone, and nitromethane, respectively, with the correlations remaining satisfactory. Additionally, five new solutes were added which improved Rohrschneider's treatment for specific types of compounds. The addition of 2-methyl-2-pentanol, for example, improved prediction of the retention indexes for branched-chain alcohols and 1-iodobutane enabled better prediction for halogenated compounds.

More than 200 liquid phases have been classified according to this system and average polarities were calculated from the sums of the ΔI's of benzene, butanol, 2-pentanone, nitropropane and pyridine at 120° C. Table 10 illustrates this type of data with a partial listing of 25 stationary phases. This type of listing for stationary phases is now readily available from manufacturers and provide a useful systematic classification of column polarity. The total ΔI for all five solutes provide an overall polarity classification and clearly demonstrates the similarities of a number of phases. In Table 10 for example, phases 4-8 show essentially identical retention behavior and are not all necessary from a separation viewpoint. The individual ΔI's are also valuable since they indicate specific retention properties. Phases 17 and 18 have essentially the same total polarity but differ rather markedly in their individual retention capabilities for specific classes of compounds and would not be considered interchangeable. From this sampling of more than 700 phases currently available, the confusion, redundancy and absence of standard reference data in the gas chromatography literature becomes understandable.

V. PRACTICAL CONSIDERATIONS

A. "Preferred" Stationary Phases

There has recently been a determined effort to limit the number of stationary phases for the reporting of retention data on new compounds. The intent is to begin to standardize the reporting of data so that comparisons with the literature can be useful to the chromatographer. While there is general agreement that there are large numbers of

TABLE 10. McReynolds Data for Selected Liquid Phases [32]

		ΔI				
		Benzene	Butanol	Pentanone	Nitropropane	Pyridine	$\sum \Delta I$
1.	Squalane	0	0	0	0	0	0
2.	Apiezon M	31	22	15	30	40	138
3.	Apiezon L	32	22	15	32	42	143
4.	SE-30	15	53	44	64	41	217
5.	E-301	15	56	44	66	40	221
6.	OV-1	16	55	44	65	42	222
7.	SE-31	16	54	45	65	43	223
8.	OV-101	17	57	45	67	43	229
9.	OV-3	44	86	81	124	88	423
10.	Dinonyl sebacate	66	166	107	178	118	635
11.	Di-isodecyl adipate	71	171	113	189	145	689
12.	DEG stearate	64	193	106	143	191	697
13.	OV-17	119	158	162	243	202	884
14.	Versamid 930	108	309	137	208	207	969
15.	Span 80	97	266	170	216	268	1017
16.	Dicyclohexyl Phthalate	146	257	206	316	245	1170
17.	Oronite NIW	185	370	242	370	327	1494
18.	QF-1	144	233	355	463	305	1500
19.	OV-210	146	238	358	468	310	1520
20.	XE-60	204	381	340	493	367	1785
21.	OV-225	228	369	338	492	386	1813
22.	Carbowax 6000	321	540	369	577	512	2299
23.	Ethylene glycol adipate	371	579	451	655	633	2689
24.	DEGS	492	733	581	833	791	3430
25.	Silar 10C	523	757	659	942	801	3682

phases which duplicate others in terms of retention behavior, the process of producing a universally acceptable list of phases is a difficult one. Among the requirements of such a list are (1) a wide range of polarities for analysis of most solutes, (2) reproducible retention characteristics, and (3) utility over a wide temperature range.

Based upon frequency of use from literature reports as well as the three characteristics listed, Mann and Preston [1] first proposed the following list of preferred phases:

1. Apiezon L or equivalent
2. Methyl silicone polymers or equivalent
3. Methylphenyl silicone polymers with 50% phenyl or equivalent
4. Trifluoropropyl silicone polymers or equivalent
5. Polyethylene glycol 20 M or equivalent
6. Diethylene glycol succinate polyester or equivalent.

The McReynolds data for these phases are included in Table 10. The authors expressed hesitancy with Apiezon L because of its uncertain composition and recommended it remain on the list only until a more suitable, very nonpolar phase evolves. The only type of phase absent from this initial listing was an extremely polar phase. Since the applications are much less frequent here, none was suggested based on minimal literature experience.

The most recent list of preferred stationary phases resulted from a distinguished committee chaired by Hawkes [33]. This group suggested a primary listing of the six phases presented in Table 11, and a secondary list of 24 phases for specific applications not satisfied by the primary list. The McReynolds data for the primary listing is included in Table 10. The secondary list of 24, spanned a greater range of

TABLE 11. Preferred Stationary Phases [33]

Phase	Example
Dimethyl silicone	OV-101, SP-2100, SE-300, SF-96
50% Phenylmethyl silicone	OV-17, SP-2750
Polyethylene glycol mol. wt. 4000	Carbowax
Diethylene glycol succinate	Several
3-Cyanopropyl silicone	Silar 10C, Apolar 10C, SP-2340
Trifluoropropylmethyl silicone	OV-210, SP-2401

McReynolds constants and provided for specific polarity, temperature, and retention requirements not satisfied by the primary listing.

This type of list appears appropriate for drug analysis. Moffat has found that the vast majority of drugs will chromatograph suitably on SE-30 [21] and he proposed that this become the standard phase for reporting retention data in the form of retention indexes for all drugs. This would have the advantage of uniformily characterizing their chromatographic properties. For those mixtures which are not adequately separated on SE-30, their retention indexes and polarities will allow another phase to be selected. Except for unusual types of specific problems, most of these separations should be possible on the 6 primary phases, and the remainder should resolve using one of the phases in the secondary group. None of these lists are intended to limit development of new stationary phases, and evaluation of phases with specific properties will result in reconsideration of preferred phases and possible replacement after suitable experience. The new cyanoethyl-cyanopropyl silicones, OV-275 and Tenax GC, have demonstrated useful retention properties for polar solutes and may eventually be included in a preferred list of phases [34].

In the specialized application of column selection for gas chromatographic-mass spectroscopic (GC/MS) analysis, the maximum temperature of the stationary phase becomes important. Because of the great increase in detector sensitivity, column bleeding must be avoided. To minimize bleed, high molecular-weight, stable liquid phases are required. While SE-30 and OV-1 are generally useful, very similar retention behavior is observed on OV-101, which produces a more stable column and avoids bleed. OV-101 appears to be the single most useful and popular nonpolar phase for GC/MS applications. Leibrand [35] has offered a preferred list for GC/MS and includes

1. Dimethyl silicone (for example, OV-101, SP-2100)
2. 50% Phenyl-50% methyl silicone (for example, OV-17, SP-2250)
3. 50% 3-Cyanopropyl-50% phenyl methyl silicone
    (for example, Silar 5CP, SP-2300)
4. 3-Cyanopropyl silicone (for example, Silar 10C, SP-2340)

Among the intermediate and special phases listed by Hawkes et al. [33], care must be used due to excessive bleed above 180°.

B. Chromatographic Supports

One assumes in all considerations of solute retention that the stationary liquid is supported on an inert solid which does not interact with the solute. Since in practice the support often does influence the chromatographic separation, a brief discussion is included. For a more detailed discussion, reference to Ottenstein [36] is suggested. In addition to being inert, an ideal support should have a large surface area and be made up of regularly shaped particles of relatively narrow size distributions to allow the stationary liquid to be coated uniformly. The structure of the support and uniformity of liquid coating governs the efficiency of the column which is associated with peak shape and peak width. Symetrical and narrow peaks are, of course, preferred. Interaction between the solute and the active polar sites on an absorbent, will result in "tailing" of the peak and this is more pronounced for polar solutes. Tailing appears related to the ability of the solute to form hydrogen bonds and becomes significant for many of the functional groups typically in drug molecules including alcohols, amines, acids, and to a lesser extent, aldehydes, ketones, and esters. While the retention time of the peak is not significantly affected, tailing tends to broaden the trailing portion of solute and affects both the separation and the sensitivity when peak-height analysis is used.

The most commonly utilized solid supports in drug analysis are the Chromosorb series of which G, W, and P are used for quantitative analysis and A is frequently used for preparative applications. In terms of inertness, these rank G, W, and P in decreasing order. For separation of polar solutes, treatment of the support with a silanizing reagent removes surface hydroxyl groups and serves to decrease tailing. Dimethyldichlorosilane (DMCS) or hexamethyldisilizane are most commonly used for this purpose. Surfaces of glass columns and the glass wool holding the packing material can also contain active -OH groups which cause tailing and therefore these are often silanized. In addition, mineral impurities in the support contribute to tailing. This effect is minimized by using acid-washed (AW) support phases.

In cases where tailing is anticipated, it is common practice to inject fairly large amounts of the sample itself onto the column prior to use. This will saturate any specific binding site for the solute molecule and it tends to reduce tailing. This phenomenon explains the gradual increase in response noted with the first few injections of a polar solute on a newly packed chromatographic column.

Although for individual separation problems, the choice of a solid support may be quite significant, in general, the Chromosorbs W or G are widely used for drug applications. For more polar molecules silanized, acid-washed solids are recommended. Moffat [21] found insignificant difference in retention indexes and peak shape for most of the 480 drug molecules reported from data obtained in several laboratories using SE-30 on Chromosorb W or G. Most of these laboratories tended to use acid-washed DMCS-treated supports. Gas Chrom supports, many of which are pretreated to reduce tailing, are also popular for drug analysis. Gas Chrom S, is acid washed and Gas Chrom Q, one of the most frequently cited supports in the literature, is presilanized. They have comparable surface characteristics to the Chromosorb series. Typical applications include steroids, pesticides, and alkaloids with a liquid-phase loading of 3% most common.

In the case of closely structured polar compounds, the interaction with the support may be as critical to the separation as the stationary liquid phase interaction. In these cases the particle size distribution of the support becomes significant with finer mesh supports such as 100/120 giving more distinct separations than those of larger particle size. In these cases, temperature programming may be required to decrease the tailing resulting from the affinity of the drug for the support. The separation of methylated derivatives of phenobarbital and diphenylhydantoin, and primidone on a 3% OV-17 on 100/120 Chromsorb W using temperature programming, is compared to the same separation using identical conditions on a 60/80-mesh Chromosorb W in Fig. 3. The slowest-eluting peak is 5-p-(methylphenyl)-5-phenyl-hydantoin, an internal standard. Changing the mesh size of the solid support significantly affected the retention times and the peak shape.

C. Stationary Phase Loading

The amount of liquid phase applied to the support depends in part on the nature of the support and also on the nature of the solutes in the mixture. Column packings are usually expressed on a weight basis and are typically 3%, 5%, 10%, and occasionally as high as 20%. Since densities of supports vary, the amount of liquid phase present will also vary and a 5% loading on a high-density support such as Chromosorb W would have about twice as much stationary phase as a 5% loading on a lower density support such as Chromosorb P. The observed retention of a solute on the higher density phase would also be expected to be about double that on a column of the same dimensions packed with a low-density support.

The efficiency of a chromatographic column is also related to the amount of liquid phase and to the capacity of the column to hold the liquid phase effectively. For analytical work, high efficiency is essential while for preparative applications, efficiency can be compromised at the expense of increased sample size while using a higher liquid-phase loading with the latter. Low-density supports have high surface areas/unit volume and can accomodate a liquid-phase loading of 15-30% depending upon the application. Higher density supports such as Chromosorbs G and W and Gas Chrom Q, are generally loaded from 2-15% depending upon the application.

The volatility and amount of the sample are other factors which influence the percentage of stationary phase. For very polar samples of low volatility, lower loaded columns (2-5%) are used since interaction with the column is favored while more volatile less polar compounds require higher loads (10-20%) to increase their residence time in the liquid phase and facilitate retention. The larger the sample size the larger the liquid load required to handle the sample. For most analytical drug applications, liquid loadings range from 2% to 5% on Chromosorb G or W, or Gas Chrom Q.

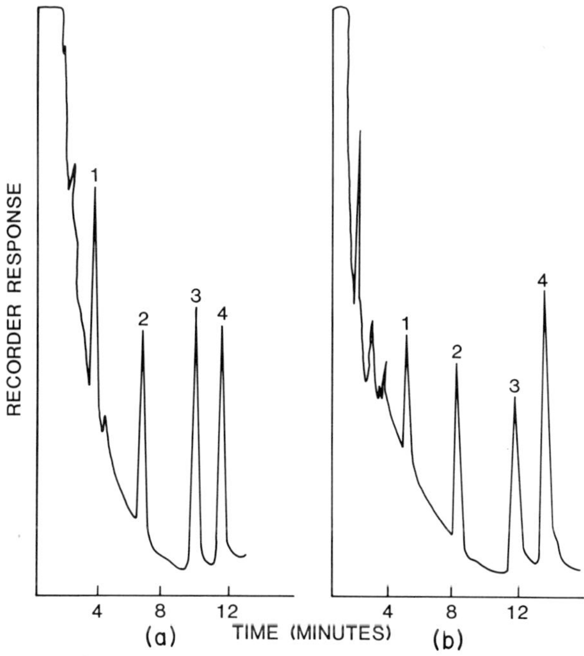

FIG. 3. Chromatogram of methylated anticonvulsants on (a) 3% OV-17 on Chromosorb W 60/80 and (b) 3% OV-17 on Chromosorb W 100/120 under identical chromatographic and programmed temperature conditions: (1) phenobarbital, (2) primidone, (3) diphenylhydantoin, and (4) internal standard. (J. L. Cohen, unpublished data.)

D. Column Packing

The packing of gas chromatographic column is critical to the success of the separation. Even if optimal stationary phases, solid supports, temperatures, and flow conditions are selected, nonuniformly packed columns can diminish column efficiency by affecting the eddy diffusion (A) term in the Van Deempter equation. In a packed column the actual flow velocity of the carrier gas will vary considerably from one point to another depending upon the particle surfaces, wall effects of the column and packing inhomogeneities. This tends to broaden a chromatographic peak and effectively reduce the number of theoretical plates in the column. In addition to column packing, the efficiency of the chromatographic system depends on how well the liquid phase is coated on the solid support.

In most cases common liquid phases are available precoated on a variety of supports in several percentage loadings. In those instances where coating is required, the two most commonly used and most effective techniques are the filtration and the slurry methods. In the filtration method, the support is soaked in a solution of the solvent of known concentration and the excess liquid is filtered off under vacuum

using a water aspirator. The wet packing is then dried while being continuously stirred to avoid nonuniform coating. The slurry method is similar and involves dissolving
an accurately weighed amount of liquid phase in a volatile solvent. The solid support
is coated by making a slurry and then evaporating the volatile solvent, while it is being agitated continuously. A rotary evaporator or heat lamp is typically used. After
the solvent is removed, the column material is mixed well on a shaker for 30-60 min.

Precoated materials or those coated specifically in the laboratory should all be
completely free flowing. These are poured through a funnel into the column which is
either metal or glass, with the latter generally preferred for drug analysis. For
U-shaped or straight columns one end of the column is plugged with silanized glass
wool and small amounts of packing are added at a time and patiently tapped down into
the column using a pencil or an equivalent tool. This is continued until no more
packing can be added and the packing is evenly settled into the column. Cautions include adding only small amounts at a time and continuous gentle tapping, rather than
vigorious vibrating which may rupture the support. For a coiled column, generally a
vacuum pump or aspirator hose is attached to the plugged end to provide a vacuum,
since gravity is not possible. The tendency to pack these too rapidly must be avoided
and gentle tapping is critical to uniform packing.

Prior to use, gas chromatographic columns should be conditioned to remove trace
quantities of absorbed moisture, residual impurities on the liquid phase or solid
support, and any solvent used to coat the support. Columns should be conditioned by
connecting the injector end and disconnecting the detector to avoid contamination.
Initial conditioning requires low carrier-gas flow (10-20 ml/min) at 100° C to remove
residual solvents. The column should then be conditioned overnight at temperatures
at least 25° above those of intended use but slightly below the recommended maximum
temperature for the stationary phase. Conditioning should continue until a stable
baseline in the desired sensitivity range is observed.

Specific columns may require special conditioning procedures which should accompany the packing from the manufacturer. In general, the methyl, methyl phenyl, and
methyl vinyl silicones, including SE-30, SE-31, SE-33, SE-51, SE-52, and the corresponding OV and SP series should be conditioned at 250° with no flow for several
hours, then conditioned at intended operating temperature with gas flow until stable.
For specific packing and conditioning details the reader is referred to respective
manufacturer data on packings and columns. An excellent practical source for the
beginning chromatographer is the book by McNair and Bonelli [29].

E. Column Selection

While no single approach to column selection is appropriate for the variety of applications of gas chromatography to therapeutic agents, a number of general schemes are

outlined in the literature which can assist the chromatographer. Littlewood [7] and McNair and Bonelli [29] outline general approaches based upon solute polarity. These considerations in conjunction with the classification of over 200 stationary phases by McReynolds [32] in terms of relative polarity based upon 10 standard solutes provide sufficient data for initial column selection for relatively simple organic molecules. In the case of drug molecules however, many of which have multiple functional groups, extension of these approaches becomes difficult. In general an empirical determination of retention of a complex solute on two phases of widely varying polarity such as OV-1, SE-30 and OV-225, or carbowax provide more rapid relevant information than a detailed theoretical treatment. The retention times and the number and shape of peaks under isothermal and temperature-programmed conditions provide the experienced chromatographer with sufficient data to utilize one of these columns or proceed to another column. This approach is widely used in conjunction with thorough literature searches to determine retention times of specific, or at least closely related compounds to solve many drug separation problems. This approach is outlined by Supina [37]. Other useful sources for drugs include a number of compilations of specific conditions utilized for the successful chromatography of various drug classes [38-40] and some retention index data for toxicological screening [20, 41].

In searching the literature for drug retention data, it is evident that most drugs can be chromatographed on a variety of columns under several conditions. It is also apparent that a minimum number of stationary phases is adequate for the vast majority of separations, consistent with the preferred-phase concept. Additional drugs, formulation components, metabolites and unknown interfering substances, however, pose significant column selection problems. The utilization of SE-30 as a standard liquid phase for drugs by Moffat [21] forms the basis for a more systematic approach for stationary phase selection. This listing of 480 drugs in terms of retention indexes provides significant comparable data for all current chemical classifications of drugs with molecular weights up to 400. While SE-30 is by no means implied to be an appropriate phase for all of these agents, the existence of comparable retention data can be extremely useful to the chromatographer with a drug separation problem. As a general rule, retention indexes should differ by more than 50 units to assure adequate separation. It should be emphasized that although retention index data are temperature independent, the absolute retentions, and therefore the resolution of the components of a mixture are a function of temperature.

Based upon retention indexes on SE-30, a general approach to column selection is presented.

1. Known Component with SE 30 Retention Data

If the compounds to be analyzed have retention data reported on SE-30, drug stability and convenience of absolute retention time in terms of solvent and interfering peaks

will dictate the polarity of the stationary phase required. Closely grouped indexes
suggest alternate conditions. If molecular weights are similar and structures vary,
a more polar phase (for example, 50% phenyl methyl silicone, OV-17) is indicated. If
structures are similar, temperature programming may be required. Widely spaced retention indexes suggest SE-30 to be appropriate unless retention times are not convenient
or drug thermal stability is a problem. Short retention times require a more polar
column or lower temperatures. Long retention times indicate either strong interactions
requiring a more polar column or very high molecular weights requiring temperature
modification.

2. Components without SE-30 Retention Data

A structural comparison to known drugs should allow prediction of a reasonable retention index range for a specific component. The retention index on SE-30 can then be
determined with reference to appropriate alkane hydrocarbons, (at least two) run under
identical isothermal chromatographic conditions. The retention index can be determined
either by calculation from Kovat's equation or from a plot of log retention time vs
carbon number of the alkanes, as illustrated in Fig. 4. After retention indexes for
all known components are determined on SE-30 the preceding general guidelines could
be applied.

Because of its low polarity (McReynolds, $\Sigma \Delta I$ = 217), SE-30 is an ideal reference
phase for drug molecules. It is unlikely that a change to a lower polarity phase than
SE-30 will separate any unresolved mixture. If possible, preferred phases of increased
polarity should be tried unless a specific phase is believed to be more satisfactory.

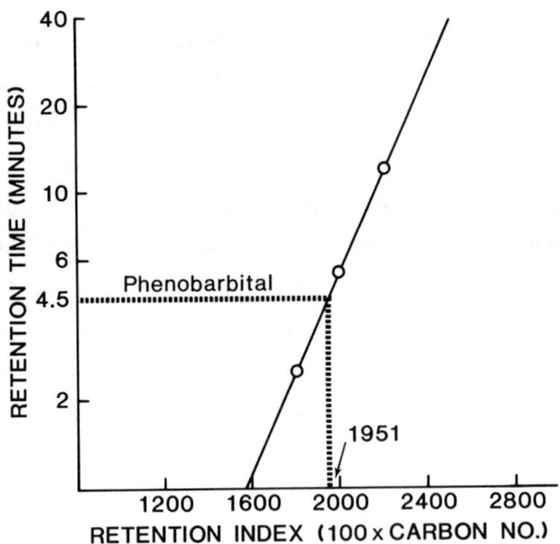

FIG. 4. Determination of retention index of phenobarbital from a retention index
plot [20]. (Reprinted with permission of copyright owner.)

This will increase the comparable data base of retention data for drugs and continue
to improve predictions of retention time for new compounds. If separation does not
occur for a given mixture, stepwise increases in stationary phase polarity should
allow resolution.

In complex mixtures containing components of unknown structure, preliminary
separation to identify the number of chromatographable components is essential. The
use of columns of varying polarity with temperature programming as well as other
chromatographic procedures is generally required to make certain that a given peak
represents a single component. If SE-30 is used as the nonpolar column, additional
retention indexes of unknown or interfering components can also be estimated. Knowing the SE-30 retention indexes of all components should then allow rapid selection
of an appropriate column and chromatographic conditions.

REFERENCES

1. J. R. Mann and S. J. Preston, *J. Chromatogr. Sci.*, *11*, 216(1973).

2. R. S. Henly, *J. Chromatogr. Sci.*, *11*, 221(1973).

3. D. H. McCloskey and S. J. Hawkes, *J. Chromatogr. Sci.*, *13*, 1(1975).

4. J. J. Leary, J. B. Justice, S. Tsuge, S. R. Lowry, and T. L. Isenhour, *J. Chromatogr. Sci.*, *11*, 201(1973).

5. C. E. Pippenger and H. W. Gillen, *Clin. Chem.*, *15*, 583(1969).

6. S. R. Lipsky and R. A. Landowne, *Anal. Chem.*, *33*, 818(1961).

7. A. B. Littlewood, *Gas Chromatography*, 2nd Ed., Academic Press, New York, pp. 16-55, 1970.

8. E. Heftman (ed.), *Chromatography*, 2nd Ed. Reinhold, New York, 1967.

9. K. A. Connors, *A Textbook of Pharmaceutical Analysis*, 2nd Ed. Wiley-Interscience, New York, 1975, p. 149.

10. A. B. Littlewood, *Anal. Chem.*, *36*, 1441(1964).

11. A. B. Littlewood, *J. Gas. Chromatogr.*, *1*, 16(1963).

12. A. T. James and A. J. P. Martin, *Biochem. J.*, *50*, 699(1952).

13. V. G. Arakelyen and K. I. Sakodoynskii, *Chromatographic Revs.*, *15*, 93(1971).

14. A. T. James and A. J. P. Martin, *Biochem. J.*, *63*, 144(1956).

15. M. B. Evans and J. E. Smith, *J. Chromatogr.*, *5*, 300(1961).

16. M. B. Evans, *J. Chromatogr.*, *12*, 2(1963).

17. M. B. Evans and J. E. Smith, *J. Chromatogr.*, *8*, 541(1962).

18. R. Kaiser, *Gas Phase Chromatography, Vol. 3*, Butterworth, New York, 1963.

19. E. Kovats, *Helv. Chim. Acta*, *41*, 1915(1958).

20. L. Kazyak and R. Permisohn, *J. Forensic Sci.*, *15*, 346(1970).

21. A. C. Moffat, *J. Chromatogr. Sci.*, *113*, 69(1975).

22. A. B. Littlewood, *Gas Chromatography*, 2nd Ed., Academic Press, New York, 1970, pp. 86-95.

23. W. O. McReynolds, *Gas Chromatographic Retention Data*, Preston Technical Abstracts Service, Evanston, Ill., 1966.

24. H. F. Martin, J. L. Driscoll, and B. J. Gudzinowicz, *Anal. Chem.*, *35*, 1901(1963).

25. F. H. Pollard, G. Nickless, and P. C. Uden, *J. Chromatogr.*, *11*, 312(1963).

26. T. L. Chang and C. Karr, *Anal. Chim. Acta.*, *26*, 410(1962).

27. H. F. Martin and J. L. Driscoll, *Anal. Chem.*, *38*, 345(1966).

28. A. B. Littlewood, C. S. G. Phillips, and D. T. Price, *J. Chem. Soc.*, 1480(1955).

29. H. M. McNair and E. J. Bonelli, *Basic Gas Chromatography*, 5th Ed., Varian Aerograph, Walnut Creek, Calif., 1969, pp. 71-79.

30. L. Rohrschneider, *J. Chromatogr.*, *22*, 6(1966).

31. W. Supina, in *Recent Advances in Gas Chromatography* (I. Domsky and J. A. Perry, eds.), Marcel Dekker, New York, 1971, pp. 99-104.

32. W. O. McReynolds, *J. Chromatogr. Sci.*, *8*, 685(1970).

33. S. Hawkes, D. Grossman, A. Hartkopf, T. Isenhour, J. Leary, J. Parcher, S. Wold, and J. Yancey, *J. Chromatogr. Sci.*, *13*, 115(1975).

34. A. Zlatkis, W. Bertsch, H. A. Lichtenstein, A. Tishbee, F. Shunbo, A. M. Cosica, H. M. Liebich, and N. Fleischer, *Anal. Chem.*, *45*, 763(1973).

35. R. J. Leibrand, *J. Chromatogr. Sci.*, *13*, 566(1975).

36. D. M. Ottenstein, *J. Chromatogr. Sci.*, *11*, 136(1973).

37. W. R. Supina, in *Theory and Application of Gas Chromatography in Industry and Medicine*, (H. S. Korman and S. R. Bender, eds.), Grune and Stratton, New York, 1968, pp. 39-42.

38. H. Kern, P. Schilling, and S. H. Miller, *Gas Chromatographic Analysis of Pharmaceuticals and Drugs*, Varian Aerograph, Walnut Creek, Calif., 1968.

39. B. J. Gudzinowicz, *Gas Chromatographic Analysis of Drugs and Pesticides*, Marcel Dekker, New York, 1967.

40. E. G. C. Clarke, *Isolation and Identification of Drugs*, Pharmaceutical Press, London, 1969.

41. A. H. Beckett, G. T. Tucker, and A. C. Moffat, *J. Pharm. Pharmacol.*, *19*, 273(1967).

Chapter 3

COLUMN SELECTION IN HPLC

Karl J. Bombaugh

Radian Corporation
Austin, Texas

I. INTRODUCTION

Column selection in high performance liquid chromatography (HPLC) involves more than choosing an appropriate packing material to fill a tube of some prejudged dimension. A single packing material may be used in any of several different chromatographic modes to make different types of separations. For example, porous silica, the most widely used LC packing, may be used for either adsorption, partition, or exclusion chromatography. Similarly ion-exchange resins may be used in either the partition, exclusion or reverse-phase adsorption mode, as well as in the conventional ion exchange mode. The change may be achieved simply by changing the composition of the carrier. The chromatogram in Fig. 1 illustrates a duality of separation in which the adsorption and exclusion modes are attained simultaneously.

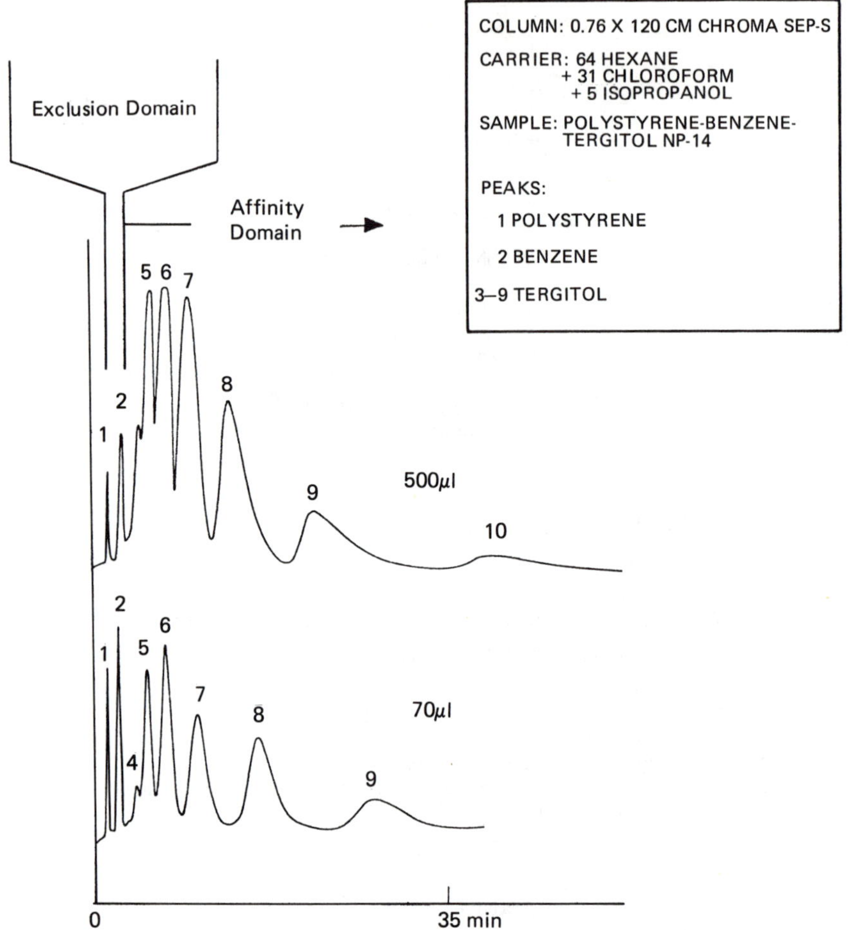

FIG. 1. Simultaneous exclusion and affinity mode separation on 60 Å pore silica [1]. (Reprinted with permission of copyright owner.)

The novice is often bewildered by the multiplicity of chromatographic supports and modes available for HPLC. To alleviate the feeling of bewilderment, this condensed overview is included in the introduction and will be expanded in the text.

There are basically four modes of chromatography and a well equipped laboratory will have columns for each mode. These include:

Adsorption

Bonded reverse-phase partition (octadecylsilane or octylsilane)

Ion exchange

Exclusion

The majority of drugs can be separated successfully on either an adsorption-phase, a bonded-phase, or a reverse-phase column. Ionic materials may be separated by ion-exchange chromatography. Recently, techniques have been developed to separate acidic and basic drugs as ion pairs on reversed phase partition columns. In the latter approach, retention is changed by adjusting the organic/aqueous solvent ratio which is preferable to preparing electrolyte solutions needed for ion-exchange chromatography. Exclusion chromatography is used primarily for high molecular-weight (>1000) components. Consequently, its application is limited in routine drug analysis but may be useful for the characterization of additives in pharmaceutical formulations.

Major advances in column packing technology have resulted in the commercial availability of high-efficiency columns prepared from totally porous microparticles (5-10 μm). These highly efficient columns with theoretical plate numbers ranging from 2000 to 12,000 plates per 15- to 30-cm column are now more widely used in drug analysis than are the older and less efficient large particle (>30μm) or pellicular packings.

Since a majority of drugs can be separated by normal adsorption (silica gel) or reverse-phase (octadecylsilane) microparticle columns, the choice is made primarily on the basis of whether or not the compounds to be separated are structurally related. When contaminants are not closely related to the drug, either type of column may suffice. However, when closely related compounds must be separated, column selection should be facilitated by structural considerations since silica gel and reverse-phase columns do not perform equally well on all types of compounds. Silica gel columns often do not separate homologous compounds as well as reverse-phase columns. Conversely, silica gel is superior to reverse-phase columns for the separation of structural isomers.

This simplistic approach to column selection may be useful to the novice, but a deeper understanding of the separation process is required for the intelligent selection of columns and conditions essential to the effective practice of HPLC.

The objective of this chapter is to acquaint the reader with the steps needed to select the right column and carrier to make a separation. This procedure includes selecting a chromatographic mode, choosing an appropriate column packing, selecting a suitable carrier, and then optimizing the separation to achieve the needed resolution. Since optimization can best be approached on a fundamental basis, it will be necessary to define basic terms and relationships in simple terms using basic relationships. This chapter will provide a description of the respective chromatographic modes, the role of packing and the role of the carrier in the separation process. It will also provide a basis for selecting the right conditions for a given sample and for optimizing the separation using basic relationships.

II. THE SEPARATION PROCESS

The chromatographic separation is based upon the differential rate of migration of
solute molecules swept by a carrier through a bed of retardant. The separating process is based upon the selective retardation of one solute relative to another by a
stationary phase. When the transfer is between the two bulk phases, it is an absorption process. When the exchange is between the surface of the retardant and the bulk
carrier phase it is an adsorption process. When a species transfers between the carrier and retardant rapidly and uniformly, high-speed separations are attainable. If
the exchange is slow and non-uniform, high-speed separations cannot be made regardless
of the carrier flow rate, carrier head pressure, or packing efficiency.

The extent of solute retardation is governed by the distribution coefficient K_d
defined as

$$K_d = \frac{[s]}{[m]} \tag{1}$$

where

    $[s]$ = solute concentration in the stationary phase

    $[m]$ = solute concentration in the mobile phase

An empirical classification has been developed in HPLC that recognizes the four
modes of separation, namely partition, adsorption, exclusion, and ion-exchange chromatography. The classification depends more on the origin of their development than
on physical principles. (In fact, a fifth mode termed affinity chromatography is now
being developed by a particular segment of practitioners that is in reality a special
form of adsorption chromatography.) A classification based upon fundamental phenomena
would be preferred to the one in use since it could help the novice gain insight into
the retardation process involved. Unfortunately the four-mode classification is deeply
entrenched in the vocabulary of the practitioners, and the best that can be done here
is to relate the respective modes to the fundamental principles of the separation
process.

Since K_d is a concentration ratio, it is independent of the retarding mechanism.
It is equally applicable to adsorption and absorption processes, where a concentration
gradient between phases exists.

The currently used four-mode classification is summarized in Table 1. Apparent
inconsistencies emanate from the diverse origin of the nomenclature such as between
gel permeation chromatography (GPC) and gel filtration chromatography (GFC) where each
technique is now used with both aqueous and organic carriers and the only real difference is the orientation of the practitioner. An additional deviation has resulted from
the introduction of chemically bonded stationary phases which perform the function of

TABLE 1. Relationship between Types of Sorption and the Commonly Used Modes of Retardation

Adsorption processes (Bulk-to-surface transfer)	Absorption processes (Bulk-to-bulk transfer)
Liquid solid chromatography (LSC), nonionic adsorption: Normal phase LSC (polar retardant) Reverse phase LSC (nonpolar retardant)	Liquid-liquid partition chromatography (LLC): Normal phase LLC (polar retardant) Reverse phase LLC (nonpolar retardant)
Ion-exchange chromatography, ionic adsorption: Cation-exchange chromatography (retards cations) Anion-exchange chromatography (retards anions)	Liquid-liquid exclusion chromatography (LEC): Gel permeation chromatography (GPC) (Organic Carrier) Gel Filtration Chromatography (GFC) (Aqueous Carrier)
Affinity chromatography: Specific adsorption chromatography using immobilized enzymes.	

liquid partition phases but which may be surface active retardants. Since solute
transfer between bulk phases poses problems different from surface exchange, the difference should be recognized. However, rigorous adherence to the classification is
impractical since mixed mechanisms are a common occurrence in modern HPLC.

A. Retardation Processes

A brief description of the four commonly used modes of chromatography is presented to
acquaint the reader with the commonly used retardation processes.

1. Liquid-solid (Adsorption) Chromatography (LSC)

Separations by LSC are carried out on porous adsorbents offering a relatively high
specific surface area (A_s > 200 m^2/g) and a small and uniform particle size. Particles may be either fully porous or superficially porous. Polar adsorbents such as
silica or alumina are most generally used in the so-called normal-phase mode with
carriers of moderate polarity. Such carriers are prepared by adding a small quantity
of a polar modifier (e.g., isopropanol) to a hydrocarbon solvent (e.g., *n*-hexane).
The nature of the functional group is a primary factor affecting the relative
retardation of a solute molecule in adsorption chromatography. Relative adsorption
increases as the polarity of, and number of polar groups increases. However, adsorptive retardation is a competitive process in which the polar functional groups of
the solute molecule and the polar groups of the carrier compete for the active sites,
rigidly fixed on the adsorbent surface. Retardation is therefore a function of the
total interaction between the polar adsorbent surface and the net polarity of the
solute molecule in the carrier. While solute molecules in a liquid phase are free
to align themselves with the solvent molecules for maximum interaction between their
functional groups, they lack this freedom in the adsorbed phase where interaction
with the spatially fixed sites will vary with the geometry of the solute molecule.
Consequently, a uniqueness in selectivity can arise from the multiple interactions
between the functional groups on the solute molecule and the corresponding site on
the rigid adsorbent surface. Spatially oriented geometric isomers with similar
total polarity can often be separated by LSC when other methods fail. In contrast,
weak interaction of alkyl groups in the adsorption process renders it less effective
for separating a homologous series with a similar total polarity [2].

2. Ion-Exchange Chromatography

Ion-exchange chromatography is an adsorption process in which a reversible exchange
of ions occurs between the aqueous carrier and the charged surface of the ion exchange
resin. The ion-exchange resin consists of an insoluble polymeric matrix that is perme-

able to ionic solutes. The matrix contains fixed charge groups and mobile counter
ions, i.e., ions of opposite charge. These counter ions can be exchanged for other
similarly charged ions in the carrier via chemical equilibrium. The charge of the
counter ion identifies the type of resin. Resins which attract negative counter ions
are called *anion-exchange resins* while those attracting positively charged counter ions
are called *cation-exchange resins*. Anion resins may be either strongly or weakly basic
while cation-exchange resins may be either strongly or weakly acidic.

The most commonly used cation-exchange resins contain a sulfonic acid ($-SO_3^-H^+$)
group at the active site where the H^+ counter ion can be exchanged with positively
charged ions in the carrier. Weakly acidic cation-exchange resins contain ($-COO^-H^+$).
The most commonly used strongly basic anion-exchange resins contain a quarternary
ammonium exchange group $[-CH_2N^+(CH_3)_3Cl^-]$ while weakly basic exchangers have protonated
amines such as $[-N^+H(R_2)Cl^-]$. Traditional porous resins are usually prepared from
cross-linked polystyrene-divinyl benzene copolymer. Pellicular resins are prepared
by coating soda lime glass beads with a physically bonded layer of polystyrene. Ionic
groups are then introduced into the polymer matrix chemically. The newer bonded-
phase resins are prepared by introducing the ionic groups onto organic matrices which
were chemically bonded to a silica base.

Solute retardation in ion-exchange chromatography is controlled by four factors:
(1) charge strength of the solute ion, (2) ionic strength of the carrier, (3) charge
strength of the counter ion on the resin, and (4) the pH of the carrier. Retardation
can be decreased by increasing ionic strength of the carrier, decreasing the charge
strength of the counter ion and by adjusting carrier pH in a manner to decrease dissociation of either the solute, the counter ion on the packing, or both.

3. Partition Chromatography

Liquid-liquid partition chromatography is an absorption process, since the mass transfer is between the two bulk phases. The solute molecules are free to diffuse within
both the carrier and the stationary phase. For this reason, both the thickness of the
phase layer and the viscosity of the stationary phase are relevant properties affecting
the speed of separation. Thick pools of viscous stationary liquid must therefore be
avoided to acheive high-speed separations.
Normally a nonpolar carrier is used with a hydrophilic adsorbent as previously described. In this form the technique is commonly called *normal phase chromatography*.
When the system is reversed and a polar carrier is used with a lipophilic adsorbent, the technique is commonly termed *reversed phase chromatography* (RPC). RPC is
most useful for very nonpolar solutes where the distinguishing characteristic is the
differential solubility of the species in the carrier.

In liquid-liquid chromatography, as in gas-liquid chromatography, an inert solid support is coated with a liquid capable of retarding the solute, i.e., one in which the sample shows a degree of solubility. For the phase to remain stationary it must be maintained at equilibrium with the carrier. This is best accomplished by using a material that is immiscible with the carrier and by saturating the carrier with the stationary phase. Unfortunately the selection of solvent pairs becomes limited as the following requirements are met:

1. Immiscibility between the carrier and the stationary phase
2. A favorable polarity relationship between the carrier, the solute, and the stationary phase
3. A differential distribution ratio between sample components

Some partition systems which meet these requirements are listed in Table 2.

a. Blended-Solvent Phase Systems

A major difficulty in classical partition chromatography is the limited number of phase systems which meet the listed requirements and still afford a broad applicability. Many solvent pairs which meet the immiscibility requirements provide extreme partition coefficients. Thus, many components favor the carrier and are unretained while others favor the stationary phase and are not eluted. This difficulty can be overcome by use of a blended-phase system. A ternary mixture such as chloroform, isopropanol, and water will form two liquid phases, one rich in water and the other rich in chloroform. Isopropanol or ethanol acts as a moderator and adjusts the interphase composition. An increase in alcohol increases the water content of the carrier which in turn alters the solute distribution between phases. Phase diagrams can be constructed to define

TABLE 2. Typical Binary Partition Systems

Stationary phase system	Solid support	Carrier
Ethylene glycol	Silica (4 m^2/g)	Light hydrocarbon
Triethylene glycol	Silica (4 m^2/g)	Isooctane
ββ'-Oxydipropanitrile	Silica (4 m^2/g)	Isooctane
Dimethyl sulfoxide	Silica (4 m^2/g)	Isooctane
Ethylene diamine	Silica (4 m^2/g)	Isooctane
Methanol	Silica (4 m^2/g)	Isooctane
Dimethyl sulfoxide	Silica (4 m^2/g)	Chloroform-benzene-nitromethane
Dimethyl sulfoxide	Silica (50-200 m^2/g)	Chloroform-benzene-nitromethane
Hydrocarbon polymer	Silica (4 m^2/g)	Aqueous methanol
Isooctane	Silanized silica	Methanol

TABLE 3. Solvents for Use in Ternary Systems for HPLC

Nonpolar members	Modifiers	Polar members[a]
Fluorocarbon	Ethyl acetate	Water
Isooctane	Tetrahydrofuran	Formamide
Hexane	Dioxane	Ethylene glycol
Cyclohexane	Propanol	Propylene carbonate
Chloroform	Ethanol	Dimethyl sulfoxide
Benzene	Dimethyl sulfoxide	Formamide
Benzene	MeOH	Formamide

[a]Components listed in order of polarity (δ).

the composition and quantity of the respective phases [3]. The blended phase system
provides a distinct advantage over pure solvent systems in that a continuum of solubilities exists between the limiting solubilities of the solute in the respective
phases. Some solvents for use in ternary systems are listed in Table 3. To prepare
a ternary, one component from each column is combined to form two phases. The amount
of each component can be varied to control polarity and the amount of each phase
formed.

Indeed any binary system can be converted to a ternary system by the addition of
a modifier that serves as a mutual solvent for the polar and nonpolar members.
In common practice the water-rich phase is held stationary by a silica support
while the organic-rich phase is used as the carrier. (If the organic-rich phase is
to be held stationary, silanized silica should be used as the support.) Since phases
are produced as an equilibrium mixture they are intrinsically presaturated with each
other. A precolumn packed with a support comparable to that used in the column assures phase stability. The columns can be coated in situ by pumping a plug of the
stationary phase through the column.

Liquid-liquid partition systems have been somewhat neglected in favor of the
newer chemically bonded packings because of the presumed ease of using the latter.
However, ternary systems are worthy of consideration since they are easy to use and
offer a wide range of applicability [4, 5].

b. Nonequilibrium Partition Systems

A useful modification of the ternary system is attained by adding sufficient modifier
to produce a single phase that is used in conjunction with an active support. In this
state the active support retains a disproportionate amount of the polar elements producing a stationary phase which is more polar than the carrier; partition occurs between the more polar and less polar phases. A highly useful system can be prepared

from a mixture of chloroform, isopropanol, and water for use with conventional silica-
type packings. Coarse adjustments in retardation are made by varying the isopropanol,
while fine adjustments can be made by varying the chloroform or by adding a less polar
element such as hexane.

The nonequilibrium phase system was employed by Samuelson and others to separate
saccharides using cation-exchange resins as the active support with mixtures of ethanol
and water as the carrier [6-8]. The sulfonate groups on the cation-exchange resin selectively retained water producing a water-rich stationary phase. The ethanol limited
the solubility of the solutes in the carrier producing favorable partition ratios
needed for the separation.

c. Ion-Pair Partition Systems

A useful variation of classical partition chromatography is a technique known as ion-
pair partition chromatography [9, 10]. By this technique a water soluble solute is
combined with an appropriate counter ion to make it lipophilic [11]. Sample pH is
adjusted to induce sample ionization and to facilitate its combination with the counter ion. The ion pair is then partitioned between an aqueous and an organic phase,
either of which may be held stationary. In normal phase chromatography a conventional
high-efficiency silica packing is used as the support for the stationary aqueous phase
while a stable reverse-phase packing may be used as the stationary organic phase. The
sample is added to an aqueous solution of the counter ion to form the ion pair which
is then extracted into a portion of carrier solvent for injection into the chromatographic column. Carrier pH is adjusted to minimize dissociation of the ion pair. By
this technique Karger et al. [12, 13] separated a variety of materials, e.g., carboxylic acids, sulfa drugs, thyroid hormones, and biogenic amines. They used tetrame-
thylammonium hydrogen sulfate (TMAHS) or perchloric acid/sodium perchlorate to form
the ion pair with the solute and as the stationary phase. Mixtures of butanol and
methylene chloride were used as the carrier. Figure 2 shows a typical separation
of sulfonamides [14].

Operation in the reverse phase permits the use of stable bonded phases as supports. The counter ion is added directly to the aqueous mobile phase thereby eliminating the need for an extraction step [15].

d. Bonded Phases

Phases can be bonded to the support to hold them stationary [16-19]. The retarding
mechanism of the bonded phase may be either partition or adsorption. The nature of
the phase and the manner of use determine whether the solute transfer is into the bulk
stationary phase or is limited to surface activity.

Several types of bonded phases are commercially available. Since phase stability
differs between types, a review of the most widely used structures is in order.

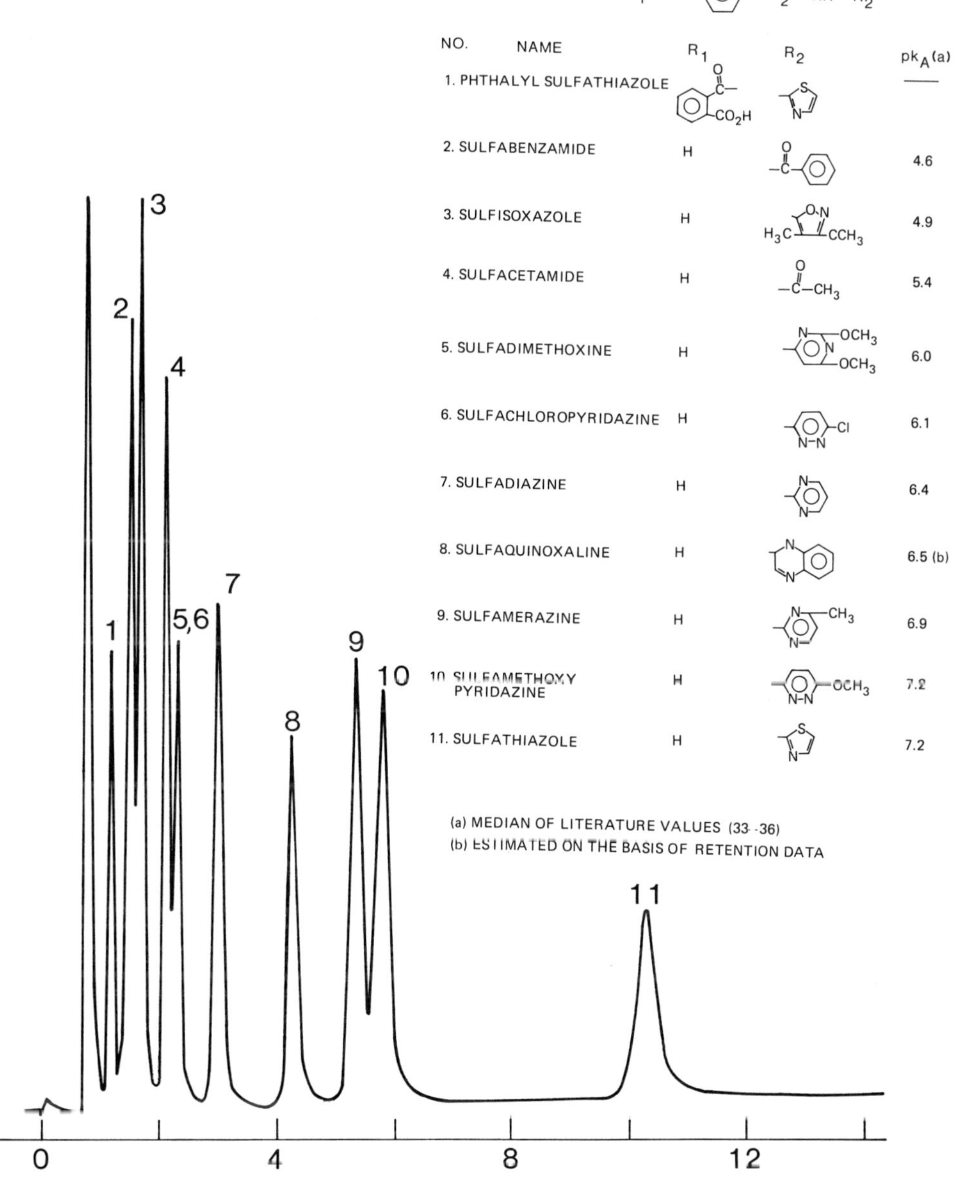

FIG. 2.   Separation of sulfa drugs by ion-pair partition chromatography: column: spherical silica, 10 μm (LiChrospher), 25 cm x 3.2 mm; stationary phase 0.04 *M* tetra-butylammonium hydrogen sulfate 0.10 *M* borate buffer (pH 8.4); mobile phase, *n*-butanol-heptane (25:75).   Compound: (1) phthalylsulfathiazole, (2) sulfabenzamide, (3) sulfisoxazole, (4) sulfacetamide, (5) sulfadimethoxine, (6) sulfachloropyridazine, (7) sulfadiazine, (8) sulfaquinoxaline, (9) sulfamerazaine, (10) sulfamethoxypyridazine, (11) sulfathiazole [14].   (Reprinted with permission of copyright owner.)

Basically, a chemically bonded packing consists of an organic species coval-
ently bonded to a silica particle bearing either a random surface [18, 19] or controlled surface porosity [20]. The particle usually consists of either a glass bead
($\bar{d}_p$ 25-30 μm) coated with a layer of silica or a porous microparticle ($\bar{d}_p$ 5-10 μm) of
silica. Porous particles may be either spherical or irregular in shape. The chemically bonded phase may consist of either a polymeric or monomolecular layer. Polymeric layers offer greater solute capacity but show a higher resistance to mass
transfer [21].

The commonly used bond structures are listed in Table 4. All of these types of
bonded-phase packings may be needed to cover a broad range of selectivity.

e. Silane Derivatives (Nonpolar Bonded Phases)

The organosilyl derivatives of silica are the most commonly used chemically bonded
phases. They are prepared by reacting octadecyl or phenyl silane with the silica
particle to obtain a nonpolar packing that is stable to virtually all conventional
carriers even at elevated temperatures. They are commonly used with carrier of aqueous
methanol or aqueous acetonitrile. (When attempting a separation of an unknown substance one should start with a 1:1 mixture of water and methanol, and then increase
the water content to increase retention or decrease water to decrease retention.)
Since retardation is controlled by limiting the solubility of the solute in the carrier, sample capacity is limited as compared to comparable normal phase adsorptive
packings. These materials offer a limited range of selectivities. While several
different phases of this general type are commercially available from several pro-
ducers, their retention characteristics are similar because solute retardation depends
more on solubility in the carrier than on the nature of the retarding force of the
packing.

These reverse-phase packings are most applicable to relatively nonpolar solutes
such as hydrocarbons and other materials which, though similar in polarity show a
significant difference in lipophilic character.

f. Polar Bonded Phases

Polar bonded phases in which polar groups such as ether or nitrile groups are coupled
to the silica by means of an -Si-C- bond afford an increased range of selectivity
[20, 22-25]. These materials can be used with nonpolar carriers such as hexane modified with chloroform or isopropyl chloride. It appears that selective adsorption
rather than reverse solubility is a dominant influence in retardation with the polar
bonded packings.

g. Silica Esters: "Brushes"

"Brushes" [19, 23] offer a wide range of possible derivatives of silica. Virtually
any product with an active hydrogen, e.g., polyols, nitriles, or aliphatic alcohols,

TABLE 4. Commonly Used Methods of Bonding Stationary Phases

Type	Reported structure	References	Remarks
Hydrocarbon derivatives (octadecyl silane)	$\begin{array}{c} \text{O-Si} \\ \text{R-Si-O-Si} \\ \text{O-Si} \end{array}$	[17, 18]	Nonpolar
Multifunctional derivatives (amine, ether, nitrile)	$-Si-O-\left(-Si-O-\right)_n$ ROR, O	[20, 22]	Moderately polar
Esterified silica	$-Si-O-R$	[23, 24, 19]	Hydrolytically unstable (wide polarity range)
Nitrogen derivatives	$-Si-N\big\langle$	[25, 26]	Limited pH stability range (5-7)

will esterify activated silica. The disadvantage of silica esters in liquid chromatography is that they are readily hydrolyzed by water and suffer alcohol exchange. They, therefore, have a limited shelf life unless well protected from moisture and they cannot be used with carriers containing either water or lower alcohols. Brushes are normally used with hydrocarbon carriers which may be modified as needed with chloroform or acetonitrile. A carrier more polar than chloroform is seldom necessary even for quite polar solutes.

4. Exclusion Chromatography

Exclusion chromatography, also known by the names of gel permeation chromatography (GPC) and gel filtration chromatography (GFC), is a method of separating molecules by size in solution. The method is applicable to virtually any soluble species from small molecules with an average molecular weight $(\overline{M})$ of less than 100 to very large ones exceeding 10 million. To make a size separation, the sample is passed through a column that has been tightly packed with a porous permeable gel in a carrier which fills the gel pore and interparticle spaces. Sample molecules are free to diffuse into any liquid available to them. Small molecules can diffuse into all of the gel's pores while large molecules are excluded from the pores and must remain in the interstices. Intermediate molecules can penetrate some fraction of the pores (K_0) in accordance with their size in solution. As the carrier passes through the column it carries with it the molecules contained in the interstitial carrier. The molecules inside the pores are left behind until they can diffuse back out of the pore and into the moving carrier. Large molecules that cannot enter the pores are swept along with the carrier and therefore are eluted first. Small molecules that spend the greatest amount of time in the pore are eluted last. Thus, any species is eluted at a volume exactly equal to the pore volume available to it in the column $(K_0 K_{pa})$. Since the exchange is between the two bulk phases, i.e., the trapped carrier and moving carrier, liquid exclusion chromatography (LEC) is a form of absorption chromatography.

Because the carrier and the stationary phase are of the same composition the classical partition coefficient $K_d = [s]/[m] = 1$. In this mode, the retardation is controlled by the permeation coefficient K_0 defined as

$$K_0 = K_d \frac{V_{pa}}{V_{pg}} \tag{2}$$

where

V_{pa} = pore volume available to any volume species

V_{pg} = pore volume of the gel

K_d = partition coefficient

The separation is therefore described by a modification of the classical elution equation [27]:

$$V_r = V_m + K_0 V_s \tag{3}$$

where

V_r = retention volume

V_m = volume moving phase

K_0 = pore fraction available

V_s = volume of trapped carrier

Since $K_d = 1$ it vanishes from the equation. When $K_0 = 0$ the species is excluded. When $K_0 = 1$ the species is retained to the extent possible which is one column volume $(V_s + V_m)$. When $1 > K_0 > 0$ the molecules are retained in accordance with their size which may be measured by means of a calibration curve of the type illustrated in Fig. 3.

a. Packing Selection in Exclusion Chromatography

Packing selection is based on the permeation range of the gel. Since the pore size distribution of a given gel is confined to certain limits, the permeation range must be matched to the hydrodynamic size of the solute as illustrated diagramatically in Fig. 3. The permeation range of the system may be increased by coupling columns packed with gels of different porosities. Since coupling order does not seem to affect performance at normal sample level, it is a common practice to arrange gels in order of decreasing porosity. Gels of different porosity may be intermixed in the same column to expand the permeation range and to linearize the calibration curve. This practice tends to reduce the separating efficiency for a given column size and must be compensated for by increasing column length.

Since permeation range is the property of interest, representative calibration curves should be obtained to aid in the selection. Table 5 lists exclusion packings from commercial sources. Caution is required when comparing packings from different suppliers since different units have been used to describe the various packings. Pore size may be given in one of the following dimensions.

1. Molecular weight of the calibration standard, e.g., polystyrene, polyglycol, polydextran, polypeptide, or a protein. (The dimension may describe either the exclusion limit or the permeation range.)

2. Average diameter of the gel pore as determined by mercury porosimetry (pore diameter).

3. Extended chain length of a polystyrene molecule excluded from the gel pore.

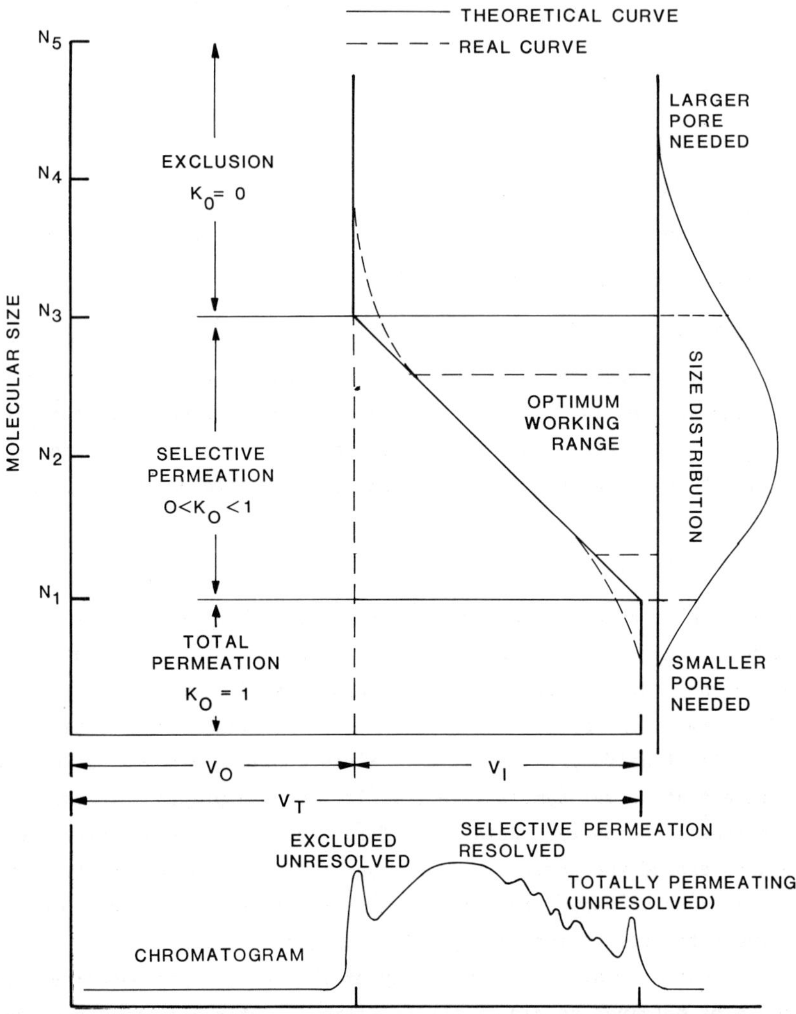

FIG. 3. Diagramatic representation of the interrelationship between the molecular-size distribution of the sample (top right), the calibration curve (top left), and the chromatogram (bottom) [27]. (Reprinted with permission of copyright owner.)

Exclusion packings can be reduced to a common base by an empirical calibration. However, the curve can shift with changes in operating conditions because molecules can change size with change in solvating power of the carrier and the pore size of nonrigid gels can change with the degree of swelling.

b. Carrier Selection for Exclusion Chromatography

A carrier for exclusion chromatography should dissolve the sample and be sufficiently similar to the packing to suppress sorption. Ideally, it should be low in viscosity

TABLE 5. Types of Exclusion Packings from Commercial Sources

Type[b]	Trade name[d]	Particle shape[c]	Use[a]	Vendor source
Silica	LiChospher	SP	B	E-M Laboratories
Silica	EM Si gels	IR	B	E-M Laboratories
Silica	Spherosil	SP	B	Supelco, Inc.
Silica	Zorbax	SP	B	E. I. DuPont
Silica	Porasil	SP	B	Waters Associates
Modified Silica	Sil-X	IR	B	Perkin-Elmer Corporation
Porous Glass	CPG	IR	B	Electronucleonics, Inc.
Porous Glass	Corning glass	IR	B	Corning Scientific Instruments
Porous Glass	Bioglass	IR	B	Bio-Rad Laboratories
Polystyrene	Styragel	SP	O	Waters Associates
Polydextran	Sephadex	IR	A	Pharmacia Fine Chemicals
Polydextran	Sephadex LH-20	IR	O	Pharmacia Fine Chemicals
Polyacrylamide	Biogel	IR	A	Bio-Rad Laboratories
Agarose	Biogel (A)	IR	A	Bio-Rad Laboratories
Agarose	Sephadex	IR	A	Pharmacia Fine Chemicals

[a] O = organic carrier, A = aqueous, and B = either aqueous or organic.

[b] Unmodified silica-type packings may require a polar modifier when used with polar solute to suppress adsorption.

[c] SP = spherical; IR = irregular shape.

[d] All names are registered trademarks of the listed company.

and compatible with the detector. Solvents commonly used as LEC carriers are listed in Table 6.

Under most circumstances, carrier composition does not affect solute retention in exclusion chromotography, as it does in affinity chromatography. The carrier can influence retention either by altering the size of the solvated molecule [28] or by altering pore dimension by swelling the gel. Organic molecules dissolved in a theta solvent will be smaller and will therefore be retained slightly longer than in a fully relaxed state [29]. This effect is much greater with polyelectrolytes and certain biopolymers which require the addition of an electrolyte to the aqueous carrier to prevent relaxation in solution [27]. Modifiers may be added to exclusion carriers to suppress adsorption particularly with glass and silica type packings. A moderate amount of an appropriate glycol is often added to either the aqueous or organic carrier to suppress adsorption.

TABLE 6. Physical Properties of Solvents Commonly Used for GPC at or Near Ambient Temperature

Solvent	Boiling point ($^{\circ}$C)	Density (20°)	Viscosity (25°)	Refractive index (20°)
Tetrahydrofuran	66.0	0.8892	0.51	1.4070
Ethylene dichloride	84.0	1.2569	$0.84^{20^{\circ}}$	1.4443
Toluene[a]	110.6	0.866	0.52	$1.4893^{24^{\circ}}$
Chloroform	61.2	1.489	$0.56^{20^{\circ}}$	1.4476
N,N'-Dimethylformamide	153.0	$0.9445^{25^{\circ}}$	0.90	$1.42803^{25^{\circ}}$
1,1,2,2-Tetrachloroethane[b]	146.5	$1.58658^{25^{\circ}}$		1.49419
Dimethylsulfoxide	189.0	1.100	2.2	$1.4787^{21^{\circ}}$
Water[c]	100.0	0.9999	$1.0^{20^{\circ}}$	1.3330

[a]Generally used at 80°C.

[b]Generally used at > 135°C for polyolefins.

[c]Aqueous buffers may be required to stabilize pH, molecular composition, or molecular size.

B. Theoretical Treatment of the Retardation Process

Solute retention volume (V_r) is determined by three factors as defined in Eq. (4).

$$V_r = V_m + K_d V_s \tag{4}$$

where

V_m = volume of the moving phase

V_s = volume of the stationary phase

Since V_r is a function of column volume and phase load, as well as K_d, the normalized term capacity factor (K') is generally preferred because it is independent of column dimension. Capacity factor (K') also known as capacity ratio and phase ratio is defined in Eq. (5) as:

$$K' = \frac{K_d V_s}{V_m} = \frac{V_r - V_m}{V_m} = \frac{t_s}{t_m} \tag{5}$$

where

t_s = time solute in stationary phase

t_m = time solute in mobile phase

Since $K_d = [S]/[M]$, it is evident that

$$K' = \frac{[S]V_s}{[M]V_m} = \frac{M_s}{M_m} \tag{6}$$

where

M_s = mass of solute in the stationary phase

M_m = mass of solute in the total mobile phase.

Equation (6) shows that solute retention is a function of the mass distribution of the solute between the stationary phase and the mobile phase. K', therefore, is an indication of how well the stationary phase is used. A K_d of 2 indicates that the solute spends twice as much time in the stationary phase as in the mobile phase and requires (2 + 1) column void volumes to transport the solute through the column. It further indicates that twice as much of the solute mass is contained in the stationary phase as in the mobile phase. It is therefore a most basic parameter of the chromatographic process.

To achieve a chromatographic separation between two solutes requires that the retardation of one solute species be greater than that of the other as they are swept through the column by a carrier, that is, $K'_2/K'_1 > 1$.

In practice, all molecules of the same species do not pass through the column at the same rate. Consequently a solute band spreads as it passes through the column while subjected to the migration and retardation process. Two bands which migrate through the column at different rates will not be resolved unless their respective widths are sufficiently narrow as to prevent overlap.

Two components may be considered to be resolved when their peak centers are separated by one peak width that is, when R = 1. R is defined as

$$R = \frac{V_2 - V_1}{\overline{W}} \tag{7}$$

where

V_2, V_1 = retention volume of components 1 and 2

$\overline{W}$ = average peak width

The separation is complete when one solute elutes from the column in a different volume of carrier than the other, i.e., when $R \geq 1.5$. Figure 4 illustrates how two columns providing the same separation of peak center (i.e., the same retardation) can differ significantly in resolution because of differences in w. (To see the effect of doubling R with w constant, compare the center curve with the top, and with a constant, compare the center with the bottom curve.)

Solute band width is a reciprocal function in the separation process that must be minimized since its influence can nullify the effect of the selective retardation

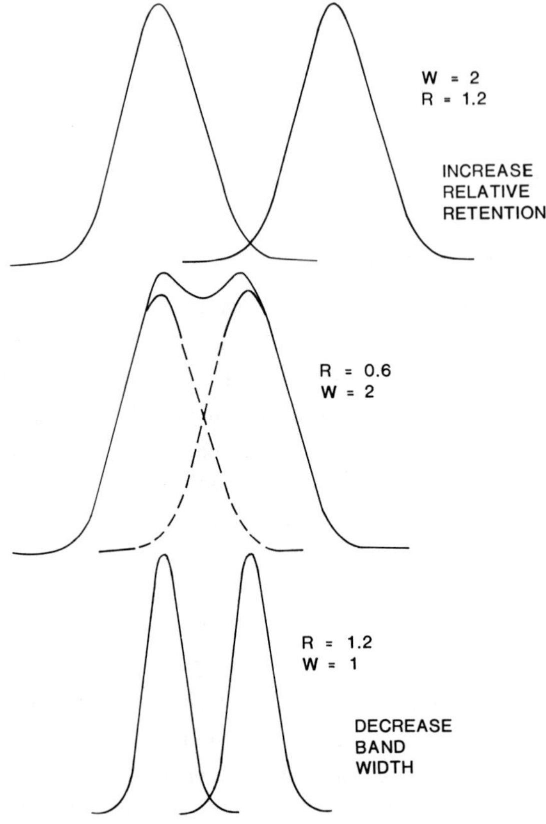

FIG. 4. Effect of band width and band position on resolution of gaussian peaks.

process. Since W varies with column conditions as well as column performance the term HETP, or height equivalent to a theoretical plate, is the preferred term to describe column efficiency because theoretically H is independent of column dimension and solute retention.

$$H = \frac{L}{N} = \frac{LW^2}{16V_r^2} \tag{8}$$

where

L = column length

$N = 16V_r^2/W^2$

    = number of theoretical plates

V_r = solute retention volume in ml

W = peak width in ml

It is worth noting that in practice H is not independent of either column length or retention volume. Within limits, H generally decreases with increasing retention and increases with increasing column length.

If under a given set of test conditions column length L and retention volume V_r are held constant, Eq. (8) becomes

$$H = aW^2 \tag{9}$$

Equation (8) shows that H is proportional to W^2 and may therefore be treated statistically as a variance, i.e.,

$$H_{total\ column} = \sum (H_1 + H_2 + H_3 + \cdots + H_n)$$

but this condition holds only when L and V_r are held constant for all members of the set. However, peak widths may be added as variances, i.e.,

$$W_T^2 + W_1^2 + W_2^2 + \cdots + W_n^2 \tag{10}$$

This principle is an important consideration in the utilization of coupled columns where one bad column can destroy the efficiency of several good ones [30]. The peak width W_i for an individual column is measured directly or is calculated using Eq. (11).

$$W_i = \frac{4V_r}{\sqrt{N}} \tag{11}$$

The sum of (W_T) of n columns is then calculated for the coupled column with Eq. (10) from which the plate number of the sum can be calculated using Eq. (12).

$$N_{total} = \frac{16V_{r\ total}^2}{W_T^2} \tag{12}$$

Any other method for summing column efficiencies can lead to erroneous results.

1. Theoretical Plates

The term theoretical plate N defined in the preceding section is commonly used to define a column's performance (where a large value for N signifies a good column and a small N signifies a poor column). It is a common practice to increase column length to increase N. However, plate numbers are not additive. For example, a 4-ml column of 4000 plates coupled to a column of similar size but yielding only 1000 plates would not yield 5000 plates. By Eq. (10), the coupled set would yield only 3200 plates. Plate numbers may be added only when the coupled segments are identical in size and performance. Otherwise, peak widths must be added as variances and the resultant N determined with the plate equation. Coupling is advantageous only when segments of nearly equal quality are combined.

Much emphasis has been placed on column efficiency H and plate number N since a band broadening caused by a poorly packed column can prevent a possible separation. Plate number alone, however, does not indicate the columns separational capability. It merely indicates the solute band width relative to the distance traveled by the carrier front.

2. Resolution

The measure of band separation is resolution R defined by Eq. (13). Three factors determine R: (A) differential retardation, commonly called the *selectivity term*, (B) fraction of the solute mass retarded, commonly called the *capacity term*, and (C) transfer number, commonly called the *plate term*. Thus, the two equivalent expressions for resolution are

$$\text{(A)\,(B)\,(C)}$$

$$R = \left(\frac{\Delta V}{V'_r}\right)\left(\frac{V'_r}{V_r}\right)\left(\frac{V_r}{W}\right) = \frac{\Delta V}{W} \tag{13}$$

$$\text{(A)\ (B)\ (C)}$$

$$R = \left(\frac{\Delta t_r}{t'_r}\right)\left(\frac{t'_r}{t_r}\right)\left(\frac{t_r}{W}\right) = \frac{\Delta t_r}{W} \tag{13a}$$

The respective parameters are defined diagramatically in Fig. 5. The factors are defined in classical terms as follows:

(A) Column selectivity $\alpha = K^2/K^1$ relates to A as

$$A = \frac{\alpha - 1}{\alpha} = \frac{K^2/K^1 - 1}{K^2/K^1}$$

$$= \frac{(V^2 - V_M)/(V^1 - V_M) - 1}{(V_2 - V_M)/(V_1 - V_M)} = \frac{\Delta V}{V'_r} \tag{14}$$

(B) The capacity factor K' defined in Eq. (5) relates to B as

$$B = \frac{K'}{1 + K'} = \frac{V_2 - V_m}{V_2} = \frac{\Delta V}{V} \tag{15}$$

When K' is defined in terms of solute mass distribution as per Eq. (6), the capacity term of the resolution equation becomes

$$\frac{K'}{1 + K'} = \frac{M_s/M_m}{1 + M_s/M_m} = \frac{M_s}{M_s + M_m}$$

$$= \frac{\text{Mass of solute in the stationary phase}}{\text{Mass of solute transported}} \tag{16}$$

In this form it is evident that the capacity term of the resolution equation defines the fraction of the solute mass retarded per peak width.

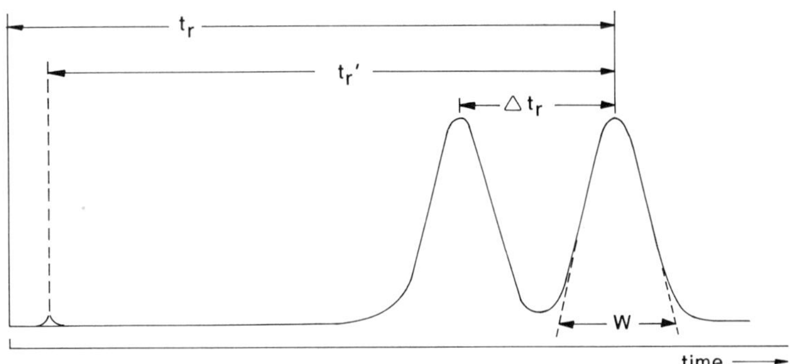

FIG. 5. Diagramatic representation of the expanded resolution equation expressed in terms of retention time and peak width [34]. (Reprinted with permission of copyright owner.)

(C) Column plate number N relates to C as

$$C = \frac{1}{4}\sqrt{N} = \frac{1}{4}\sqrt{\frac{16V_r^2}{W^2}}$$

$$= \frac{V_r}{W} = \text{Band capacity}^{\dagger} \tag{17}$$

indicating that the band capacity is a basic parameter of the chromatographic process.

The resolution equation is defined in traditional terms as [32, 33]

$$R = \left(\frac{\alpha - 1}{\alpha}\right) \left(\frac{K'}{1 + K'}\right) \left(\frac{1}{4}\sqrt{N}\right) \tag{18}$$

R can also be defined in terms of normalized volume K' as

$$R = \left(\frac{\Delta K'}{K'}\right) \left(\frac{K'}{1 + K'}\right) \left(\frac{1 + K'}{W/V_m}\right) \tag{19}$$

where V_m, in this case, is the total volume moved or V_r. The various expressions for the resolution equation are included here to further illustrate analytically the concept of resolution as a differential retardation process and should be helpful to understanding the mechanism of the process. When combined into a single statement, R is the product of the difference in migration rate, the fraction of time the solutes spend in the stationary phase, and the reciprocal band width or column band number. A difficult separation may require maximum optimization of the three factors.

---

†The term *band capacity* should not be confused with the term *peak capacity* defined by Gidding as $n = 1 + C \log (V_n/V_1)$ [31]. The band capacity represents the number of band turnovers and is somewhat analogous to the number of stages in a counter current extraction. For example, a four-stage extraction would be equivalent to a separation on a column offering 256 plates.

TABLE 7. Relationship between Basic Units and Term Values of the Expanded
Resolution Equation [34]

α	A	K'	B	N	C
1.05	0.05	1	0.50	200	3.5
1.10	0.09	2	0.67	500	5.6
1.15	0.13	3	0.75	1,000	7.9
1.20	0.17	4	0.80	2,000	11.2
1.30	0.23	5	0.83	4,000	15.9
1.50	0.33	6	0.86	6,000	19.4
2.0	0.50	7	0.88	8,000	22.4
5.0	0.80	8	0.89	10,000	25.0
50.0	0.98	50	0.98	40,000	50.0

When flow rate is held constant, time (expressed by chart travel) becomes a
valid indication of volume, and Eq. (17) becomes

$$R = \left(\frac{\Delta t}{t_r'}\right) \left(\frac{t_r'}{t_r}\right) \left(\frac{t_r}{W}\right) \tag{20}$$

In this form the equation permits four simple distance measurements taken directly from the chromatogram, and used in three ratios, to provide a numerical analysis of the separation [34]. Each term contributes directly to resolution. It is
evident from Table 7 that terms A and B each approach unity as a limit. Term C is
simply the elution volume measured in peak widths which as a reciprocal of band width
becomes the *band number* of the column. At constant H, the band number is independent
of V_r or K'. In practice, the A term should be greater than 0.14 (α = 1.16) to achieve
a good separation on a typical column. [As an illustration in the use of the data in
Table 7, assume that the selectivity term A = 0.06 and the capacity term B is adjusted to 0.89. Since R = ABC, a band number (C) of 25 would be required to provide
baseline resolution. Since $C = (1/4)\sqrt{N}$ a column of 10,000 plates would be required
to provide a band number of 25 required to make the separation.]

III. APPROACHING A SEPARATION PROBLEM

A systematic method of approaching a new separation problem is defined by the following steps

   1. Select a mode of separation,

   2. Select a column representing the mode,

   3. Choose a carrier of moderate strength,

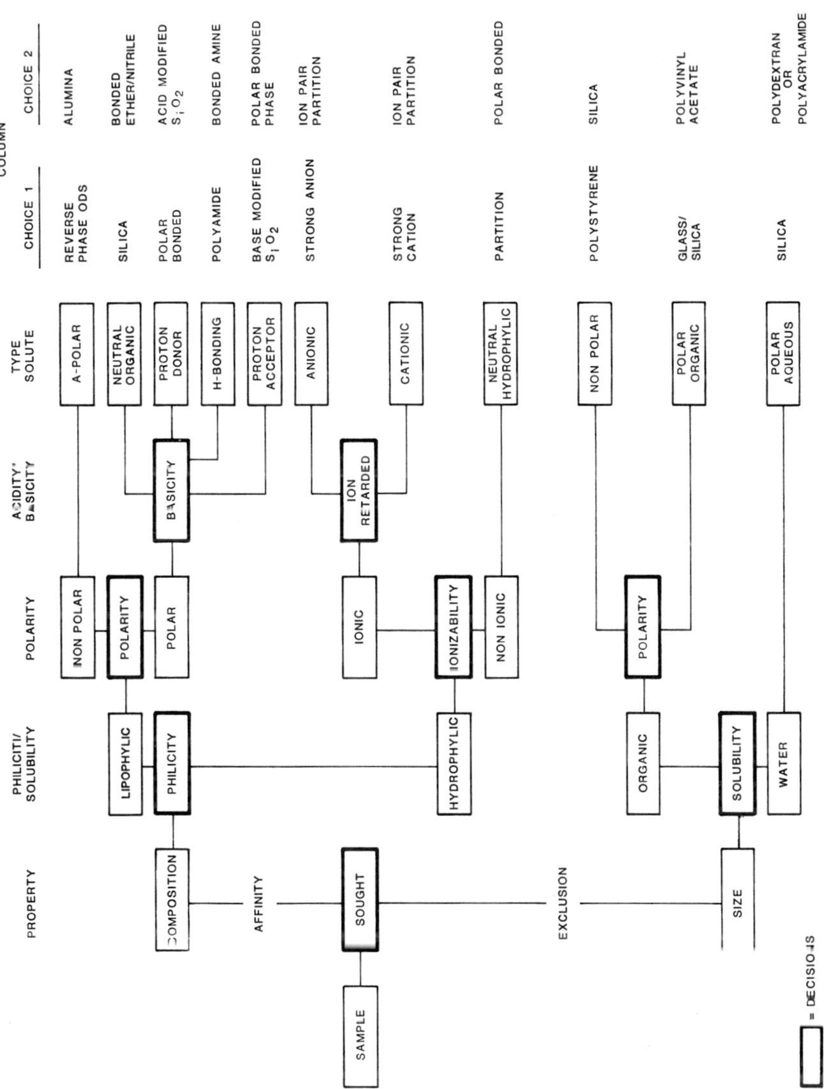

FIG. 6. Column selection chart for HPLC. To use the scheme, information is entered into the decision diagram at the left and advanced to the right. Bold-faced blocks indicate decisions; regular blocks denote solute properties.

4. Make a trial separation,

5. Optimize the separation.

A. Column Selection Scheme

The schematic diagram shown in Fig. 6 provides guidelines for making selections. The
dominant factors in the selections are

1. The information sought (size vs composition)

2. The solubility of the solute

3. The chemical nature of the solute

1. Property of Interest

The first choice in the scheme is based on the property of interest. When molecular
size is the property of interest, exclusion chromatography is the mode of choice ir-
respective of the size of the solute molecule. Exclusion chromatography is often used
for materials with $\overline{M} > 2000$ because small differences in composition are inadequate
to provide a basis for a separation between related species; also, because the effect
of these differences are masked by the effects of differences in size.

If molecular composition is the property of interest, the preferred mode is the
one that will produce the best selectivity. While a problem may be encountered where
the greater selectivity is provided by the exclusion mode, such cases are usually the
exception to the rule that the affinity mode is preferred to separate components which
differ in composition.

2. Philicity

The second choice in the scheme is based on the lipophilic/hydrophilic character of
the solute. The term *philicity* is used to denote "preference" since many substances
are soluble in both organic and aqueous solvents and indeed might under appropriate
circumstances be handled in either medium. Where multiple routes are available, the
first choice should follow the route of greatest philicity. The choice may then be
altered if the desired selectivity is not obtained by the first choice.

3. Polarity

While solute polarity is related to solubility, it is a separate consideration in
mode selection. Chromatographic polarity is an empirical concept that cannot be defined in absolute terms and must be related to the system of measurement. Many factors affect solute polarity, a discussion of which is beyond the scope of this chapter.
However, the following example illustrates the concept of polarity as related to functional group activity without regard for molecular interactions [35, 36].

$$-NH_3^+X^- > R\overset{\displaystyle O}{\overset{\displaystyle \|}{C}}OH > -OH > -NH_2 > -OCOCH_3 > -NR_2$$

$$> -NO_2 > -OR > aryl > alkyl > -Cl-C_xF_y$$

4. Basicity

The basicity of a solute relates to its ability to act as either an acid or a base, and to form hydrogen bonds in various chromatographic media. Solute basicity is a significant consideration in selecting a workable (tailing-free) chromatographic system, particularly for strongly polar solutes such as are encountered in biologically active materials.

5. Type of solute

The diverging scheme (Fig. 6) culminates in listing 11 types of solutes for which columns and carriers must be selected. Table 8 lists the first and second choice of columns and carriers suggested for the trial separation for the respective problem types.

B. Porous Versus Porous Layer Packings

A choice must still be made between totally porous and porous layer packings. Following are some criteria to aid in the selection.

Totally porous packings are usually preferred when

1. High efficiency is needed over a narrow range of K' as with a mixture of isomers,
2. A heavy sample load is needed to detect a trace component,
3. Both analytical and preparative separation are required of the same column,
4. The component of interest is a minor component similar in structure to a major component that comprises more than 90% of the mixture.

Porous layer packings are preferred when
1. Components in the sample show a wide range of polarity or composition differences and can therefore be expected to show a wide range of relative retentions.
2. Components are present in similar amounts; no component sought is present at a level below 0.1%.

Thus, when attempting to isolate an impurity which is an isomer of a major component in a sample, the fully porous microparticle packing is probably preferred. However, when the sample is made up of a wide-range mixture of nonpolar and polar materials at similar concentration levels, the porous-layer bead is the wiser choice.

TABLE 8. Column Systems for Various Types of Solutes

Type solute	First Choice			Alternative		
	Column	Carrier	Modifier	Column	Carrier	Modifier
Affinity:						
Apolar	Reverse phase	Water	Methanol or acetonitrile	Al_2O_3	Hexane → $MeCl_2$	Alcohol, H_2O
Neutral organic	Silica	Hydrocarbon	$CHCl_3$ → Alcohol → H_2O	Bonded ether or nitrile	Hydrocarbon	Alcohol
Proton donor (weakly acidic)	Polar bonded Phase	Hydrocarbon	Alcohol	Acid-modified SiO_2	Organic	Alcohol, H_2O, HAc
H Bonding	Polyamide	Hydrocarbon	Alcohol HAc	Polar bonded phase	Hydrocarbon $CHCl_3$	Alcohol
Proton acceptor (weakly basic)	Silica	Hydrocarbon	Alcohol NH_3 or Amine	Polar bonded phase	Hydrocarbon	Polar organic, alcohol
Anionic	Anionic resin	Water	[Anion] pH	Silica aq/ $HClO_4$, $NaClO_4$	$MeCl_2$	Butanol
Cationic	Cationic Resin	Water	[Cation] pH	Silica/TBA, HSO_4	Hexane	Butanol
Neutral polar	Silica, Water	Hydrocarbon	Alcohol or dioxane ternary	Bonded amine	Organic	Water, alcohol, or acetonitrile
Exclusion:						
Nonpolar	Polystyrene	THF toluene		Silica	THF or toluene	Triethylene glycol
Polar organic	PVA or silica	THF		Silica or glass	THF or toluene	Triethylene glycol
Polar aqueous	Silica	Water	Buffer salts, triethylene glycol	Polydextrous or polyacrylamide	Water	

Because the porous-layer packing can elute a wider range of material at a single solvent strength, it is the preferred packing for scouting the new sample [37]. When little is known about a sample or if the choice is not clear, the porous layer packing should be the first choice. If experimental data indicate that higher efficiency or greater capacity is needed, then a fully porous packing can be employed. Since optimum carrier strength is not as critical with porous layer packings, it is often desirable to use the porous layer packing for initial screening even when preliminary results indicate that a fully porous packing will ultimately be required. Porous microparticles are preferred where high resolution is required within a narrow range of K´, while porous layer packings are generally preferred for applications where the components elute over a wide range of K´.

C. Phase Selection

Phase selection is a highly empirical art because selectivity is to a degree unpredictable. Guidelines are presented here to aid the practitioner. Table 9 shows an arbitrary numerical rating for commonly used stationary phases; however, these guidelines are not a substitute for empirical data. Therefore when attacking a new problem, first consideration should be given to the available data. The questions to be answered here are:

> Has the material or similar material been separated previously?
> Under what conditions?
> Is the information applicable to this problem?

If similar materials have been separated, then use that information to help solve your problem. Usable information may be in the form of published chromatograms,

TABLE 9. Sensitivity[a] of Packing Type to Structural Differences [38]

	Adsorptive	Partition	Polar Bonded	Nonpolar Bonded
Molecular size	2	4	1	4
Structural isomers	2	1	3	
Steric isomers	3	3	3	
Optical isomers	1	1		
Unsaturation	4	1		2
Number of nonpolar substitutions	2	2	3	3
Number of polar substitutions	5	5	4	2
Polarity of substitutions	5	5	4	1

[a]Sensitivity 5 > 1.

R_f values from TLC separations, or relative retention data from classical LC separations. Such data are useful in answering the three questions that are fundamentals to LC.

1. What packing retards the solute?
2. What carrier moves the solute?
3. Is there a difference in the solute migration rates?

1. Adsorptive Stationary Phases

When published data indicate that a separation is feasible, materials comparable to those used previously can generally be applied to the high-pressure system. Because of the abundance of experience on silica and the ease of running a preliminary separation by TLC, silica is an ideal starting material. If published R_f values are lacking, a preliminary trial by TLC can easily be made. Valuable information can be gained not only on the choice of column and carrier but on the complexity of the sample as well. R_f values can be related to K' and subsequently to α by the equation

$$R_f = \frac{1}{1 + K'} \tag{21}$$

where $K'_{LC} \leq K'_{TLC}$ and $\alpha = K'_2/K'_1$. The relationship plotted in Fig. 7 shows that R_f values between 0.1 and 0.5 are most applicable to column chromatography. While the TLC developer may require a slight adjustment for optimum performance, its composition can be used to select the starting composition of the carrier.

Retention values at a given carrier strength are frequently lower in column chromatography than in thin layer chromatography, requiring a strength adjustment to achieve the same retention. With a little experience the proper adjustment can usually be made with a minimum of trial and error.

Although numerous adsorbents have been used for modern HPLC, siliceous packings are most commonly used. Table 10 lists some typical high-efficiency packings which are available from commercial sources. Most of these are sized particles with a 60- to 150-Å pore diameter. Zorbax and Zipax are produced by a patented process to provide a controlled surface porosity while Sil-X and Vydac have chemically modified surfaces to limit the surface energy. Conventional 60-Å silica gel has a "specific" surface area of 200-400 m^2/g. This applies to both the porous gels and the porous layer gels. Thus, while the porous-layer packing (e.g., Perisorb and Corasil) report lower surface areas due to the solid core, the effective surface is just as active as a high-area adsorbent and should be treated accordingly.

A range of lower surface area siliceous packings, sold for use in exclusion chromatography, can be used as low-energy adsorbents. These materials are produced in a range of pore diameters and surface areas. The user should consult the vendor's

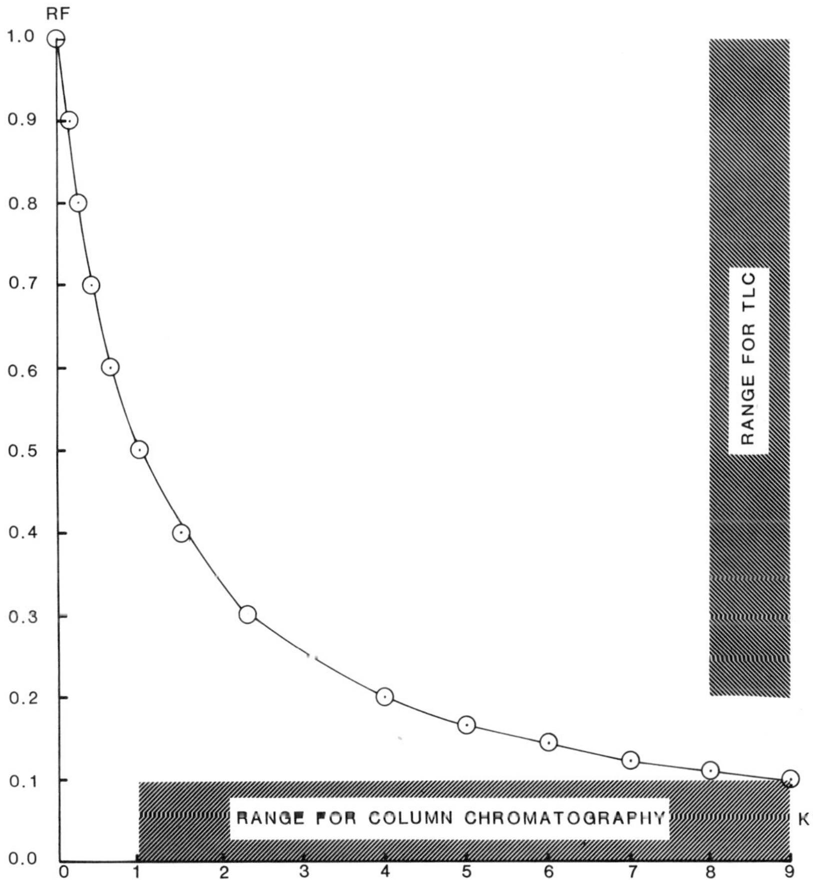

FIG. 7. Relationship between R_f and K' in liquid chromatography.

catalogs for the specifications of these materials. They include E. M. SiGel, Spherasil, and Porasil. The lower surface area adsorbents normally require less modifier to effect elution at a given K' than do 60-Å gels, and they offer a more narrow range of retardation. When molecular size is not a problem and solvent programming is available, the high-area packings offer the broadest capability. Low surface-area adsorbents may therefore be considered special purpose adsorbents but they should not be overlooked for high-polarity solutes.

Silica gels with pore diameters greater than 150 Å are useful with high molecular-weight solutes because the larger pores minimize adverse effects from restricted diffusion within the pores. Thus, a 150 Å silica may be preferred to a 60-Å silica for solutes where $\overline{M} > 1000$. Use of wide-pore adsorbents may be used to extend the practical range of affinity chromatography into the higher molecular-weight region.

TABLE 10. Some Typical Silica Adsorbents for HPLC from Commercial Sources

Vendor[a]	Spherical porous particles (5 and 10 μm)	Irregular porous particles (5, 10, 20, 30, 40 μm)	Spherical porous layer (∿35 μm)
E. M. Laboratories	LiChrospher	Lichrosorb	Perisorb
E. I. DuPont	Zorbax		Zipax
The Separations Group		Vydac	Vydac
Waters Associates, Inc.	Porasil	Porasil	Corasil
Whatman, Inc.		Partisil	Pellosil
Spectra-Physics	Spherisorb		
Perkin-Elmer Corp.		Sil-X	
Varian		MicroPak	

[a]All names are registered trademarks of the listed company.

a. Chemically Modified Silica

The chemically modified adsorbents, such as Vydac and Sil-X, do not require the addition of water to block their high-energy sites. They may be used with polar solutes with alcohol modified carrier without tailing. They are, however, incapable of discriminating between paraffins and aromatics and other nonpolar species.

While many nonpolar materials can be separated on fully active silica with a nonpolar carrier, e.g., fluorocarbon or dry hexane, they can usually be separated more readily in a reverse-phase system.

2. Alumina

Alumina adsorbents are available as both porous and porous-layer particles from a number of the vendors listed in Table 10. The selectivity of alumina is slightly different from silica and may therefore afford an advantage even though the efficiency of alumina colums generally runs significantly lower than that of silica. The ability of alumina to display acid-base characteristics, plus its unique selectivity to aromatic compounds, does justify its consideration under appropriate circumstances.

3. Bonded Phases

Bonded phases for both direct and reverse phase applications are experiencing increased use. Since retention data from outside sources is limited, these packings except for the popular octadecylsilanes, often receive secondary consideration to silica. Bonded phases should receive broader consideration since they are applicable to a wide range of solutes and offer unique selectivities. As illustrated by Fig. 8,

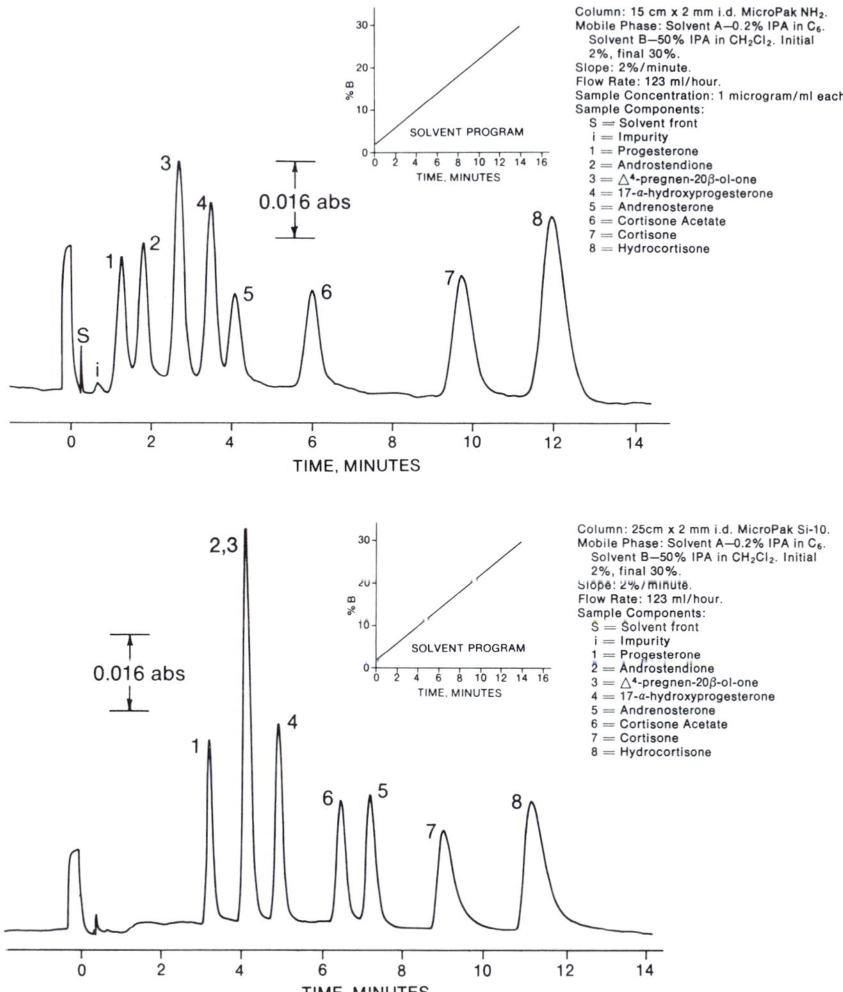

FIG. 8. Comparison of the selectivity of a MicroPak NH₂-type bonded phase with that of a conventional microparticle silica column for steroids, some of which contain functional groups exhibiting Bronsted basicity. (Courtesy Varian Instrument Division.)

they sometimes afford separations not provided by siliceous adsorbents. If the rule "like seeks like" is applied, then bonded phases containing functional groups such as ethers, nitriles, and amines should be considered for use with solutes having similar functionality. The criteria defined for partition systems can also be applied to bonded-phase systems.

Polar bonded phases are particularly desirable for use with strongly polar materials which normally would require highly modified carriers to elute them from

silica with symmetrical peaks. Ideally these phases are selected according to functional group, basicity and polarity in order to attain favorable retention (1 < K´ < 10)
with a simple carrier.

The nonpolar bonded phases such as the octadecyl silane derivatives of silica are
the packing of choice for solutes less polar than benzene using a reversed phase system. Support-coated mechanically bonded phases consisting of hydrocarbon polymers that
are stabilized by either cross-linking or by solubility are also available for use
where selectivity warrants.

Bonded phases containing a variety of functional groups are available from vendors
listed in Table 10. Functional groups include ethers, amine, and nitrile, as well as
the commonly used hydrocarbon polymeric phase. Polar phases are bonded by means of
silicon to carbon or silicon to nitrogen bonds as either polymeric or monomolecular
layers. Vendors' literature should be consulted for a description of their respective
materials. Monomolecular layers are prepared to afford faster mass transfer, however,
relevant data validating this claim are limited.

4.   Ion-Exchange Phases

Packings for ion-exchange chromatography are either cationic or anionic. Cationic
packings are used to retard cations while anionic resins retard anions. Since some
organic solutes are amphoteric, they may be used with either type of packing.

Cationic resins are received from vendors in the stable sodium form while anionic
resins are received in the chloride form. Prior to use they should be equilibrated in
the carrier ions since activity depends upon the charge strength or affinity of the
counter ion attached to the resin's surface. The counter ion must be selected to provide the activity desired. For example, a cation resin is most active in the lithium
form and least active in the silver form. The sequence of affinities for univalent
counter ions is as follows: $Li^+ < H^+ < Na^+ < NH_4^+ < K^+ < Rb^+ < Cs^+ < Tl^+ < Ag^+$. Affinity increases with molecular weight and with valency. For bivalent ions the affinity sequence is $Be^{2+} < Mg^{2+} < Ca^{2+} < Sr^{2+} < Ba^{2+}$. The affinity sequence for anions is:
$F^- < HCO_3^- < OH^- < Cl^- < BrO_3^- < HSO_3^- < CN^- < Br^- < NO_3^- < I^-$. Consequently, resins in
the nitrate form would be less active than in the chloride form. Because this sequence may vary somewhat with various types of resins, the vendors' catalogs should
be consulted for guidance.

Both pellicular and porous resins are commercially available. The fully porous
resins ($d_p \simeq$ 5-10 μm) afford highest efficiency and capacity but require closely controlled gradients to elute solutes having a wide range of charge strength (polarities).
Pellicular resins are easy to pack and elute a wider range of solutes at a given carrier strength. They require comparatively low ionic-strength carriers. As with conventional adsorptive packing, pellicular resins are generally simpler to use and

desirable for survey work while fully porous resins are required for highest performance. Columns packed with porous resins of small uniform particle size have successfully resolved samples containing at least 150 components [40, 41].

Ion-exchange packings can enter into nonionic interactions with both the solute and the carrier which can influence solute retention. When such interactions are stable they can be used to advantage. However, when these interactions are concentration dependent and interfere with the separation, they must be suppressed by maximizing the net dissociation of the solute and resin.

D. Stationary Partitioning Phases

The stationary phase in partition chromatography is normally a highly polar or a nonpolar substance which is capable of retarding the solute. To achieve solute retardation the stationary phase must exhibit favorable solubility characteristics which can best be defined in terms of solubility parameters. Since the stationary phase must be chosen as an integral part of the partitioning system (i.e., with the carrier in mind), further consideration of the topic is reserved for the discussion of mobile phase selection for partition chromatography.

E. Stationary Phases for Use with Carrier Programming

Carrier programming in HPLC is the analog of temperature programming in GC and is used when the sample components cannot be eluted in a reasonable time at any given carrier strength. Carrier programming is, therefore, not used with exclusion chromatography. Stable packings such as silica and chemically bonded phases are essential to successful carrier programmed operations. Therefore, when a sample is known to contain components with a wide range of K values (or R_f values from 0 to 1), first consideration should be given to the selection of a stable stationary phase to permit carrier programmed operation.

F. Mobile Phase Selection

It is the carrier's role to impart motion to the solute by counteracting the retarding action of the packing. The migration rate is governed by the net interaction between the carriers, the solute, and the packing. As illustrated diagramatically in Fig. 9, selection of column activity depends on solute polarity and determines carrier strength. Movement of the triple-arrowed pointer locates the three parameters on a given line.

1. Carrier Selection in Adsorption Chromatography

In adsorption chromatography the carrier competes with the solute for the active sites on the adsorbent. Carrier strength is increased by increasing the affinity of the

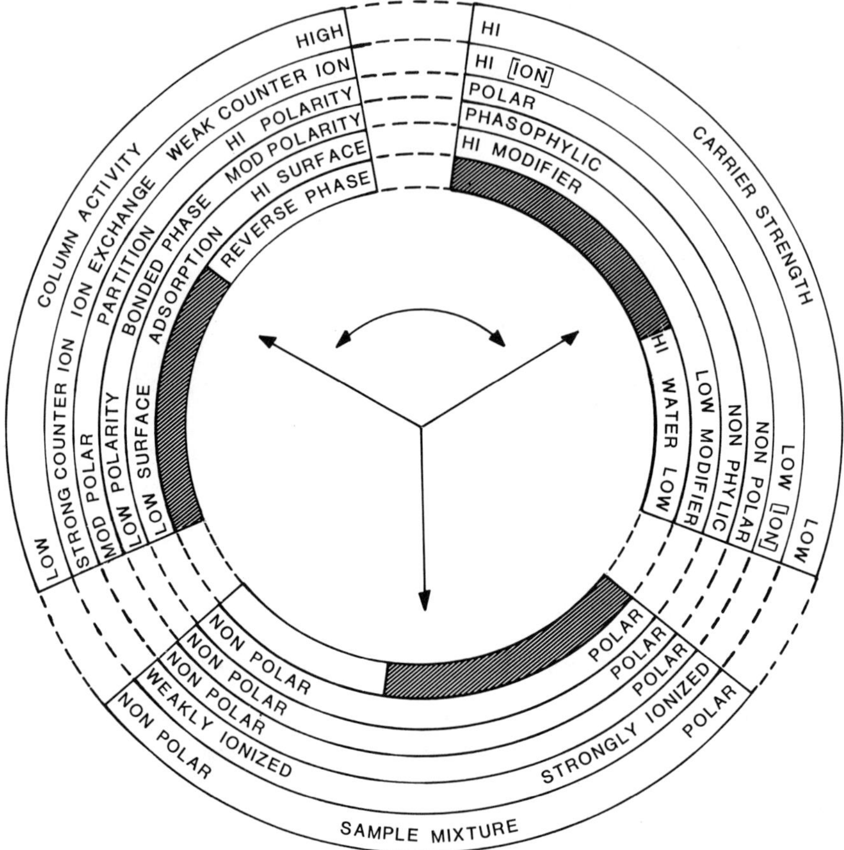

FIG. 9. Phase-strength selection diagram for sample mixture to be separated.

carrier for the active sites on the adsorbent. A table which lists solvents in order
of chromatographic solvent strength is called an *elutropic series*. The series in
Table 11 provides a variety of physical properties for chromatographic solvents along
with solvent-strength parameter E_0. Although the numerical values of solvent strength
were determined for alumina, the relative order is essentially the same when applied
to silica.

Ideally, the carrier should dissolve the sample, be low in viscosity, and transparent to the detector. Carrier strength for the trial separation should be chosen
to elute the sample somewhere within the optimum range of K´ (for example $1 \leq K´ \leq 10$).
If the doubt factor is large, it is preferred to start with a higher strength carrier
since the sample components will elute more quickly and the most strongly retarded
components will not hang up on the column to cause base line problems in subsequent

injections. Once the elution characteristics are determined, carrier strength can be reduced to increase retention and resolution.

Rather than seek a solvent that provides the precisely required strength, one can use solvent admixtures to constitute a carrier from nonviscous to transparent solvents. It is a common practice to add a small quantity of a strongly polar modifier such as an alcohol to a nonpolar hydrocarbon to compose a carrier rather than to blend nearly equal amounts of more similar materials. This practice permits a wider range of carrier strengths to be prepared from a single set of components but demands precise blends to achieve reproducible results. Empirical judgment is required in preparing blends from widely differing components because the solvent-strength scale is not a linear function of binary composition. Snyder demonstrated that the addition of 20% of a polar modifier achieved 80% of the strength of the polar component while 5% achieved 50% of the maximum. The nonlinearity was less pronounced for more similar binaries but the trend was the same [43].

Rather than attempt to make small adjustments in carrier strength with strongly polar modifiers, a preferred method is to use ternary mixtures containing a secondary modifier of intermediate strength such as chloroform in conjunction with the polar modifier. In a mixture such as hexane, chloroform, and isopropanol, the chloroform can be used to make fine adjustments in retention since a large change in composition imposes a relatively small change in strength. An additional benefit from this type of carrier mixture is that the chloroform provides an increase in the solubility range of the mixture.

The wide range of solubility and polarity offered by the ternary blend makes it the nearest approach to a universal carrier for modern HPLC. Particularly since the system is often usable in both the adsorption and partition modes.

In reverse-phase chromatography using chemically bonded phases, the adsorptive force of the packing is minimal. Retardation is enhanced by forcing the solute out of the carrier onto the retardant by a partitioning mechanism. Carrier strength is increased here as in partition chromatography by increasing the solubility of the solute in the carrier. This is usually done by decreasing the amount of water in the carrier which is, in effect, decreasing the "insolubility" of the solute in the aqueous carrier. Reverse-phase carriers need not be limited to water-based systems. Indeed any solvent-nonsolvent binary may be used which provides the desired retardation by the packing. The organic components most commonly used in reverse-phase carriers are methanol, acetonitrile, and tetrahydrofuran in admixture with water. Materials such as THF can also be added to aqueous methanol to increase the solubility of high molecular-weight solutes in the carrier or to alter selectivity.

TABLE 11. Solvent Properties of Chromatographic Interest [42, 43, 45]

	E_o	Solubility parameter δ	Dispersion δ_d	Orientation δ_o	Proton acceptor δ_a	Proton donor δ_b	Viscosity η	RI20^o	UV cutoff	BP
Fluoroalkane	-0.25	6.0	6.0	0	1.0	0		1.25		
Pentane	0.00	7.1	7.1	0	0	0	0.23	1.3577	210	36
Hexane	0.01	7.3	7.3	0	0	0	0.32	1.37536	210	69
Isooctane	0.01	7.0	7.0	0	0	0	0.50	1.3915	210	99
Cyclohexane	0.04	8.2	8.2	0	0	0	1.00	1.4264	210	81
Carbon tetrachloride	0.18	8.6	8.6	0	0.5	0	0.97	1.4607	265	77
Isopropyl ether	0.28	7.3					0.37	1.3678	220	69
Toluene	0.29	8.9	8.9	0	0.5	0	0.59	1.4893	285	111
Benzene	0.32	9.2	9.2	0	0.5	0	0.65	1.50142	280	80
Ethyl ether	0.38	7.4	6.7	2.0	2.0	0	0.23	1.3497$^{24.8^o}$	220	35
Chloroform	0.40	9.1	8.1	3.0	0.5	0	0.57	1.4476	245	61
Methylene chloride	0.42	9.6	6.4	5.5	0.5	0	0.44	1.3348$^{15^o}$	245	40
Methyl isobutyl ketone	0.43							1.3959	330	119
Tetrahydrofuran	0.45	9.1	7.6	4.0	3.0	0	0.51	1.4040$^{25^o}$	220	65
Ethylene dichloride	0.49	9.7	8.2	4.0	0	0	0.79	1.41655	230	84
Methyl ethyl ketone	0.51	9.3						1.38071$^{15.9^o}$	330	80

Nitropropane	0.53							1.4003	380	131
Acetone	0.56	9.4	6.8	5.0	2.5	0				
Dioxane	0.56	9.8	7.8	4.0	3.0	0	1.54	1.4232	220	101
Ethyl Acetate	0.58	8.6	7.0	3.0	2.0	0	0.45	1.3728	260	77
Methyl Acetate	0.60	9.2	6.8	4.5	2.0	0	0.37	1.35935	260	58
Dimethyl sulfoxide	0.60	12.8	8.4	7.5	5.0	0	2.20	$1.4787^{21°}$		189
Nitromethane	0.64	11.0	7.3	8.0	1.0	0	0.67	1.3818	380	101
Acetonitrile	0.65	11.8	6.5	8.0	2.5	0	0.37	1.3460	210	80
Porpanol	0.82	10.2	7.2	2.5	4.0	4.0	2.30	1.38543	210	82
Ethanol (anhydrous)	0.88	11.2	6.8	4.0	5.0	5.0	1.20	$1.36242^{18.4°}$	210	78
Methanol	0.95	12.9	6.2	5.0	7.5	7.5	0.60	1.3276^{25}	210	64
Water	Large	21.	Large	Large	Large	Large		1.3330	200	
Ethylene glycol		14.7	8.0	Large	Large	Large	19.90	1.430		
Acetic acid		12.4						1.37182		
Dimethyl formamide		11.5	7.9				0.09	1.4280		153
Propylene carbonate		13.3								
Formamide		17.9	8.3	Large	Large	Large		1.4472		109

2. Carriers for Ion-Exchange Chromatography

Aqueous solutions of buffers and neutral salts are most commonly used as carriers for
ion-exchange chromatography. Since solute retention is inversely proportional to
ionic concentration, neutral salts with a counter ion of proper affinity are used to
control carrier strength. Buffers are used to maintain the carrier at the desired
pH where retention is independent of the solute concentration.

The effect of pH on solute retention is difficult to predict accurately since
numerous interactions are involved. Since the exchange capacity of the resin is pH
dependent, a titration curve of the packing may be useful to determine the pH range
of the carrier required to provide maximum dissociation of the resin.

Weak bases are normally protonated and treated as cations where dissociation
can be increased by decreasing the [OH$^-$] concentration (i.e., increasing pOH).[†] It
must be remembered that the exchange capacity of some ion-exchange resins decrease as
pH is increased. When Δ capacity/Δ pH at the inflection point in the curve is small,
the range of stable exchange capacity is narrow the freedom to maximize solute dissociation independently of the resin is limited.

Carrier pH also affects the dissociation of the solute. To maximize retardation
one must find the point of net maximum dissociation between the resin and the solute.
Since salts of weak acids experience greater dissociation than the acid, pH should be
increased from the pK_a by one or two units as predicted by the following.

$$K_a = \frac{[H^+][\text{Anion}^-]}{[\text{Acid}]} \tag{22}$$

which can be expressed as

$$pH - pK_a = \log \frac{[\text{Anion}]}{[\text{Undissociated acid}]} \tag{23}$$

Dissociation of weak acids should be 50% complete when pH = pK, and 99% when
$pH - pK_a = 2$.

Nonaqueous carriers may be used to influence dissociation in a manner similar
to that used in nonaqueous titrations. For example, a weak acid may appear as a
base in a methanol solution of glacial acetic acid while a weak base may appear
acidic in the presence of an amine. Such techniques may be tried when conventional
techniques do not provide the required retention or selectivity.

3. Partition Chromatography

In the partition chromatography, solute interaction between the carrier and the retardant determines retention. Carrier strength is increased by increasing the affinity

---

[†] $pH + pOH = P_w = 14$; $K = [\text{cation}^+][\text{OH}^-]/[\text{base}]$; and $pOH - pK = \log([\text{cation}]/$

[undissociated base]).

of the carrier for the solute. This usually means increasing the solvating power of the carrier for the solute. Similarly, increasing retardation is accomplished by increasing the solvating power of the stationary phase. Because many factors affect solubility, it is impossible to predict partition behavior reliably.

When selecting solvents for a partition system, the respective solvents must be taken from near the opposite ends of the polarity scale in order to maintain separate phases (that is, $\delta_s - \delta_m > 4$). Normally solvents with high solubility-parameter (δ) values are used as the stationary phase and held in place on a semi-inert silica support while solvents with low δ values are selected for use as the carrier. When solvents of the precisely desired polarity cannot be found, solvents can be blended to achieve the desired strength. Since the δ values for mixtures equal the arithmetic average of the volume fraction of the insoluble components blended, the δ values listed in Table 11 can be a useful aid in selecting components and proportions to produce carriers with the desired characteristics.

The solubility parameter can be used to calculate partition coefficients using Eq. (24).

$$\log K_d = \frac{\overline{V}(\delta_x - \delta_m)^2 - (\delta_s - \delta_x)^2}{2.3RT} \tag{24}$$

where

$\overline{V}$ = molar volume

   = formula weight/density

R = 1.987 cal degree^{-1} mole^{-1}

T = absolute temperature

TABLE 12. Relationship between Solute Solubility Parameter (δ_x) and Partition Coefficient (K_d) for Several Phase System Values[a]

	Partition coefficients (K_d)		
δ_x	$\delta_s = 13.4$ $\delta_m = 8.0$	$\delta_s = 14.7$ $\delta_m = 7.3$	$\delta_s = 16.0$ $\delta_m = 6.0$
9	0.03	4.29×10^{-3}	6.3×10^{-4}
10	0.17	6.5×10^{-2}	2.56×10^{-2}
11	1.0	1.0	1.0
12	5.9	15.28	39.0
13	34.0	233.0	1585.0

[a] $\overline{V}$ in Eq. (24) assumed to equal 100.

Although δ values are not readily available for most solutes and calculated K_d values may be less than accurate, this treatment can be used to gain insight into partition phenomena.

The values for several hypothetical phase systems listed in Table 12 show the relationship between δ and K as predicted by Eq. (24). It is evident from these data that an increase in the difference in polarity between phases is accompanied by an increase in α. Compare, for example, K_d values for the $\delta_x = 12$ columns and note that $\Delta\delta$ (or $\delta_s - \delta_m$) is increased from 4.8 to 10, α increases from 5.9 to 39. According to Eq. (24), any solute with a δ_x midway between the phase values (that is, $\delta_m - \delta_x = \delta_s - \delta_x$) would distribute equally between phases and show a K_d of one.

When δ_x is greater than the midpoint value, K_d will be greater than one. When δ_x is less than the midpoint value, K_d will be less than one. When K < 1, the phases can be reversed since $(K_{normal\ phase} = 1/K_{reversed\ phase})$. This principle is evident in Table 12 where the reciprocal of K_d at $\delta = 10$ is equal to the value of K_d at $\delta = 12$. This relationship may be used to decide when to reverse phases in a partition system.

The effect of the polarity difference between phases on relative retention in an equitropic system is illustrated in Table 13. A comparison of the underlined values show that doubling the incremental difference between phase parameters reduced proportionately the required difference between solute-solubility parameters in order to maintain α constant.

In these calculations molar volume $\overline{V}$ was held constant at an assumed value of 100 to illustrate the mathematical relationship between δ and K. However $\overline{V}$ is a significant variable in the relationships that cannot be neglected as is illustrated by the data in Table 14. For example, ethanol which has a larger δ than nitrobenzene would elute from Systems 1 and 2 before nitrobenzene because of its smaller $\overline{V}$. These data illustrate that relative retention is a function of both relative $\overline{V}$ and the relative position of δ_x between phase values. It is also evident that retention can be decreased by increasing δ_m or decreasing δ_s. In simplistic terms, in order to increase a separation, one should increase the range between the phases and simultaneously reduce the difference between the δ of the solute and δ of the stationary phase.

As stated earlier, no single parameter can accurately define all of the interactions affecting solubility. The Hildebrand solubility parameter has been broken

TABLE 13. $\Delta\delta_x$ Required to Maintain Constant Relative Retention (α)

δ_m	δ_s	$\delta_s - \delta_m$	δ_x (K = 1)	δ_x (K = 20)	$\Delta\delta_x$
8.6	13.4	4.8	11.0	12.6	1.6
7.3	14.7	7.4	11.0	12.0	1.0
6.0	16.0	10.0	11.0	11.8	0.8

TABLE 14. Relationship between Solute Solubility Parameter (δ_x) and Partition Coefficient (K_d) of Various Materials with Several Phase Systems

Solute	δ	$\overline{V}$	K_1	K_2	K_3	K_4
Benzonitrile	10.7	103	0.9	3.6	0.11	0.4
Nitromethane	11.0	54	1.2	2.7	0.45	1.0
Nitrobenzene	11.1	102	1.8	8.2	0.28	1.3
Ethanol	11.2	58	1.5	3.7	0.56	1.4
Phenol	11.4	88	2.5	10.6	0.6	2.6
Dimethyl sulfonate	12.8	58	6.5	25.0	4.2	17.3
Methanol	12.9	40			3.0	7.9

		δ_m	δ_s
System 1:	chloroform-methanol	9.1	12.9
System 2:	hexane-methanol	7.3	12.9
System 3:	chloroform-ethylene glycol	9.1	14.7
System 4:	hexane-ethylene glycol	7.3	14.7

down into components as listed in Table 15 [43]. The components may be related to the Hildebrand solubility parameter by Eq. (25) [44].

$$\delta^2 = \delta_d^2 + \delta_o^2 + \delta_a^2 + \delta_b^2 \tag{25}$$

Each parameter defines a different aspect of solvent strength which can be related to solvent selectivity. For example, methanol and dimethyl sulfoxide are very similar in total strength, but DMSO would be more selective to nitro compounds while methanol would be more selective to alcohols.

When a phase with the desired physical properties cannot be found, a blend can be prepared which incorporates the best interactions. Since the component parameters are a linear function of composition, solvents can be blended for component strength as well as for total strength.

TABLE 15. Solubility Parameter; Component Selectivity[a]

Solvent parameter	Symbol	Solute type	Functional group
Dispersion	δ_d	Nonselective	Aromatic Cl or S
Orientation	δ_o	Large dipole	Nitrile, nitro
Proton acceptor	δ_a	Proton donor	Alcohols, phenols
Proton donor	δ_b	Proton acceptor	Ethers, amines

[a]A comprehensive treatment of the solubility parameter is provided by the following: B.L. Karger, L.R. Snyder and C. Eon *J. Chromatogr.*, *125* 71 (1976); A.F. Barton *Chemical Reviews*, *75* (No. 6), 731 (1975); D.M. Koenhen and C.A. Smolders, *J. Applied Polymer Sci.*, *19* 1163 (1975); J.H. Hildebrand and R.L. Scott *The Solubility of Non-Electrolytes*, Dover Publications, New York 3rd Ed. (1964).

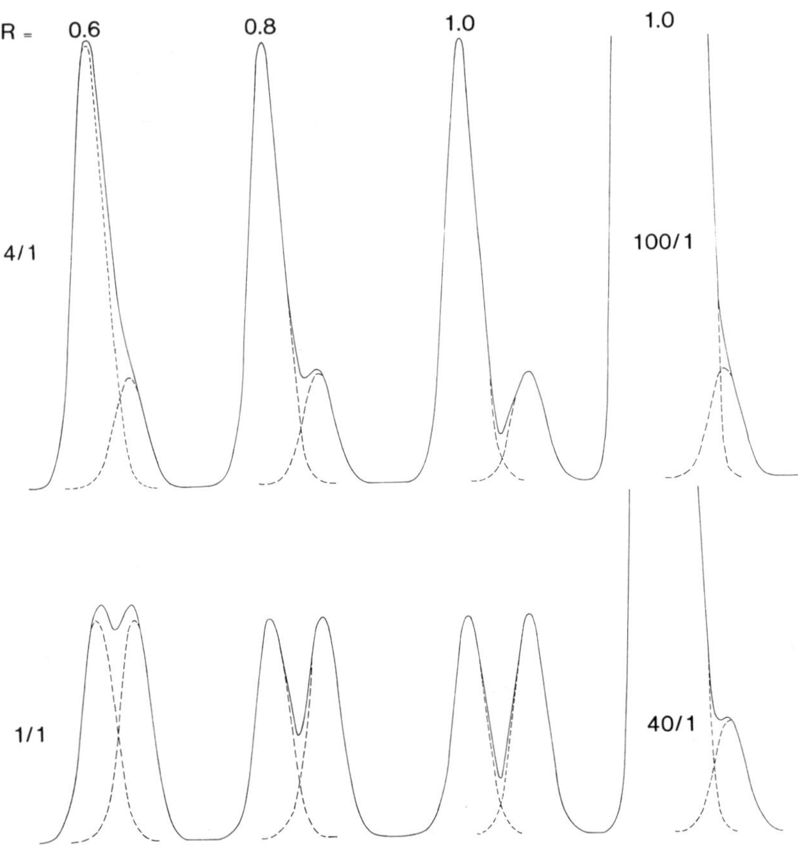

FIG. 10. Effect of relative peak size on peak overlap for values of R as indicated. Peaks defined by $y = ae^{-x^2}$ and a = 5, 20, 200, and 500 [46].

IV. OPTIMIZING THE SEPARATION

When faced with a new problem, the column systems must be selected for a trial separation using the concepts previously described.

Following the trial separation, a visual inspection of the chromatogram will often show whether the resolution R is adequate. Although an R of 1 is often considered adequate for quantitative work (since peak centers are free of contaminations from adjacent peaks), the resolution required for peak characterization varies with the relative amounts of the respective components present. Figure 10 illustrates how the relative amount of solute affects peak overlap at several values of R [46]. Note that when amounts are equal, peak centers are identifiable at R = 0.6, while at a 100/1, the trace component is masked at an R = 1. Since chromatographic peaks are usually asymmetrical, the masking effect of the second peak would be greater than that shown by a purely gaussian curve ($y = ae^{-x^2}$).

When additional resolution is required the chromatogram should be evaluated to determine what is needed before an attempt is made to increase the separation.

The steps in optimizing a separation are summarized as follows:

1. Make a trial separation using a column and carrier chosen by the procedure described previously,
2. Measure t_r, t_r', and W for the unresolved components of interest,
3. Calculate ratios for A, B, and C per Eq. (13a),
4. Adjust conditions starting with B, followed by C and A.

If the peaks of interest are not fused, peak centers can usually be identified and values for t and t´ measured. When peaks are fused into a single envelope, a numerical analysis of the separation cannot be clearly achieved and numerical values must be obtained by other means,[†] If standards of the unresolved components are available, they can be run individually to determine the values of t_r, t_r', and W from which values of A, B, and C can be calculated. If standards are not available, peak widths can be computed from associated resolved peaks or from reference material of comparable K´ and used numerically to estimate the separation between peak centers. A reasonable estimate of A is useful information since it affords a ready means of assessing the difficulty of the separation and helps to identify the choice of practical alternatives.

First effort in the analysis should be given to factor B in Eq. (13a) which depends on carrier strength since that adjustment requires least effort. Figure 11 illustrates the effect of retention on resolution. In many cases a carrier-strength adjustment would be adequate to achieve the desired resolution. Should other components (eluting after the components of interest) be retained excessively, carrier programming may be necessary to obtain an acceptable chromatogram of the entire spectrum of components.

In those cases where the unresolved components are already retained sufficiently (0.67 < B < 0.9), column length must be increased or the carrier-retardant system changed to increase selectivity. Since the search for an optimum carrier-retardant system may become a long and involved process, the choice must be weighed in terms of need to effort. If R must be increased by a factor of 1.7 or less, the increase can be achieved by increasing column length to increase C. Since the increase in C is proportional to $(L_2/L_1)^2$, an increase of at least 3.5 is needed to achieve an R of one. Column length is generally increased in LC by coupling columns of equal dimension and quality. However, efficiency is usually lost in the union. It is not uncommon to lose 25-50% of the theoretical plates as a result of the union. Therefore

---

[†]A visual method of estimating R as reported by Snyder provides a series of gaussian curves with several values of R and several relative peak sizes [46, 47]. A copy of these curves is a valuable aid in chromatogram evaluation.

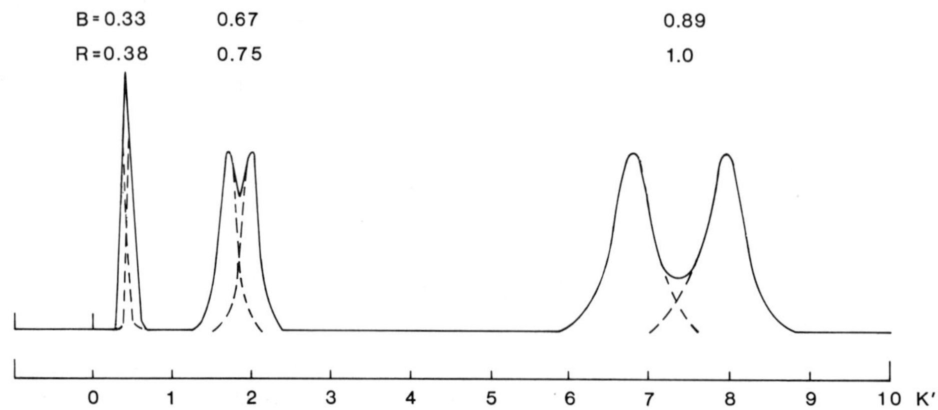

FIG. 11.   Illustration of the effect of B on resolution at constants A and C [34].   (Reprinted with permission of copyright owner.)

to increase R by 1/0.6 may require a sevenfold increase in column length and simultaneously require a sevenfold increase in column head pressure to maintain a constant carrier velocity.   A fourfold increase in column length is a practical consideration and should increase resolution by a factor of at least 1.4.   A sevenfold increase in length would require a greater investment and should be weighed against other alternatives such as recycle chromatography.

In those cases where resolution cannot be achieved by the direct means described, a more selective carrier-packing system should be sought to increase A.   In fact, attention should be given to A from the start since any difficiency in A must ultimately be made up by C, and any increase in C requires a significant increase in L.

$$\frac{L_2}{L_1} \simeq 2\left(\frac{C_2}{C_1}\right)^2 \tag{26}$$

While vendors of commercial small-particle (5-10 μm) columns report theoretical plate numbers of 6000 plates per unit length as determined with selected test components and low-viscosity carriers, comparable plate numbers are generally not achieved at practical working conditions using polar modifiers.   Instead, plate numbers on the order of 1000 to 4000 plates are typically obtained.   If we assume a plate number of 2000 plates per length as representative of a typical component eluting at K' = 0.8 from a small-particle column and we consider a four-column system as a convenient practical limit of column length, we can determine that an A of 0.05 is needed to provide an R = 1.   When efficiencies are less than 2000 plates per length or if more efficiency is lost by the union of the coupled sections, the limiting value of A will be greater.   An A of 0.09 (α = 1.1) is required to separate a pair of components on a 1000 plate column to an R = 0.63, where peak centers will be separated by 0.6 μm.

At this resolution symmetrical peaks of equal height will have just begun to show a valley and be identifiable as individual peaks (see Fig. 11). It is evident, therefore, that when A is less than 0.09 consideration should be given to improving selectivity, and when A is less than 0.05 selectivity must be increased.

A. Improving Selectivity

The initial selection for the trial separation is made by judgment. If the choice turned out to be a poor one as indicated by A values of less than 0.05, effort must be directed to increasing selectivity. First consideration should be given to selection of a different phase system. Options include a change from a silica adsorbent to a bonded phase containing a selective group, as illustrated in Fig. 6. Perhaps a change from normal phase to reverse phase or a change in carrier (e.g., from alcohol modifier to an ether modifier such as dioxane), or a change from methanol in a reverse-phase system to acetonitrile or THF. All that can be said at this writing is that the choices are highly empirical. In the author's own work, visual solubility tests are generally run to seek a solvent in which the solute is sparsely soluble and then vary the moderator to adjust K'. In this way small differences in the solubility of the unresolved solute pair are maximized. If a partition system is used, the difference in polarity between the phases may be increased so that the difference in the solubility parameters is maximized. The same principles can be applied to the functionality of a chemically bonded packing.

If such variations fail to yield a separation, a variety of packings must be evaluated. As a practical consideration, a search of this type can best be done when the unresolved pair are free of other materials. If they cannot be obtained as individual components, the unresolved pair should be isolated as a fraction from preparative LC for use in the search. In this way the pair can be studied with least delay and least uncertainty from background interferences.

TLC can be used to screen a variety of adsorbents, e.g., alumina and Florisil in place of silica. A variety of selective solvent systems can be screened to find a partition system that will enhance selectivity. If the search fails, a chemical derivative of the solutes may alter selectivity and provide a viable alternative. If indeed all efforts fail - the only practical alternative is to increase L until the separation is made by brute force.

It can be assumed that a separation warranting the previously described effort is indeed needed and would justify the increase in L to achieve the needed increase in C. A practical alternative to coupling columns in HPLC is to use recycle [48, 49]. Since retarded peaks are less susceptible to extra-column band broadening, a four- to eight-column system used with recycle can achieve a significant increase in resolution. While the benefits of recycle are usually greatest in the exclusion mode where K' is

limited to approximately 1, significant benefits can be realized in the affinity mode
at $K' = 5$ by collecting the outer extremities of the peaks prior to overlap. By this
technique seemingly unresolvable materials may be separated at high purity.

B. The Role of Temperature

Although temperature is less significant to LC than to GC it deserves consideration
as a variable to be optimized for difficult separations. Since increasing temperature
reduces carrier viscosity and increases diffusion rate, column efficiency can be increased by increasing temperature. The benefit is most predictable in exclusion chromatography where K_0 is not affected by temperature.

Temperature effects are greatest in partition chromatography where K_d is temperature dependent. Since $\Delta K_d/T$ varies with solutes and phase systems, an increase in
temperature may help or hinder a separation. Consequently the optimum temperature
must be determined empirically. Elevated temperatures are commonly used in ion exchange chromatography to improve mass transfer and facilitate faster separation.

C. Flow Rate

One remaining consideration to achieving improved resolution is to reduce flow rate.
This proposal may seem to be in conflict with current thought. With the advent of
porous layer and small porous particles, as well as improved column packing procedures,
much higher carrier velocities may be used to achieve fast analysis [50-58] with only
minor band broadening. Some workers have demonstrated H versus U curves, which were
virtually flat [57]. This would imply that peak width was independent of flow rate.
While this condition may occur with high diffusion-coefficient solutes in a nonviscous
carrier eluting at a relatively low K', it is not representative of the complete range
or normal practice. In any separation where H increases with an increase in K', a
considerable increase in resolution can be gained from a decrease in flow rate.

Although the consideration of flow rate is discussed last in this chapter, it
should not be considered as a last resort. Rather, the effect of flow rate on resolution for a given sample in a particular system should be scanned early in the
development program.

GLOSSARY OF TERMS

A   Selectivity term of the resolution equation defined by Eq. (13)

a   A proportionality constant

α Net relative retention

B   Capacity term of the resolution equation defined by Eq. (13)

[ ]   Concentration

C   Efficiency term of the resolution equation defined by Eq. (13)

$\overline{d}_p$ Average particle diameter in micrometers (μm)

δ Hildebrand solubility parameter

δ_a Proton-acceptor solubility parameter

δ_b Proton-donor solubility parameter

δ_d Dispersion solubility parameter

δ_0 Orientation-dipole solubility parameter

δ_m Solubility parameter of mobile phase

δ_s Solubility parameter of stationary phase

δ_x Solubility parameter of the solute

Δ Difference between two values

e   Base of natural logarithm

f   A function of H

H   Height equivalent to a theoretical plate (HETP) = L/N

K_a Acidic dissociation constant

K_d Distribution coefficient defined in Eq. (1)

K_0 Permeation coefficient defined in Eq. (2)

K' Capacity factor defined in Eq. (5)

K'_{LC} Capacity factor in column chromatography

K'_{TLC} Capacity factor on an equivalent system in thin-layer chromatography calculated from Eq. (21)

L   Column length

M_s Mass of solute in the stationary phase

[M]   Solute concentration in the mobile phase

M_m	Mass of solute in the mobile phase
$\overline{M}$	Average molecular weight
N	Number of theoretical plates
N_T	Ttoal number of theoretical plates obtained from coupled colums
n	Column peak capacity
pH	Classical definition of hydrogen ion concentration
pK_a	Classical definition of acidity as defined by Eq. (23)
R	Resolution
R^o	Gas constant expressed in calories per degree per mole
Rf	Solute migration rate relative to the developer front
t_m	Time solute spends in the mobile phase
t_r	Retention time
t_s	Time solute spends in the stationary phase.
Δt	Retention time difference between two peaks.
t'	Net retention time.
T	Absolute temperature.
U	Carrier velocity in cm/sec.
V_r	Retention volume.
V_r'	Net retention volume.
V_m	Volume of the mobile phase.
V_s	Volume of the stationary phase.
V_{pa}	Volume of gel pores available to a size species.
V_{pg}	Pore volume of a gel.
ΔV	$V_2 - V_1$ difference between solute retention volumes
$\overline{V}$	Molar volume = formula weight/density
V_1, V_2, V_n	Retention volumes of various components
W	Peak width or solute band width
$\overline{W}$	Average peak width
W_T	Total peak width = $(\Sigma W_i^2)^{1/2}$
W_i	Peak width on a single column

REFERENCES

1. K. J. Bombaugh and P. W. Almquist, *Chromatographia*, *8*, 109(1975).

2. L. R. Snyder, in *Modern Practice of Liquid Chromatography* (J. J. Kirkland, ed.), Wiley, New York, 1971, pp. 205-236.

3. J. F. K. Huber, *J. Chromatogr. Sci.*, *9*, 72(1971).

4. J. F. K. Huber, J. C. Kraak, and H. Veening, *Anal. Chem.*, *44*, 1554(1972).

5. J. F. K. Huber, C. A. M. Meijers, and J. A. R. Hulsman, *Anal. Chem.*, *44*, 111(1972).

6. L. I. Larson and O. Samuelson, *Acta Chem. Scand.*, *19*, 1357(1965).

7. E. Martinsson and O. Samuelson, *J. Chromatogr.*, *50*, 429(1970).

8. J. S. Hobbs and J. G. Lawrence, *J. Chromatogr.*, *72*, 311(1972).

9. S. Eksborg and G. Schill, *Anal. Chem.*, *45*, 2092(1973).

10. S. Eksborg, P. O. Lagerstrom, R. Modin, and G. J. Schill, *J. Chromatogr.*, *83*, 99(1973).

11. G. Schill, in *Advances in Ion Exchange and Solvent Extraction* (J. A. Marinsky and Y. Marcus, eds.), Vol. 6, Marcel Dekker, New York, 1974.

12. B. L. Karger, S. C. Su, S. Marchese, and B. A. Persson, *J. Chromatogr. Sci.*, *12*, 678(1974).

13. B. A. Persson and B. L. Karger, *J. Chromatogr. Sci.*, *12*, 521(1974).

14. S. C. Su, A. V. Hartkopf, and B. L. Karger, *J. Chromatogr.*, *119*, 523(1976).

15. D. P. Wittmer, N. O. Nuessel, and W. G. Haney, *Anal. Chem.*, *47*, 1422(1975).

16. J. J. Kirkland, in *Gas Chromatography* (L. Fowler, ed.), Instrument Society of America, Pittsburgh, 1963.

17. E. Able, Z. H. Pollard, P. C. Uden, G. Nickless, *J. Chromatogr.*, *22*, 23(1966).

18. W. A. Aue and C. R. Hastings, *J. Chromatogr.*, *42*, 319(1969).

19. I. Halasz and I. Sebastian, *Int. Ed. in Eng.*, *8*, 453(1969).

20. J. J. Kirkland and J. J. DeStefano, *J. Chromatogr. Sci.*, *8*, 309(1970).

21. J. H. Knox and G. Vasuari, *J. Chromatogr.*, *82*, 181(1973).

22. J. Wartman and H. Deuel, *Chimia*, *12*, 455(1958).

23. J. J. Kirkland, in *Gas Chromatography* (L. Fowler, ed.), Instrument Society of America, Pittsburgh, 1963, p. 77.

24. C. Rossi, S. Munari, G. Angaile, and G. F. Tealdo, *Chim. Ind. (Milan)*, *42*, 724(1960).

25. O. E. Brust, I. Sebastian, and I. Halasz, *J. Chromatogr.*, *83*, 15(1973).

26. R. V. Vivilecchia, R. L. Cotter, R. J. Limpert, N. Z. Thimot, and J. N. Little, *J. Chromatogr.*, *99*, 407(1974).

27. K. J. Bombaugh, in *Modern Practice of Liquid Chromatography* (J. J. Kirkland, ed.), John Wiley and Sons, New York, 1971, pp. 237-285.

28. J. G. Hendrickson, *Anal. Chem.*, *40*, 49(1968).

29. J. C. Moore and M. C. Arrington, in *Proceedings: Third International Seminar on Gel Permeation Chromatography*, Geneva, Switzerland, 1966.

30. K. J. Bombaugh, The Cost of a Bad Column in GPC, Technical Information Bulletin, GPC-T1B1068/1, Waters Associates, Milford, Mass.

31. J. C. Giddings, *Anal. Chem. 39*, 1027(1967).

32. J. H. Purnell, *J. Chem. Soc., 1960*, 1268(1960).

33. L. R. Snyder, *Principles of Adsorption Chromatography*, Marcel Dekker, New York, 1968, p. 18.

34. K. J. Bombaugh, in *Progress in Analytical Chemistry* (I. L. Simmons and G. W. Ewing, eds.), Vol. 6, Plenum Press, New York, 1973, p. 208.

35. F. C. Saville, in *Comprehensive Analytical Chemistry Analysis* (C. L. Wilson and D. W. Wilson, eds.), Vol. 2, Part B, Elsevier, Amsterdam, 1968, pp. 212-348.

36. W. R. Wilcox, in *An Introduction to Separation Science* (B. L. Karger, L. R. Snyder, C. Horvath, eds.), John Wiley and Sons, New York, 1973, p. 345.

37. R. P. W. Scott and P. Kucera, *J. Chromatogr. Sci., 12*, 473(1974).

38. K. Randerath, *Verlag Chemie GMBH*, Academic Press, New York, 1964, p. 19.

39. Varian Instrument Division, MicroPak Columns for High Performance Liquid Chromatography.

40. P. B. Hamilton, in *Handbook of Biochemistry: Selected Data for Molecular Biology*, The Chemical Rubber Co., Cleveland, 1968, pp. 43-55.

41. C. D. Scott and R. L. Jolley, W. W. Pitt, and W. F. Johnson, *Amer J. Clin. Path., 53*, 701(1970).

42. L. R. Snyder, *Principles of Adsorption Chromatography*, Marcel Dekker, New York, 1968, p. 194-195.

43. L. R. Snyder, in *Modern Practice of Liquid Chromatography*, (J. J. Kirkland, ed.), John Wiley and Sons, New York, 1971, pp. 125-157.

44. C. M. Hansen, *J. Paint Technol., 39*, 104(1967).

45. John A. Dean, in *Lange's Handbook of Chemistry*, 11th. Ed., McGraw-Hill, New York, 1973.

46. L. R. Snyder, *J. Chromatogr. Sci., 10*, 200(1972).

47. L. R. Snyder, *J. Chromatogr. Sci., 10*, 369(1972).

48. K. J. Bombaugh, W. A. Dark, and R. F. Levangie, *J. Chromatogr. Sci., 7*, 42(1969).

49. K. J. Bombaugh, and R. F. Levangie, *Sep. Sci., 5*, 6(1970).

50. L. R. Snyder, *J. Chromatogr., 7*, 352(1969).

51. J. F. K. Huber, *J. Chromatogr., 7*, 85(1969).

52. J. H. Knox and M. J. Saleem, *J. Chromatogr. Sci., 7*, 614(1969).

53. C. D. Scott, *Anal. Biochem., 24*, 292(1968).

54. C. G. Horvath, B. A. Preiss, and S. R. Lipsky, *Anal. Chem., 39*, 1422(1967).

55. J. J. Kirkland, *Anal. Chem., 41*, 218(1969).

56. R. E. Majors, *Anal. Chem.*, *44*, 1722(1972).

57. R. E. Majors, *J. Chromatogr. Sci.*, *11*, 88(1973).

58. J. J. Kirkland, *J. Chromatogr. Sci.*, *10*, 593(1972).

Chapter 4

DERIVATIZATION TECHNIQUES IN GAS-LIQUID CHROMATOGRAPHY

John A. Perry

Consultant
Chicago, Illinois

Charles A. Feit

Regis Chemical Company
Morton Grove, Illinois

I. INTRODUCTION

Compared to its parent compound, the gas chromatographic derivative should be more
volatile, separable, stable, and detectable.

In an ideal derivatization, a given compound is derivatized quantitatively and
quickly (in less than 10 min). Then, without further processing, the reaction mixture can be injected directly into the gas chromatograph. The derivative, stable and
preservable in the reaction mixture and also stable during gas chromatography, yields
a symmetrical peak, and can be separated from similar derivatives.

If a given substance contains two or more functional groups that no single reagent can successfully derivatize, a sequenced derivatization is usually used. The
characteristics of an ideal sequenced derivatization, first developed by Horning
et al. [1a] for the catecholamines, are described in Sec. VI. C.1.

This chapter is a highly selective review of current optimum practice in derivatization for gas chromatography. Described in it are the reagents and procedures
that are both the most nearly ideal according to the definitions just given for ide-
ality, and the most generally usable. This survey is limited to compound classes
which are of biomedicinal interest. In the course of preparing this chapter review
articles on derivatization in gas chromatography appeared by Drozd [1b], Ahuja [1c],
Nanbara and Goto [1d], and a book by Blan and King [1e].

A. Abbreviations

The following abbreviations are used in the text and tables of this chapter, and coin-
cide with generally accepted usage in the literature.

Ac_2O acetic anhydride

BMDCS	bromomethyldimethylchlorosilane
BSA	N,O-bis(trimethylsilyl)acetamide
BSTFA	bis(trimethylsilyl)trifluoroacetamide
BTFA	bis(trifluoroacetamide)
CMDMCS	chloromethyldimethylchlorosilane
DA	dopamine
DAE	diazoethane
DAM	diazomethane (CH_2N_2)
DEA	diethylamine
DIMAP	dimethylamino-dimethyl phosphine
DMF	N,N-dimethylformamide
DMF-DMA	N,N-dimethylformamide dimethylacetal
DMSO	dimethylsulfoxide
DNBS	2,4-dinitrobenzenesulfonic acid
DNFB	2,4-dinitrofluorobenzene
DNPH	2,4-dinitrophenylhydrazine
E	epinephrine
HAc	acetic acid
HFDA	heptafluorobutyric anhydride
HFBI	heptafluorobutyrylimidazole
HMDS	hexamethyldisilazane
HVA	homovanillic acid
MBTFA	N-methyl-bis(trifluoroacetamide)
MN	metanephrine
MSTFA	N-methyl-N-trimethylsilyl-trifluoroacetamide
MSA	N-methyl-N-trimethylsilyl-acetamide
NBBA	*n*-butylboronic acid
NE	norepinephrine
NMN	normetanephrine
PDAM	phenyldiazomethane
PEA	phenethylamine
PEOA	phenethanolamine
PF	pentafluoro-
PFB	pentafluorobenzyl-
PFBA	perfluorobenzaldehyde
PFBBr	pentafluorobenzyl bromide
PFBHA·HCl	O-(2,3,4,5,6-pentafluorobenzyl)hydroxylamine hydrochloride
PFPA	pentafluoropropionic anhydride
PHI	phenylhydrazine
py	pyridine
$RB(OH)_2$	a boronic acid with radical R

TCE	2,2,2-trichloroethanol
TFAA	trifluoroacetic anhydride
TFAAc	trifluoroacetic acid
THF	tetrahydrofuran
TMA	trimethylamine
TMAH	trimethylanilinium hydroxide
TMCS	trimethylchlorosilane
TMH	tetramethylammonium hydroxide
TMSDEA	trimethylsilyldiethylamine
TSIM	trimethylsilylimidazole
VMA	vanillyl mandelic acid

II. DERIVATIZATION REAGENTS

Presented here for ease of reference, the following comparisons among and comments on
the most common reagents, can be and have been gleaned from the works reviewed.

A. Silylation Reagents

If it is applicable, silylation is the preferred method of derivatization.

At this writing, the silylating reagents of choice are BSA, TSIM, and BSTFA.
BSA and BSTFA are usually used with TMCS as catalyst. Of these, BSTFA is by far the
most effective. BSTFA, for instance, silylates acids 10 times faster than BSA [2, 3].
BSTFA is usuable where BSA is not, for instance, in the silylation of homovanillic
acid and vanillyl mandelic acid [4], purines and pyrimidines [5], intact glucuronides
[6], shikimic and quinic acids [3], and neuraminic acid [7].

TSIM is generally less effective than either BSA [1] or BSTFA, but excels in
silylating highly hindered hydroxyls [8-10]. TSIM is also the reagent of choice for
silylation in the presence of water [11, 12], for instance, in the silylation of
sugar syrups.

BSA-TMCS is much more effective than HMDS-TMCS, but may be erratic [11]. For
example, keto groups that are not affected by either TSIM or BSTFA may be enolized
and silylated by BSA [13]. Although BSA has been found less effective than TSIM for
silylation in the presence of water [11], trace water can catalyze silylation by
BSA [1, 14].

The most widely employed trimethylsilylating reagents, all of which are commercially available, are shown in Fig. 1.

B. Perfluoro-Acylating Reagents

The groups most commonly used for sensitizing substances to detection by electron
capture are the perfluoro groups: the trifluoroacetyl (TFA), the pentafluoropropionyl

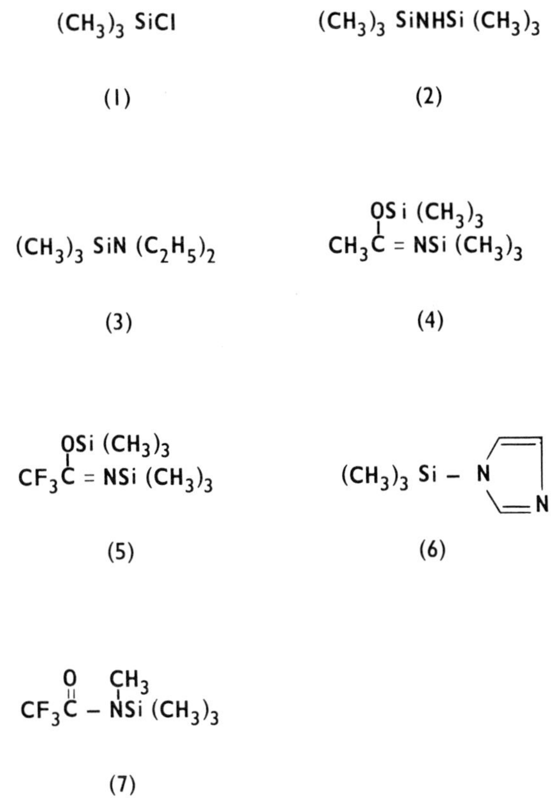

$(CH_3)_3\ SiCl$

(1)

$(CH_3)_3\ SiNHSi\ (CH_3)_3$

(2)

$(CH_3)_3\ SiN\ (C_2H_5)_2$

(3)

$$CH_3\overset{\overset{\displaystyle OSi\ (CH_3)_3}{|}}{C} = NSi\ (CH_3)_3$$

(4)

$$CF_3\overset{\overset{\displaystyle OSi\ (CH_3)_3}{|}}{C} = NSi\ (CH_3)_3$$

(5)

$(CH_3)_3\ Si - N$ (imidazole ring)

(6)

$$CF_3\overset{\overset{\displaystyle O}{||}}{C} - \overset{\overset{\displaystyle CH_3}{|}}{N}Si\ (CH_3)_3$$

(7)

FIG. 1. Commonly used trimethylsilylation reagents. 1. Trimethylchlorosilane (TMCS), 2. Hexamethyldisilazane (HMDS), 3. Trimethylsilyldiethylamine (TMSDEA), 4. N,O-Bis(Trimethylsilyl)-acetamide (BSA), 5. N,O-Bis(Trimethylsilyl)-trifluoro-acetamide (BSTFA)[+], 6. N-Trimethylsilylimidazole (TSIM), 7. N-Methyl-N-Trimethylsilyltrifluoroacetamide (MSTFA)

(PFP), the pentafluorobenzoyl, and the heptafluorobutyryl (HFB). They have repeatedly been compared [15-31]. On grounds of stability to hydrolysis and high temperatures, ease of overall use, and the imparted degree of sensitivity to detection by electron capture, the pentafluorobenzoyl derivative seems most preferable [27].

The heptafluorobutyryl group is stable toward water at pH 6, but decomposes with aqueous ammonia [18]. HFBA is not easily removed from a reaction mixture by evaporation and although washing the mixture is quite effective, it is less convenient [28, 32]. As described in detail in Sec. VI.B.2, the heptafluorobutyryl group can (with favorable structures) impart more sensitivity to detection by electron capture than either the trifluoroacetyl or the pentafluoropropionyl; however, the sensitivity is highly dependent on the structure of the compound derivatized [33].

[+]The exact structure of BSTFA continues to be the object of study.

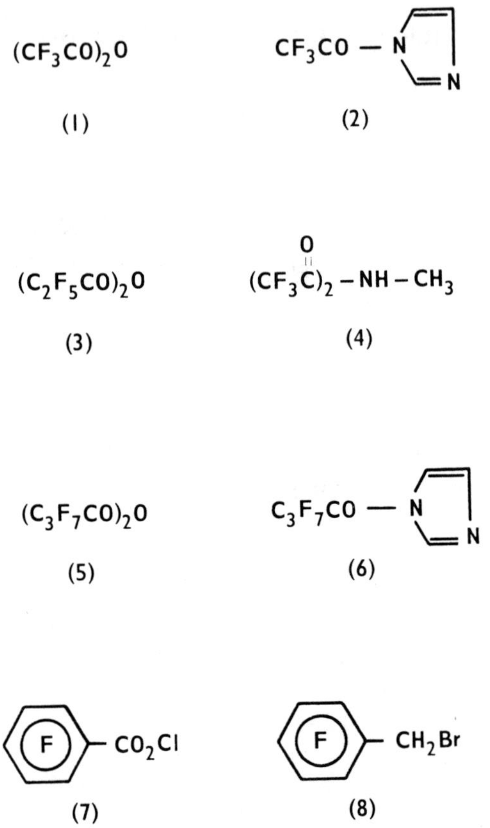

$(CF_3CO)_2O$

(1)

CF_3CO-N (imidazole ring)

(2)

$(C_2F_5CO)_2O$

(3)

$(CF_3\overset{O}{\overset{\|}{C}})_2-NH-CH_3$

(4)

$(C_3F_7CO)_2O$

(5)

C_3F_7CO-N (imidazole ring)

(6)

(pentafluorophenyl)$-CO_2Cl$

(7)

(pentafluorophenyl)$-CH_2Br$

(8)

FIG. 2. Commonly used ECD reagents. 1. Trifluoroacetic anhydride (TFAA), 2. N-Trifluoroacetylimidazole (TFAI), 3. Pentafluoropropionic anhydride (PFPA), 4. N-Methyl-bis (trifluoracetamide) (MBTFA), 5. Heptafluorobutyric anhydride (HFBA), 6. N-Heptafluorobutyrylimidazole (HFBI), 7. Pentafluorobenzoyl chloride, 8. Penta-fluorobenzyl bromide (PFB-Br)

A listing of the most frequently used reagents for EC sensitization is shown in Fig. 2.

III. HYDROXYLS

A. Alcohols

1. Flame Ionization Detection

Typically fast, mild, and convenient, trimethylsilylation is the most widely used method of derivatizing alcohols. For example, BSTFA quantitatively silylates vita-min D_2 (secondary alcohol) in 30 min at room temperature [34]. The reaction mixture is injectable. A general equation for the silylation of alcohols is shown in Scheme 1.

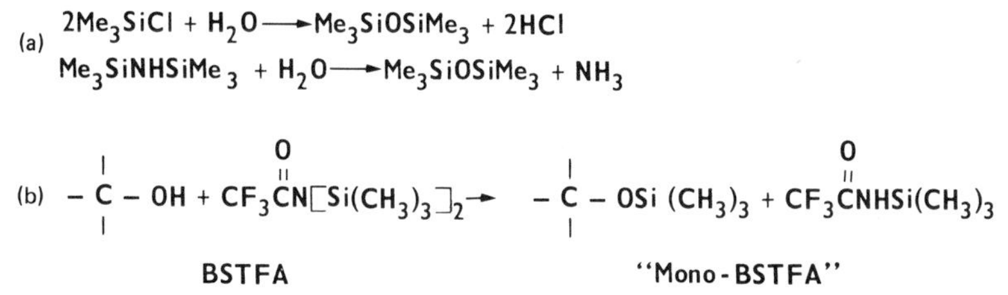

(a)
$$2Me_3SiCl + H_2O \longrightarrow Me_3SiOSiMe_3 + 2HCl$$
$$Me_3SiNHSiMe_3 + H_2O \longrightarrow Me_3SiOSiMe_3 + NH_3$$

(b)
$$\underset{|}{\overset{|}{-C-}}OH + CF_3\overset{O}{\overset{||}{C}}N[Si(CH_3)_3]_2 \longrightarrow \underset{|}{\overset{|}{-C-}}OSi(CH_3)_3 + CF_3\overset{O}{\overset{||}{C}}NHSi(CH_3)_3$$

BSTFA   "Mono-BSTFA"

SCHEME 1

Typical silylation reactions (a) with water and (b) with alcohol.

Either in aqueous solution or with the dried samples, the alkylboronic acids, $RB(OH)_2$, (usually with *n*-butyl as R) form cyclic boronates rapidly at room temperature with pairs of 1,2 and 1,3 hydroxyls (e.g., the *n*-butylboronates of sorbitol, mannitol, and galactitol [35]). The reaction mixture is injectable. For long storage, the derivatives should be dried.

2. Electron Capture Detection

For high sensitivity to detection by electron capture and for ease of preparation, Braestrup [36] chose the pentafluoropropionyl over the trifluoroacetyl and the heptafluorobutyryl derivatives (see also Refs. 15 and 16). A model alcohol, 3-methoxy-4-hydroxyphenylglycol, was derivatized with PFPA (15 min, room temperature; dry with nitrogen). The redissolved derivative was injected.

B. Phenols

1. Flame Ionization Detection

Unless hindered, phenolic hydroxyls are readily silylated by any silylating reagent. Hindered hydroxyls are best silylated by TSIM [9, 10] or, slightly better, by TSIM-TFAAc [9].

2. Electron Capture Detection

For sensitization to detection by electron capture, McCallum and Armstrong judged the pentafluoroaryl derivatives of thymol (phenolic hydroxy) the best of seven on grounds of reactivity, sensitivity (seven times better than the heptafluorobutyryl), and cost [17]. (The thymol pentafluorobenzoate is about 10,000 times more sensitive to detection by electron capture than the underivatized thymol is to detection by flame ionization.)

Ehrsson et al. [18] found that the heptafluorobutyryl ester of p-tert-butylphenol decomposes either in aqueous ammonia or with hydroxylated stationary phases (e.g., Carbowax 20M, Amine 220). The heptafluorobutyryl group is stable in an aqueous acid (pH $\leq$ 6.0) and over a polyester stationary phase such as neopentylglycol succinate.

Seiber et al. [37] found ester derivatives unstable to hydrolysis. Ethers, however, particularly the pentafluorobenzyl (PFB) ethers, are stable, easily formed, and highly sensitive to detection by electron capture, and show *relatively short retention times*. Similarly, Kawahara [19] found that the water-stable phenyl pentafluorobenzyl ether shows 1,000 times better sensitivity to electron capture detection than the water-unstable trifluoroacetate ester. Johnson [23] removes the PFBBr from the trace pentafluorobenzoyl ester and pentafluorobenzyl ether derivatives of acids and phenols, respectively, by silica gel column cleanup. Brotell et al. [20] could detect as little as 3 pg of the water-stable pentafluorobenzyl ether of the drug pentazocine.

C. Antibiotics

Margosis reviewed antibiotic derivatization for GC [38]. Detected by flame ionization, most gas chromatographable antibiotics are trimethylsilylated. For example, to determine erythromycin (polyhydroxyl) and its derivatives, Tsuji and Robertson silylated 10 mg of the powdered sample with BSA-TMCS-TSIM (5:5:2) mixed in that order and then added to an equal volume of pyridine containing an internal standard, 1,3-dimyristin (24 h, 75°) [39].

The functional complexity of antibiotics requires that the wide variation in silylation efficiency possible with different reagents and reaction conditions be fully exploited. Spectinomycin, for instance, has two secondary amines, two secondary hydroxyls, one tertiary hydroxyl, and one enolizable carbonyl, nevertheless conditions yielding a single derivative were developed using HMDS for 1 h at room temperature [40].

D. Prostaglandins

Functional-group complexity and trace concentrations principally complicate applying gas chromatography to prostaglandin analysis [41]. Almost as prerequisites, multiple derivatization, ultra-selective and ultra-sensitive detectors, and internal standards characterize such analyses.

1. Flame Ionization Detection and Mass Spectrometric Detection

In prostaglandin (PG) analysis, flame ionization detection is used only as an adjunct to gas chromatography-mass spectrometry.

In preparing derivatives of the A, B, E, and F prostaglandins, Vane and Horning subsequentially derivatized first the keto groups as methoximes (CH_3ONH_2 in pyridine,

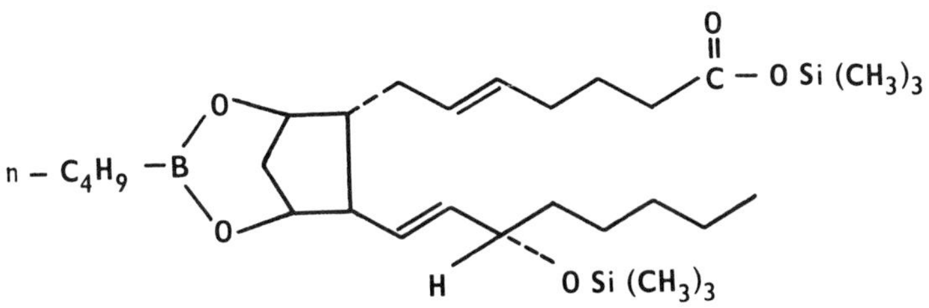

FIG. 3. *n*-Butylboronate trimethylsilyl derivative of $PGF_{2\alpha}$.

2-3 h, 60°) and then the hydroxyl and carboxylic acid groups as trimethylsilyl ethers and esters, respectively (BSTFA, 1-2 h, room temperature) [42].

With a mixture of the E_1, E_2, $F_{1\alpha}$, and $F_{2\alpha}$ prostaglandins, Pace-Asciak and Wolfe [43] also first formed the methoximes, thus derivatizing the E prostaglandin but not the F. In the presence of these stable methoximes, and using dimethoxypropane as the dehydration solvent, they then formed the PGF 9,11-*n*-butylboronates (NBBA, 2 min, 60°). (The dimethoxypropane reacts endothermically [44] with water to form acetone and methanol. The easily hydrolyzable trimethylsilyl groups are thus kept free of moisture, and the derivatives are easily dried by nitrogen.) Finally, the remaining hydroxyl and carboxyl groups were silylated (TSIM, 5 min, 60°). The 9,11-*n*-butyl-boronate derivative of $PGF_{2\alpha}$ is shown in Fig. 3.

2. Electron Capture Detection

Unless the prostaglandin can be either converted to or already is a B prostaglandin, and thus inherently sensitive to detection by electron capture, it must be derivative-sensitized to such detection. The thermally stable pentafluorobenzyl ester is first formed at the C-1 carboxyl (PFB-Br in CH_3CN, 5 min, 40°) and then the hydroxyl groups of prostaglandins are silylated (BSA, 5 min, room temperature). The pentafluoro-benzyl trimethylsilyl derivative of $PGF_{2\alpha}$ can be determined from 0.2 - 5.6 ng [22].

3. Mass Fragmentographic Detection

a. Methyl-trimethylsilyl

Nicasia and Gaui [248] prepared prostaglandin methyl esters with diazomethane, silyl-ated (TSIM) the nitrogen-dried esters, and after a few seconds injected the reaction mixture. The alkali sensitive derivatives required completely silanized equipment. Nanogram levels were detectable by mass fragmentography.

b. Methyl-*n*-butylboronate-trimethylsilyl

Kelly [45] methyl-esterified (DAM, 10 min, room temperature) prostaglandin $F_{2\alpha}$, con-verted the dried methyl esters to *n*-butylboronates (NBBA, 20 min, 60°), and again

dried the products under vacuum. The products were finally silylated (BSTFA, 2 h,
room temperature) and dried again under vacuum. For stability (10 days, room temperature), BSTFA was added again and the glass reaction tube sealed in a flame. The
tube contents were injected. The n-butylboronate esters distinguished the cis from
the trans diols and exhibited a relatively high m/e value for the derivatives.

c. Deuterated Internal Standard

Using CD_3ONH_2, Samuelson et al. [46] prepared and mixed a trideuterated prostaglandin
methoxime with an unaltered prostaglandin methoxime. When injected as the trimethylsilyl derivatives and monitored as the normal and 3-mass-unit heavier ions, 3 ng of
the PGE_1 derivative could be determined.

IV. CARBONYLS

The vast majority of carbonyl-containing compounds do not require derivatization.
When the derivatization is required, the most widely employed carbonyl derivatives
are oximes and hydrazones. Oximes increase thermal stability; hydrazones may decrease
polarity or increase stability and sensitivity.

Papa and Turner [47] reviewed carbonyl derivatization via DNPH. In 1972, Kallio
et al. [48] applied the approach to flavor carbonyls, using both flame ionization detection (10-10,000 ng) and electron capture detection (20-500 pg).

Fales and Luukkainen [49] criticized the dinitrophenylhydrazone derivatives as
potentially unstable to light and air. They suggested instead the easily crystallized
methoxime, stable to Ac_2O or HMDS-TMCS. Methoxime formation: CH_3ONH_2-pyridine, overnight, room temperature; dry. The benzene-redissolved methoxime derivative is injectable.

Methoxime derivatives are stable to further derivatization, including silylation.
For example, Butts silylated methoxime-cytidine and methoxime deoxycytidine (BSTFA-1%
TMCS, 3 h, 75°) [50]. In contrast to the methoxime, simple oximes from hydroxylamine
(NH_2OH) can be silylated.

Enolizable keto groups can be silylated by prolonged reaction [13]. At room
temperature, either BSA or HMDS-TMCS (2:1) silylated the phenolic hydroxyls of acetovanillone (4-hydroxy-3-methoxyacetophenone) in 12 min, but also enolized and silylated the acetovanillone keto group partly in 10 h and completely in 24 h. This enoli-
zation and silylation of the keto group proceeded without regard to excess reagent,
heating, or solvent catalyst. Neither TSIM nor BSTFA was found to produce this keto
silylation with acetovanillone.

In 1974, Korolczuk et al. [52] substituted the phenylhydrazones for the 2,4-
dinitrophenylhydrazones. They stated that preparation of the derivatives is simpler,
[faster] ... separation from aqueous solution by extraction with diethylether is also
very easy, rapid, and quantitative. The phenylhydrazones do not require glass columns,
as the 2,4-dinitrophenylhydrazones do, nor such high column temperatures.

V. ACIDS

A. Carboxylic Acids

1. Flame Ionization Detection

The usual carboxylic acid derivative for GC is the methyl ester.

At $-60°$ with dioxane as solvent, diazomethane (DAM) methylates both mono- and dibasic acids [53, 54].

To butylate stearic acid, Greeley [55] first dissolved it in a solution of an organic base (phenyltrimethylammonium hydroxide, TMAH) in an anhydrous polar solvent (80% N,N-dimethyl acetamide, 20% methanol). Then, on the further addition of 1-iodobutane, the stearic acid became 99.4% butylated in 10 min at room temperature. Methylation is similarly accomplished [56].

Carboxylic acids can be silylated in 3 min at room temperature by BSTFA used in the mole ratio, 100 mole BSTFA/mole acid. Such silylation is 10 times faster with BSTFA than with BSA [2, 3].

2. Electron Capture Detection

Smith and Tsai [57] quantitatively esterified eight carboxylic acids with 10% trichloroethanol (TCE) in TFAA. After the TFAA was evaporated, the residual trichloroethanol was removed by preparing a solution in ethyl acetate (100 ml) and passing it over a short silica gel column (20g). Derivatization: 10 min, $100°$; 0.1-1.0 ng detectable.

B. Fatty Acid Mixtures

1. Flame Ionization Detection

Of the numerous methods [58] recommended in 1975 for the esterification of fatty acids of higher [59-62], high and low [63], and low and medium [64] molecular weights, the standardized BF_3-methanol method (10 min) [61] is probably the most widely accepted and certified procedure. A modification of this method cleaves cerebrosides to give the fatty acid methyl esters without at the same time producing undesired artifacts [65].

Alternatively, fatty acids that are extracted into aqueous trimethyl(α,α,α-trifluoro-*m*-tolyl)ammonium hydroxide and mixed with methyl propionate within a syringe, become methylated on injection of the mixture. Fatty acid determinations that depend on such methylations are statistically indistinguishable from similar determinations that depend on other procedures [66].

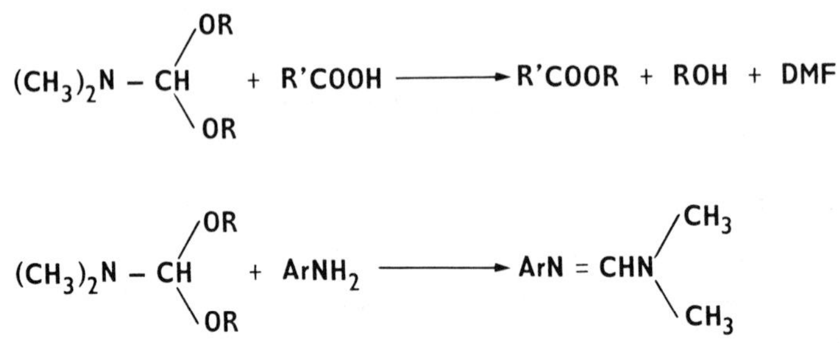

$$(CH_3)_2N-CH \begin{smallmatrix} \diagup OR \\ \diagdown OR \end{smallmatrix} + R'COOH \longrightarrow R'COOR + ROH + DMF$$

$$(CH_3)_2N-CH \begin{smallmatrix} \diagup OR \\ \diagdown OR \end{smallmatrix} + ArNH_2 \longrightarrow ArN=CHN \begin{smallmatrix} \diagup CH_3 \\ \diagdown CH_3 \end{smallmatrix}$$

SCHEME 2

Reations of DMF-dimethyl acetal.

Mixed with acids, N,N-dimethylformamide dialkylacetal quantitatively converts them to esters either within 10 min at 60° or at once on injection [67]. The reaction of DMF-acetal with an acid is shown in Scheme 2.

Free fatty acids react with N,N-carbonyldiimidazole and methanol (or other alcohols) to form methyl (or other) esters (a few minutes, room temperature) under conditions gentle enough not to transesterify triglycerides and cholesteryl esters [68].

Less volatile than the methyl, C_1-C_{20} benzyl esters can be quickly, easily, and quantitatively prepared by reaction with phenyldiazomethane [69, 70]. They furnish distinctive fragments for GC-MS.

2. Electron Capture Detection

Using TCE with HFBA as the catalyst, Alley et al. [71] esterified and sensitized C_2-C_8 aliphatic acids to detection by electron capture. They removed excess TCE by addition of and esterification with palmitic acid.

C. Phenolic, Hydroxy, and Keto Acids

1. Flame Ionization Detection

Excess BSTFA-1% TMCS silylates, for instance, primary and secondary hydroxyls of a galactocerebroside (30 min, 25°) [72] and the carboxyl, phenol, and hydroxyl groups of homovanillic acid and vanillyl mandelic acid (20 min, 70°) [73]. As usual, the reaction mixture is injectable and the derivatives are stable in closed containers with excess BSTFA. Others found that similar acids are immediately silylated at room temperature by BSTFA [74]. Still others first converted keto acids to the ethoxime ($C_2H_5ONH_2$), then in one step quantitatively silylated (BSA) all the carboxylic acid, alcoholic, phenolic, and enolic keto functional groups (15-30 min, 70°). Formation of the oxime under aqueous conditions stabilizes the keto acid during drying and subsequent silylation [75].

2. Electron Capture Detection

Wilk and co-workers [25, 76] developed a two-step sequential procedure that comes
close to being a "universal" derivatization technique. The technique has been applied to samples containing compounds with amino, hydroxyl, enolic keto, and carboxylic acid functional groups. Samples are treated first with 20% pentafluoropropanol
in PFPA (75°, 15 min) and excess reagents are removed in a stream of dry N_2, then
treated with PFPA alone (75°, 5 min), and the excess PFPA reagent is evaporated.
After the residue is dissolved in ethyl acetate, the sample is ready for injection.
Pentafluoro reagents have the double advantage of being easily evaporated (easier
than HFBA) as well as yielding derivatives with good (nearly equivalent to HFB-)
electron-capture and chromatographic characteristics. As little as 5 pg of the
derivatives were detectable [25].

In a similar vein, Brooks and co-workers [76b] derivatize polyfunctional compounds using a reagent mixture comprised of HFBA, ethanol, and pyridine. Carboxyl
groups react to form the ethyl ester, while hydroxyl and amino groups form the N-
heptafluorobutyryl derivative. Although derivatization is reproducibly obtained in
only 3 min without heat, the manipulative steps involved appear to be a drawback.

D. Sulfonic Acids

Both hydroxyl and sulfonate groups of exemplary aromatic and aliphatic sulfonic
acids, or their salts, were silylated by shaking with TSIM for 10-15 min at room
temperature [77].

Aminobenzenesulfonic acids were easily derivatized by heating with PCl_5-$POCl_3$
(45 min, 95°). The benzene- and toluene-soluble products ($-SO_2Cl$, and $-NH-POCl_2$)
are directly injectable [78].

VI. AMINES

A. Introduction

Primary, secondary, and mono derivatized amines can be gas chromatographed. Secondary
amines are usually derivatized only for sensitization to detection by electron capture.
Tertiary amines, which for derivatization require a radical change of molecular structure, are derivatized either for more certain identification or sensitive detection.
sensitive detection.

Hydroxyamines are increasingly handled by *ideal sequenced derivatization* (see
Sec. VI.C.1).

B. Primary Amines

1. Flame Ionization Detection

In contact with DMF-DMA, one primary amine group is quickly and quantitatively converted to one N-dimethylaminomethylene group either within 10 min at 60° (80) or instantaneously upon injection [79]. Scheme 2 illustrates this reaction.

The difficulty of amine derivatization increases in the order, primary-N, di-primary-N, and secondary-N. Derivatization of only the primary-N derivative is usually enough. For instance, only the N-trifluoroacetyl derivatives were used in the derivatization of mixtures of the polyamines putrescine (primary) and spermidine and spermine (both contain primary and secondary amino groups) for GC analysis (TFAA-CH_3CN, 5 min, 100°; dry with nitrogen) [21].

In another study, primary amines were fully converted to the N,N-di(trimethylsilyl) derivatives. The secondary amines present could have been silylated by harsher conditions than those used (BSA-TMCS-CH_3CN, 2 h, 60°), but instead were allowed to remain unsilylated [1]. Usually, tertiary amines are derivatized specifically for increased sensitivity to detection by electron capture. However, the Hofmann degradation has been used to assist identification by producing one or more derivatives (usually related olefins) having retention times different from, and usually shorter than, that of the parent compound [81].

2. Electron Capture Detection

Cummins (33) found wide variation in sensitivity to detection by electron capture among the pentafluorobenzoyl and heptafluorobutyryl derivatives of six amines (see Table 1).

The sensitivity of ECD for HFB-cyclohexylamine is no higher than that of FID for cyclohexylamine. Aliphatic and nonaromatic amine heptafluorobutyryl derivatives are not ECD sensitive.

TABLE 1. Sensitivity of ECD Derivatives [33]

Amine	PF benzamide	HFB amide
Cyclopropylamine	1000	15
Cyclohexylamine	500	1
Aniline	133	670
Benzylamine	153	25
Dibenzylamine	23	83
Heptylamine	670	5

Of ten perfluorobenzene reagents tested for the ECD-sensitizing derivatization of phenethylamine and N-methyl-phenethylamine, pentafluorobenzoyl chloride was found to give best overall ECD/FID sensitivity ratio of about 2,000. Down to 10 pg of the amines were detectable [27].

In a thorough study, Matin and Rowland [87] compared a variety of ECD reagents for sensitivity versus a group of primary and secondary amines developing a hypothesis for understanding the electron capture mechanism. Pentafluorobenzoyl chloride was found suitable for primary amines and some secondary amines. Where primary and secdary amines occur together, and the former might interfere, the HFB derivative is probably the choice.

Highly ECD-sensitive trifluoroacetophenone derivatives, detectable to less than 1 pg, can be formed from tertiary aromatic amines having a suitably activated ring (TFAA in DMF, 60 min, 50°) [82].

An N,N-dimethyl tertiary amine can be sensitized to detection by electron capture *via* a quaternary ammonium intermediate. In each example to follow, the initial reactant was a chloroformate. In the first example, the tertiary amine DTM (N,N-dimethyl-dibenze [b,f] thiepin-10-methylamine) was deaminated using ethyl chloroformate (30 min, 60°); the chloroformate chlorine sensitized the bulky residue to detection by electron capture [83].

Tertiary amines were reacted with an alkyl chloroformate (e.g., ethyl chloroformate) and the resultant carbamate converted by hydrolysis to a secondary amine which is sensitized to electron capture detection by a pentafluorobenzoyl [84-86] or a heptafluorobutyryl group [87]. Alternately, tertiary amines can be determined directly as a carbamate by coupling with the strong electrophore reagent such as pentafluorobenzyl chloroformate [88]. It is important to remove excess reagent with 0.5 *M* KOH in 75% methanol prior to GC [89]. Not all tertiary amines react favorably to this approach, for example, methadone does not, nor are reaction conditions without complications. The general reaction for derivatization of tertiary amines through carbamate formation is shown in Scheme 3.

C. Catecholamines and Indoleamines

1. Flame Ionization Detection and Mass Fragmentographic Detection

Ideal sequenced derivatization began with a method [1] in which the catecholamine hydroxyl but not the amino groups were silylated by TSIM (CH$_3$CN, 2-3 h, 60°). BSA-TMCS was then added directly to the reaction mixutre (2-3 h, 60°), affecting neither the O-trimethylsilyl ethers resulting from the first derivatization nor the secondary amines present, but fully converting the primary amines to the N,N-di(trimethylsilyl) derivatives. The final reaction products were stable for several days and were separable by gas chromatography.

$$(a)$$

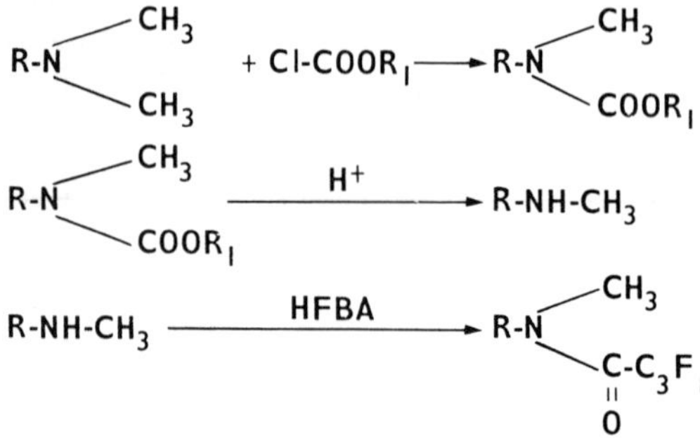

$$R\text{-}N(CH_3)_2 + Cl\text{-}COOR_1 \longrightarrow R\text{-}N(CH_3)(COOR_1)$$

$$R\text{-}N(CH_3)(COOR_1) \xrightarrow{\;H^+\;} R\text{-}NH\text{-}CH_3$$

$$R\text{-}NH\text{-}CH_3 \xrightarrow{\;HFBA\;} R\text{-}N(CH_3)\!\!\left(\underset{\underset{O}{\|}}{C}\text{-}C_3F_1\right)$$

$$(b)\quad F\text{-}C_6H_4\text{-}CH_2OCOCl + R\text{-}N(R_1)(R_2) \longrightarrow F\text{-}C_6H_4\text{-}CH_2\text{-}O\text{-}\underset{\underset{O}{\|}}{C}\text{-}N(R_1)(R_2)$$

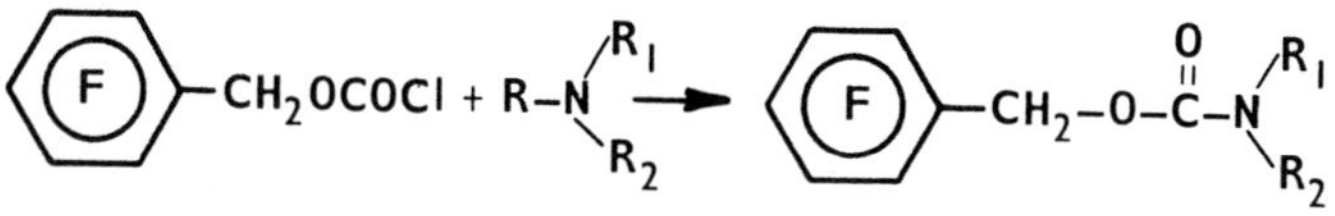

SCHEME 3

Derivatization of tertiary amines through carbamate formation [84-88]: (a) through
secondary amine formation and (b) direct coupling.

An *ideal sequenced derivatization* thus has the following six characteristics:

1. The different groups to be derivatized require more than one reagent and/or
   derivatization.
2. Each reagent yields one clean derivatization, affecting neither prior nor
   subsequent derivatizations.
3. Each successive reagent is added directly to the reaction mixture, which is
   not processed between successive derivatizations.
4. The reaction products are stable and preservable in the reaction mixture.
5. The final reaction mixture is directly injectable.
6. The derivatives are stable during gas chromatography, yield symmetrical
   peaks without undue delay, and are mutually separable.

Because phenolalkylamines oxidize easily, it is desirable to work with the stable salt form. Donike [90, 91], working directly with the salts in an ideal sequenced
derivatization, first silylated (MSTFA) and then N-trifluoroacetylated (MBTFA) them.
When protected against moisture, the derivatives are preservable for months, and are
detectable by mass fragmentography in femtomole quantities.

In a similar, ideal sequenced derivatization, the dried amine or amine hydrochloride was silylated (BSTFA, 5 min, 80°), then, on the addition of 5 µl of MBTFA,
the sample can immediately be injected [92].

Lovelady and Foster [93] dissolved isolated catecholamine residues in 0.1 ml BSA and added 0.1 ml each of THF and TFAA (mix, 10 min, 50°). Down to 0.1 pg of the O-trimethylsilyl, N-trifluoroacetyl derivatives could be detected by flame ionization.

Using PFPA in large excess, Willner et al. [94] derivatized phenylethylamine, phenylethanolamine, and the corresponding d_4 and d_3 internal standard replicates. The excess PFPA was then evaporated by nitrogen. Concentrations down to 5 pmole/g of tissue could be determined by mass fragmentography.

2. Electron Capture Detection

The pentafluoropropionyl and heptafluorobutyryl groups have been generally preferred over the trifluoroacetyl as the perfluoroderivatives for phenolic amines, and the pentafluoropropionyl over the heptafluorobutyryl [15, 16, 95] (however, see Ref. 32 for heptafluorobutyryl derivatization of tryptamine, 5-hydroxytryptamine, and their N- and/or O-methyl derivatives by HFBI). After pentafluoropropionyl derivatization, excess PFPA and solvent are removed by drying (helium). Redissolved, the sample is injectable. The sensitivity factor (electron capture/flame ionization) increases for the tri(pentafluoropropionyl) derivatives of metanephrine and normetanephrine by 166 and 95, respectively [95].

Using, in order, DNBS and BSA, Edwards and Blau [96] made the N-dinitrophenyl, O-trimethylsilyl derivatives of certain hydroxy amines (except secondary amines and catecholamines). The sensitivity factor (electron capture/flame ionization) increase was about 20, with minimum detectable amounts of 0.1 ng of dinitrophenyl-phenylethylamines (2 ng/g tissue) and as little as 8 pg of dinitrophenyl-diethylamine.

PFBA and BSA were used to form the N-pentafluorobenzylimine, O-trimethylsilyl derivatives of catechol primary amines. Directly injectable solutions of the derivatives were shown to be preservable without loss for weeks at room temperature. The peaks resulting from electron capture detection corresponded to 1.8 ng of amine, or 180 μg/g tissue [97].

D. Nitrosoamines

Using a long, multistep reaction, Brook et al. [98] showed that some of the biologically important nitrosoamines form compounds which are detectable by an electron capture detector when reacted with fluorinated anhydrides in the presence of pyridine. Some of the derivatives were 6000 times more sensitive to electron capture than flame ionization detection; about 17 ng of the derivative is detectable.

F. Phosphoryl Amines

Karlsson [99] used BSTFA-TMCS-pyridine (10:2:5) to silylate O-phosphorylethanolamine, O-phosphorylserine, and O-phosphorylthreonine (30 min, 60°). Detection was made by mass spectrometry.

F. Guanido Compounds

Using hexafluoroacetylacetone, Erdtmansky and Roehl [100] simultaneously extract,
derivatize, and sensitize to electron capture detection, drugs and metabolites containing the guanido group (R-C(=NH)-NH$_2$). Low picogram quantities of the derivatives
are detectable.

VII. AMINO ACIDS

1. Flame Ionization Detection

A successful method for the quantitative gas chromatographic analysis of the group
of twenty natural protein amino acids should apply to any one or equally well to all
at once (this review does not treat derivatives for protein sequencing). Preferably
in a single step, all the relevant functional groups should be derivatized. Separation of the derivatives should be complete, in about one to two minutes per amino
acid, and allow automatic integration. The final figures should reflect the starting
composition reliably, accurately, and precisely (for a pre-1968 review, see Blau [101]).

Gehrke and his co-workers have, for over a decade, vigorously and inventively pur-
sured the development of such a method [102-104]. After the first five years, they
could state that the principal requirements for acceptable amino acid derivatization
are (a) quantitative (95-100%) conversion to more volatile derivatives, (b) stability
during the derivatization (no rearrangements) and subsequent storage, (c) full recovery from intermediate concentration steps, and (d) mutually separable peaks on a conventional packed column of conventional length, as well as unreactive to both the
stationary phase and support within the column [103].

Although highly usable for individual or small groups of amino acids, trimethylsilylation is not as yet (late 1975) the method of choice for protein amino acid derivatization. Repeated attempts to use silylation for this purpose (while partially
successful) have all been found wanting. For instance, BSA gave sharp single peaks
for all the amino acids except arginine, which showed signs of decomposition on the
column [105]. Similarly, BSTFA, specifically synthesized for the amino acids problem
[103, 104], was not used for that purpose [106, 107]. Silylation comparison studies
[108, 109] had not yielded a fully satisfactory amino acid silylating reagent by 1971
(see, however, Hardy and Kerrin, [110]).

These comments on silylation apply equally to other promising derivatization
methods, some much used, some very new. The 2,4-dinitrophenyl derivatives of amino
acids have been thoroughly studies and Rosmus and Deyl [111] refer to over 1000 papers; however, the applications are replete with exceptions. Similar reservations
apply to the phenyl [111] and methyl [112] thiohydantoins. Promising new derivatives
including the oxazalidinone [113, 114] and the N-heptafluorobutyryl *n*-butyl ester
[115, 116] have not undergone the necessary thorough prolonged testing.

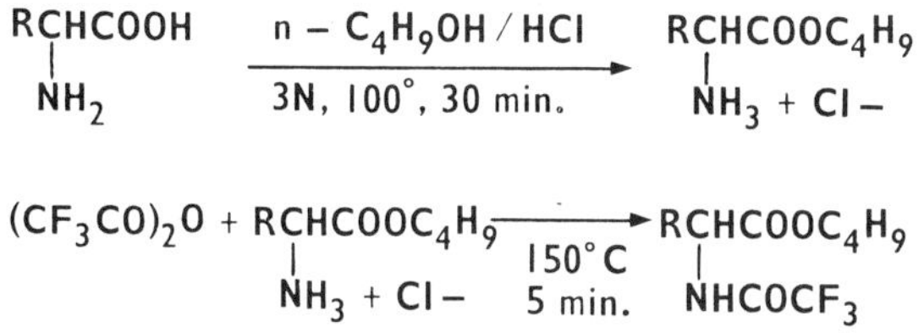

$$\underset{NH_2}{RCHCOOH} \quad \xrightarrow[\text{3N, 100}^\circ\text{, 30 min.}]{n - C_4H_9OH\,/\,HCl} \quad \underset{NH_3 + Cl-}{RCHCOOC_4H_9}$$

$$(CF_3CO)_2O + \underset{NH_3 + Cl-}{RCHCOOC_4H_9} \xrightarrow[\text{5 min.}]{150^\circ C} \underset{NHCOCF_3}{RCHCOOC_4H_9}$$

SCHEME 4

Derivatization after Gehrke for amino acids. a. Water is removed to give dry amino acids (mixture) which have been previously purified of interfering substances. b. Amino acids are converted directly to butyl ester hydrochlorides. c. The *n*-butyl ester hydrochlorides are converted to N-trifluoroacetyl *n*-butyl esters (TABs) with TFAA at elevated temperature in a sealed acylation tube. Other active functionalities are also converted to trifluoroacetyl derivatives in this process. The yields are essentially quantitative (98±3%) for the overall sequence for each of the twenty amino acids. The manipulations are routine and a number of samples may be TABed simultaneously.

In 1969 Gehrke et al. settled on the N-trifluoroacetyl *n*-butyl ester [103] and later described in detail the elaborate but effective systems of analysis involving this derivative [102, 117, 118]. Both Casagrande [119] and Raulin et al. [120] showed that this derivative allows separation of an additional 15-18 nonprotein amino acids and some amino sugars in the presence of the 20 protein amino acids, with certain modifications (Raulin used a 1.8-m (6-ft) column 0.325% ethylene glycol adipate on 80/100 Chromosorb G). The derivatization has been used [121, 122] for the analysis of nanogram and picogram quantities of the amino acids. Moodie [123] treated the N-trifluoroacetyl *n*-butyl derivatives with ethoxyformic anhydride (20 min, 150°; or 45 min at room temperature) to solve the unstable histidine N-trifluoroacetyl group "problem." N-TFA, *n*-butyl ester derivatization is illustrated in Scheme 4.

2. Electron Capture Detection

The naturally electron capture-sensitive, pg-detectable triiodothyronine and thyroxine have been converted to their methyl N,O-dipivalyl derivatives [124, 125].

For detection by electron capture, Wilk and Orlowski converted L-pyrrolidone carboxylic acid [126] and γ-glutaryl amino acids [127] to the N-pentafluoropropionyl esters with the pentafluoropropanol-PFPA combination described earlier (Sec. V.C).

Similarly, the oxazolidinone [113, 114] and N-heptafluorobutyryl *n*-butyl ester derivatives [115, 116] are used for sensitizing amino acids to electron capture detection.

VIII. AMIDES AND IMIDES

A. Barbiturates

For GC, barbiturates are usually derivatized to form dimethyl derivatives by flash heater alkylation. In this technique either of two methylating reagents TMH [130]

Uracil + $2(CH_3)_4N^+\ OH^-$ $\rightleftharpoons$ $(CH_3)_4N^+$-uracil salt + $2\ H_2O$ $\longrightarrow$ 1,3 – Dimethyluracil + $2\ (CH_3)_3N$ + $2\ H_2O$

SCHEME 5

Methylation of uracil by TMH.

or the more recently developed TMAH [128, 131-133] is injected simultaneously with
the sample. Scheme 5 illustrates the methylation by TMH of uracil. Using TMH and
three internal standards, Solow et al. [134] have analyzed thousands [129] of samples
for any of seven anti-epileptic drugs by injecting a TMH-phase serum extract at 350°.
Total processing time is 15 min.

Heavier groups, for example, ethyl or butyl, can also be derivatized by flash
heater alkylation, as for instance in the separation and determination of methylpheno-
barbitone and phenobarbitone [129, 135].

However, the results of flash heater alkylation are dependent on such critical
operational parameters [129, 136] as injection rate and port material, port-to-packing
gap, reagent-precursor concentration ratio, and pre-injection contact time (particularly the TMAH-phenobarbital combination [137]). Therefore, beginning about 1973
other approaches to barbiturate derivatization were tried.

In 1973, Dünges methylated free barbituric acids by microrefluxing [138] with
methyl iodide in acetone in the presence of K_2CO_3 for 30 min [136]. The supernatant
was injected in GC. Also in 1973, Venturella et al. [139] formed directly injectable
acetals from barbiturates heated at 100° for 10 min in DMF-DMA on a steam bath.

In 1974, using a highly polar solvent, Greeley [30] first converted an acidic
compound such as a barbiturate to a soluble salt with an organic base such as TMAH or,
preferably, TMH. An *n*-alkyl iodide of choice was then added, producing the corresponding alkyl derivative (10 min, room temperature). Xanthines are now being alkylated in this way rather than by flash-heater alkylation [140]. Although the timing
and phases differ, the chemistry of extractive methylation is essentially the same:
the acids (for instance, dimethylphenobarbital and dimethylpentobarbital) are extracted
as ion pairs from a water phase containing buffered tetrahexylammonium ion into an organic phase that contains methyl iodide. They are then methylated (10 min, room temperature [141].

B. Other Amides and Imides

Ehrsson and Mellstrom [29] reported that the pentafluorobenzoyl derivatives of the
seven secondary amides tested take much longer to prepare than the trifluoroacetates

but are far more stable, particularly to hydrolysis. These derivatives are sensitive
to detection by electron capture.

1. Saccharin

With sodium carbonate present, methylation of the imide saccharin by methyl iodide in
DMSO gives just one derivative rather than the two resulting from DAM methylation [136,
140, 142]. Alternatively, saccharin can be trimethylsilylated by BSA [143].

C. Cyclophosphamide

By treatment with TFAA-ethyl acetate (1:2,v/v) (20 min, 70°), cyclophosphamide and
isophosphamide, an internal standard, were converted to their mono-trifluoroacetates.
After dried in a nitrogen stream and redissolved in ethyl acetate, the derivatives
were injected [144].

IX. SUGARS

A. Carbohydrates

Carbohydrate derivatization for GC has been extensively reviewed by Pierce [145],
Sloneker [146], and Dutton [147].

The following silylation procedure of Sweeley et al. [148] has been very widely
used: to 10 mg carbohydrate add 1.3 ml reagent consisting of HMDS-TMCS-pyridine
(2:1:10, v/v) shake 30 sec and let stand for 5 min at room temperature and inject.

Drying prevents anomerization but water does not prevent silylation. Water
silylates rapidly at room temperature to form hexamethyldisiloxane, thus its presence
merely requires more reagent [149]. Both wet or aqueous samples have been silylated.
Using HMDS-TMCS, Langer et al. [150] silylated water-alcohol mixtures and Bentley and
Botlock [151] silylated 2% aqueous sugar solutions. For syrups, Brobst and Lott [152]
used HMDS-TFAAc, adding the TFAAc last. Investigating silylation in the presence of
water, Brittain and co-workers [11, 12] found that 21% TSIM in pyridine is excellent
and reliable for wet or dry sugars and its action is smooth and rapid.

Derivatization ideally yields one peak per component. Additional peaks may be
caused by either faulty derivatization or anomerization [151]. Alternatively, too
harsh conditions or potent silylating reagents may cleave linkages [6] or enolize
and silylate ester groups [153], making the derivatization and subsequent analysis
difficult.

A single glycose can exist in any distribution among five forms: one acyclic
form, and two stereoisomers (anomers) of each of two cyclic forms (furanose and pyra-
nose). Anomerization, the change in solution from one form toward an eventual

equilibrium distribution of all the anomeric forms, can occur during solution for
derivatization. Anomerization is the principal cause of complication in glycose
derivatization [154, 155].

The anomerization rate is essentially zero in dry DMSO, and remains low in DMSO-
water and dioxane-water, but is about 48 times higher in pyridine-water than in DMSO-
water (all solvent mixtures are 1:1) [156].

A preliminary reduction of the anomeric center (and thus of the aldose to the
alditol) eliminates the possibility of anomerization-caused multiple peaks. Acylation of the resultant alditols results in a single peak for each sugar [157]. Because of the lengthy reduction (see the end of this section for an example of the
procedure), alditol acetates tend to be more difficult to prepare than silyl derivatives, but they are more stable. The separation of the two derivative types were
found equally effective for the natural diacylglycerols [158]. The formation of
alditol acetates is found in the equations of Scheme 6.

The mass spectrum of a reduced glucose does not differentiate the end of the carbon chain from which a given ion fragment arises unless the carbonyl group is first
converted to an oxime [159]. Since such reduction will destroy the "family identity"
of components from among the aldonic, hexuronic, and hexaric acids, these compounds
are generally silylated [160]. Earlier approaches used oxidation of neutral glycoses
to the corresponding aldonic acids which were then silylated. Although the oxidation
is lengthy and multistep, the overall result is the desired one, and each glycose in
a mixture is converted to a single derivative [154].

B. Glucuronides

Direct derivatization of glucuronides circumvents the usual hydrolytic procedures,
which are time-consuming, require large amounts of the conjugate, and may be destructive to biochemically significant aglycons [6]. Ehrsson et al. [161] methylated the

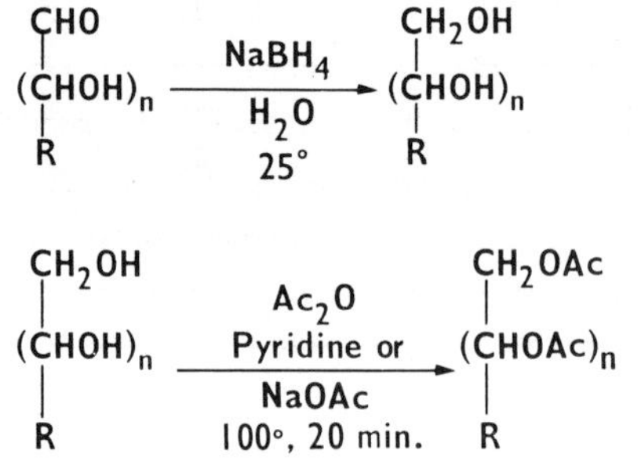

$$\begin{array}{c} CHO \\ | \\ (CHOH)_n \\ | \\ R \end{array} \xrightarrow[\substack{H_2O \\ 25°}]{NaBH_4} \begin{array}{c} CH_2OH \\ | \\ (CHOH)_n \\ | \\ R \end{array}$$

$$\begin{array}{c} CH_2OH \\ | \\ (CHOH)_n \\ | \\ R \end{array} \xrightarrow[\substack{Pyridine\ or \\ NaOAc \\ 100°,\ 20\ min.}]{Ac_2O} \begin{array}{c} CH_2OAc \\ | \\ (CHOAc)_n \\ | \\ R \end{array}$$

SCHEME 6

Preparation of alditol acetates.

carboxylic groups (DAM, 15 min, 20°; N_2 drying), then trifluoroacetylated alcoholic groups (TFAA, 15 min, 20°) of the β-D-glucuronic acid conjugates of 1-naphthol and androsterone. In the continued presence of the TFAA, the derivatives were stable. Mrochek and Rainey [6] silylated intact glucuronides (BSTFA-TMCS, then vacuum dried) that had been isolated from urine.

C. Sugar Phosphates

Sugar phosphates are usually trimethylsilylated (for example, BSA-TMCS, 1 hr, room temperature) [162]. Harvey and Horning [163] derivatized 43 sugar phosphates with BSTFA-TMCS-CH_3CN(2:1:2 v/v), using free phosphate or its metal or cyclohexylammonium salt for 10 min at 80°. The pentose and hexose-1-phosphates were derivatized (TSIM, 5 min, room temperature). In some cases, the methoximes were prepared ($CH_3ONH_2 \cdot HCl$-sugar phosphate 1:1 wt./wt., in pyridine, 1 hr, room temperature) to reduce the number of gas chromatographic peaks due to α- and β-furanose and -pyranose structures and to increase the stability.

D. Amino Sugars

Amino sugars were among the first carbohydrates to be silylated for gas chromatography [7, 148]. The primary amino groups were monoacetylated and then the hydroxyl groups were trimethylsilylated by HMDS-TMCS (15 min, room temperature) [7]. When BSTFA became available, it was used to silylate N-acetyl neuraminic acid (BSTFA-CH_3CN, 3:10 v/v, 2 hr at 125°) [164].

Porter [165], in a succinct review, suggested the resin hydrolysis of glycoprotein sugars by the $NaNO_2$ deamination of the resin-bound glucosamine and galactosamine to give the neutral 2,5-anhydrohexoses. These were then quantitated as the alditol acetates.

The derivatization of glycoses to alditol acetates is illustrated by the procedure of Porter [165]. In a water solution, the hexoses and 2,5-anhydrohexoses are reduced ($NaBH_4$, 1 hr at room temperature). The excess $NaBH_4$ is then decomposed by glacial HAc, the samples are dried, and the borate evaporated as trimethylborate using four additions of methanol-HCl, drying after each addition. To acetylate the hydroxyl group, pyridine and acetic anhydride are then added. In sealed containers, the samples are then heated at 100° for 15 min, removed, mixed, and again heated at 100° for 15 min. This final reaction mixture is injectable.

X. STEROIDS

Early studies are reviewed in *Gas Chromatographic Determination of Hormonal Steroids*, edited by Polvani et al. [166].

1. Flame Ionization Detection and Mass Spectrometric Detection

The methoxime-trimethylsilyl derivative is indicated when enolizable keto groups are
present [8, 167]. The methoxime group is formed first by reaction with methoxyamine
in pyridine for 3 hr at 60° [168] or overnight at room temperature [169]. The resulting methoxime is stable during silylation in either the methoxime reaction solution
or in a silylation solution.

The steroid methoxime [49] is relatively stable, has good GC properties, and is
the usual derivative for relatively unhindered steroid ketonic groups [167]. However, the methoxime derivative may yield syn/anti isomers [170, 171]. Also, if injected onto a surface that has received hydrogen chloride, the oximes can undergo
Beckmann fission to the nitrile [172].

The higher homologs of methoxime (e.g.) butyl [172a, 172b], pentyl [172b], and
benzyl [327] have been used to advantage. They are prepared in the same manner as
the methoxime, but possess longer retention times. In complex samples this effect
produces a simpler chromatogram and makes it possible to distinguish reactive keto
steroids from other steroids, as well as nonreactive keto steroids (11-one) [172].

Chambaz and Horning [8] suggested the following three silylation conditions
that correspond respectively to unhindered, moderately hindered, or highly hindered
hydroxyl groups: (a) BSA, room temperature; (b) BSA-TMCS or BSTFA-TMCS(2:1), room
temperature or 60°; and (c) TSIM-BSA-TMCS(3:3:2), 60°-80°. With conditions (c), the
use of methoxime-trimethylsilyl derivatives is suggested if keto groups are present.

Among silylating reagents, TSIM is atypical. TSIM reacts particularly well with
the sterically hindered 17α-hydroxyl group but not so easily with nonhindered hydroxyl
groups [10]. In better agreement with its overall silylating potential, neither TSIM
nor TSIM-MSTFA silylates the 11β-hydroxyl group, however, it can be successfully
silylated by BSA-TMCS, BSTFA-TMCS, MFTFA-TMCS, or TMSDEA-TMCS (24 hr, room temperature) [10].

Compared to trimethylsilyl ethers, the *tert*-butyldimethylsilyl ethers present
simpler mass spectra based on the M-57 peak. They are also more stable toward hydrolysis [173].

Certain steroid structures, primarily the dihydroxy acetone side chain, are
thermally unstable. Derivatives such as the methoxime-trimethylsilyl are usually
used to stabilize this side chain [174]. Alternatively, a cyclic boronate can be
formed [175, 176].

The boronic acids, $RB(OH)_2$, can form cyclic derivatives with a wide range of
functional groupings, including 1,2- and 1,3-diols, 1,2-diamines, and β- and α-
hydroxyamines. Only extremely mild and simple experimental conditions are required
and in many instances the reaction is rapid and essentially complete [175].

The derivatives show good gas chromatographic properties and under the exclusion of moisture they are stable. Also, compounds can be further derivatized in the presence of the boronate protecting group.

2. Electron Capture Detection and Mass Spectrometric Detection

Dehennin and Scholler [177] and Thomas and Walton in Polvani et al. [166] reviewed in detail the preparation and properties of steroid heptafluorobutyrates. Heptafluorobutyryl-based sensitivity to detection by electron capture can vary with the laboratory by a factor of 10 [178] and with the molecule derivatized by a factor of 100 [179]. Generally, heptafluorobutyrate detection limits are about 100 pg [179].

The important testosterone heptafluorobutyryl derivative is actually two products. These are stable in pyridine but not in benzene or acetone [180].

To increase the sensitivity to electron capture detection of derivatized testosterone and of keto steroids generally, Koshy et al. [181] prepared stable oximes with pentafluorobenzylhydroxylamine hydrochloride (30 min, 65°). With electron capture detection, as little as 5 pg of the pentafluorobenzyloxime-testosterone derivative could be detected.

Low nanogram quantities of the electron capture-sensitizing pentafluorophenyldimethylsilyl-derivatized steroids can be detected [182]. Hydroxyl groups can be derivatized in the presence of unprotected ketones and the reactivity of the system can be varied widely.

Chloromethyldimethylsilyl and iodomethyldimethylsilyl ethers, stable in both thin-layer and gas chromatography, not only enhance the sensitivity to detection by electron capture (down to 25 pg of iodomethyldimethylsilyl-testosterone are detectable) [183] but also provide intense high-mass peaks for gas chromatography-mass spectrometry [184].

XI. MISCELLANEOUS

A. Carbamates and Ureas

Cochrane [185] and Donough and Thorstenson [186] have reviewed the derivatization of insecticide and herbicide carbamates and ureas.

Carbamates and ureas can be hydrolyzed to yield amines and phenols. The original compounds as well as those formed upon hydrolysis can be acylated or alkylated.

Using TFAA, Evans [31] converted monoureas and symmetrically disubstituted ureas to their trifluoroacetates (2-3 min, room temperature). The derivatives are solid, stable for weeks if protected from moisture, and yield symmetrical peaks.

Saunders and Vanatta [187] reacted TFAA with a potential herbicide to produce a secondary N-trifluoroacetyl derivative with good gas chromatographic properties.

Using the same approach three 1,1-dimethylurea herbicides, liuron, fenuron, and monuron were studied. Only fenuron (3-phenyl-) and monuron [3-(*p*-chlorophenyl)-] yielded stable derivatives with TFAA, but not diuron [3-(3,4-dichlorophenyl)-].

However, fenuron, monuron, and diuron could be methylated to give stable and satisfactory trimethyl derivatives. Saunders and Vanatta [187] form the N-methyl compounds with methyl iodide and potassium *tert*-butoxide in THF (2-16 hr, 52°). Tanaka and Wien [188] injected a mixture of TMAH (in methanol) and the herbicide (in acetone) into a 220° injection port for flash-heater methylation. Under such conditions, a TMAH/urea mole ratio of 10 ensures reliable methylation, provided the other three nitrogen positions already have alkyl or aryl substitutions.

The phenols or amines resulting from hydrolysis are effectively derivatized and sensitized for hypersensitive detectors in ways already described in this chapter (Sec. III.C and VI.B) for such substances. For example, in one widely used approach [189-193], carbamate phenols have been converted by DNFB to their 2,4-dinitrophenyl ethers for detection by electron capture.

B. Purines, Pyrimidines, Nucleosides, and Nucleotides

Purines and pyrimidines proceed from the hydrolysis of nucleotides, as do nucleosides, which are purine- or pyrimidine-pentose combinations. Nucleoside silylations are best performed in a closed vial with a 225-fold molar excess of BSTFA for 15 min at 150° [194]. Purines and pyrimidines are also trimethylsilylated with BSTFA whereas BSA may not give complete silylation [5].

Lakings and Gehrke [195] hydrolyzed ribonucleic acid and deoxyribonucleic acid (20-µg amounts) in 100 µl TFAAc-HCOOH, 1:1 v/v, 1.5 hr for RNA or 2.0 hr for DNA at 200°. After drying under nitrogen, the resulting bases were silylated (BSTFA-CH_3CN, 150°); these silylation conditions are not critical [196], although dichloroethane-acetonitrile (1:1) is probably the best solvent [197]. Oxygen is displaced by nitrogen before the silylation to prevent its incorporation during silylation [198]. A 50-fold excess of BSTFA was used at 150° for 45 min giving 95% or better conversion with the model pyrimidines uracil and thymine [199].

Most nucleotides (30 µg in 30 µl pyridine) can be successfully trimethylsilyl-ated by BSTFA (30 µl)-TMCS (1 µl) in 30 min, at either room temperature or 150°. Adenosine 3',5'-cyclic monophosphate (cyclic AMP) required 5 hr at 100° to form the tri(trimethylsilyl) derivative [200].

C. Indoles and Imidazoles

Ehrsson [201] produced the electron capture-sensitive N-trifluoroacetyl 3-methyl-indole by reacting 100 mg of the indole in 5 ml of a 0.9 M trimethylamine solution in hexane containing 0.5 ml TFAA (20 min at room temperature). The product was extracted with water and dried under vacuum.

Most imidazoles can be quantitatively and almost instantaneously converted at room temperature to the stable (several days) acetyl derivatives by reaction with Ac_2O in dry THF [202].

D. Epoxides

If pyridine is used as solvent, epoxides react quantitatively and specifically with TMCS (30 min, $70°$) to form chloro-trimethylsilyl derivatives [14]. Trace water catalyzes the reaction.

E. Phosphoric and Phosphonic Acids

With DAM, Daemen and Dankelman [203] converted mono- and di-alkyl (C_8-C_{16}) phosphoric acids and their salts into the corresponding methyl esters.

Harvey and Horning [204] satisfactorily trimethylsilylated a number of alkyl- and aminoalkylphosphonates (BSTFA-TMCS, 2:1). However, for the chromatography of 1-aminoalkylphosphonates, conversion of the amino group into the isothiocyanate (CS_2) was found necessary subsequent to the silylation.

F. Thiols

Lofberg [205] silylated three series of homologous thiols with BSTFA (overnight at room temperature); others [206] formed the thioalkylbenzoates. Donike [227] has applied MBTFA to thioglycol (mercaptoethanol) forming the O,S-bis (TFA-) derivative.

G. Sulfonamides

Vandenheuvel and Gruber [207] quantitatively converted primary sulfonamides to the N-dimethylaminonethylene derivatives by reaction with DMF-DMA (10 min, $75-80°$). The reaction mixture was injected. With electron capture detection, 25 ppb (wt./wt.) of 3-bromo-5-cyanobenzenesulfonamide in the blood and plasma of sheep could be detected when dosed at 15 mg/kg.

XII. DERIVATIZATION TABLES

Organization of Tables 2 through 7 parallels the organization of the chapter text which is based on groupings according to functional group and compound class. The purpose of the tables is to provide the reader with more examples to consult regarding derivatization problems than the limited space for discussion in the text permits.

The tables are not intended to be comprehensive, accordingly many worthy reports were omitted (some on account of necessity, others unwittingly). The organizer apologizes in advance. Recent publications are emphasized in the Table for the reason that these papers will serve to gain the reader access to references regarding earlier reports. When many reports employing the same derivatization scheme were available, the effort was made to choose the best examples (simple reaction conditions involving "model" compounds) for reduction to practice. Early reports are listed in cases where the methodology employed continues to be used to the present substantially unchanged. Special attention is paid to recently developed derivatization techniques and reagents that hold promise for continuing use. Given the formidable monograph on silylation by Alan Pierce (over 800 references) less attention is given to trimethylsilylation techniques prior to 1968. This "omission" pertains to much of the pioneer work done using HMDS and TMCS before the flowering of the silylamides, BSA, and BSTFA.

To be readily understood, tables need to be concise, and this can be achieved through simplification which often does injury to the full truth. Simplification is practiced extensively under the table column Example. Here, in many cases, only a few compounds of many examples cited in a given report could be listed forcing the organizer to choose examples which were variously "typical," "most significant," or simply to reflect the breadth of the study, "30 aldehydes and ketones."

ACKNOWLEDGMENTS

Thanks are due to Cathy Beck for careful typing and Margaret Feit for assistance with the library work.

TABLE 2. Hydroxyls (Flame Ionization Detection)

Functional group	Example	Reagent	Derivative[e]	Reference
Alcoholic	n-Octanol	MBTFA or BTFA	O-TFA ester	[227]
	Vitamin C	MSA	O-TMS	[231]
	Vitamin A	MSA	O-TMS	[232]
	Vitamin B_6	MSA (py)	Tri(O-TMS)	[233]
	Vitamin-D_2	BSTFA, TMCS	O-TMS	[34]
Phenolic	Phenol	PFB-Br (TBHSO$_4$)^d	O-PFB	[234]
	Phenols (hindered)	Various TMS reagents plus TFAAc	O-TMS	[9]
	Stilbesterol	BSA	Di(O-TMS)	[235]
	Apomorphine	BSA	Di(O-TMS)	[236]
	Morphine and codeine (in urine and liver)	BSTFA or BSA	O-TMS	[237]
	Δ^9-Tetrahydrocannabinol	BSTFA, TFAAc	O-TMS	[238]
	Mono- and dihydroxy-cannabinoids	C_3, C_4, C_6-Trialkylchlorosilanes	O-Si(R)$_3$ ethers	[239]
	Flavinoids and glycosides	BSA, TMCS	O-TMS	[240]
Polyhydroxyl	Antibiotics	TMS various		[38]
	Mono- and diethylene glycol	BSTFA	O-TMS	[250]
	Ceramides	HMDS, TMCS	Di(O-TMS)	[242]
	Stearyl alcohol, 1,2-Glycol, Mannitol	MBTFA	TFA ester Di(TFA)ester Hexakis(TFA)ester	[227]
	1,3 and 1,2-Diacyl-glycerols	HMDS, TMCS (py)	TMS ether	[158]
	Pentaerythritol	HMDS, TMCS	O-TMS	[251]
	Styrene glycol, o-Hydroxymethylphenol	NBBA or phenyl-boronic acid	Boronate ester	[241]

TABLE 2. (Continued)

Functional group	Example	Reagent	Derivative[e]	Reference
	Sorbitol	NBBA	NBB	[35]
	1,2 and 1,3-Diols	NBBA	Butylboronate esters	[176]
	Spectinomycin	HMDS	Tetrakis TMS ether	[40]
	Lincomycin and Lincomycin B	HMDS, TMCS (py)	O-TMS	[252]
	Erythromycin and derivatives	TSIM, BSA, TMCS	O-TMS	[251]
Alcoholic and carboxylic acid	Prostaglandins (review)	Various		[41]
	$PG-F_{1\alpha}$, $F_{2\alpha}$, E_1, and E_2	CH_2N_2 TSIM (piperidine)	Me ester O-TMS	[248]
	PG-A,B,E,F	CH_3ONH_2[b] BSTFA	Methoxine O-TMS ester and ether	[42]
	$PG-E_1$[c]	CH_2N_2 CH_3ONH_2 BSTFA	Me ester Methoxime O-TMS	[46]
	$PG-E_2$ (also $F_{2\alpha}$)	$C_6H_5CH_2ONH_2$ CH_2N_2 unknown	Benzyloxime Me ester di(O-TMS)	[249]
	$PGF_{1\alpha,2\alpha}$	NBBA TSIM (py)	NBB ester O-TMS ester, ether	[43]
	$PG-F_{2\alpha}$	CH_2N_2 NBBA BSTFA	Me ester 9,11-Butylboronate 15-O-TMS ether	[45]
	Prostaglandin $F_{2\alpha}$, E_2	CH_2N_2[a]	Me ester	[194]

[a]Oximes (hydroxylamine HCl) and butyl boronates (*n*-butylboronic acid) were also utilized by the authors.

[b]BSTFA, TMCS is employed when the methoxylamine HCl is not used.

[c]GC/multiple ion analyzer.

[d]Extractive alkylation using tetrabutylammonium hydrogensulfate.

[e]For a more extensive listing of TMS-hydroxyl derivatives, see Ref. 145.

TABLE 3. Hydroxyls (Electron Capture Detection)

Functional group	Example	Reagent	Derivative	Reference
Alcoholic hydroxyl	$C_6H_5(CH_2)_nOH$ and Tropine (where n = 1, 2, 3)	HFBA	O-HFB ester	[28]
	C_2-C_7 Alcohols and phenethyl alcohol	BMDCS (NaI, DEA)	$-O-Si(CH_3)CH_2I$	[243]
Phenolic hydroxyl	40 Phenols	DNFB	DNP ether	[244]
	Phenols (from pesticides)	(a) PFB-Br DNFB	DNT ether[b] PFB ether DNP ether	[37]
	Phenols (various substituted)	HFBA, TMA	HFB ester	[18]
	Cresols, xylenols	PFB-Br	PFB ether	[19]
	1-Naphthol	PFB-Br	PFB ether	[23]
	Thymol	7 Reagents compared PF-benzoyl-Cl	PFB ester	[17]
	Thymol[d]	DiEt phosphororidate	Diethyl phosphate ester	[245]
	Δ^9-Tetrahydrocannabinol	HFBI	O-HFB ester	[246]
	Pentazocine	PFB-Br	PFB ether	[20]
Phenolic and alcoholic	MOPEG[e], free and conjugated[c]	PFPA	Tri(O-PFP)	[36]
	MOPEG, MOPEOA[e]	PFPA	PFP ester	[24]
Hydroxy acid	$PFG_{2\alpha}$	PFB-Br BSA	PFB ester O-TMS ether	[22]
Hydroxyamines	Mono, di, tri-ethanolamine	BSA	N,O-TMS	[247]

[a]Reagent is 4-chloro-α,α,α-trifluoromethyl-3,5-dinitrobenzene.   [d]Flame photometric detection.

[b]Dinitrotrifluoromethylphenyl ether.   [e]3-Methoxy-4-hydroxyphenethylene glycol and 3-methoxy-4-hydroxyphenethanolamine.

[c]Detection GC/MS.

TABLE 4. Carbonyls (Flame Ionization Detection)

Functional group	Example	Reagent	Derivative	Reference
	(Also see keto acids, Kreb's cycle acids)			
	C_{1-12} Aldehydes and C_{3-11} ketones	DNPH	DNP hydrazone	[357]
	C_{1-4}, C_{6-9} Aldehydes	DNPH	DNP hydrazone	[48]
	C_{1-5} and C_6H_5-Aldehydes	DNPH	DNP hydrazone	[47]
	36 Aldehydes 32 ketones	NH_2OH	Oxime	[51]
	27 Aldehydes and ketones	$C_6H_5NHNH_2$	Phenylhydrazone	[52]
	Ketols (e.g., cycloheximide)	NBBA	Butylboronate ester	[176]
	(Also see keto steroids)			
	Cholestane-3-one	N,N-Dimethylhydrazine	Dimethylhydrazone	[337]
	Ketosteroids (various)	CH_3ONH_2[a]	Methoxime	[49]
	Ketones, phenolic (e.g., acetonvanillone)	BSA	Mono- and di-OTMS	[13]
	Cytidine and deoxycytidine	CH_3ONH_2(py) BSTFA, TMCS	Methoxime O-TMS ether	[50]

[a]Combined with the use of silylation reagent when hydroxyl groups are present.

TABLE 5. Carbonyls (Electron Capture Detection)

Functional group	Example	Reagent	Derivative	Reference
	C_{1-4}, and C_{6-9} Aldehydes	DNPH	DNP hydrazone	[48]
	Ketosteroids	PFBHA	Perfluorobenzyl-oxime	[181]

TABLE 5. Carboxylic Acids (Flame Ionization Detection)

Functional group	Example	Reagent	Derivative	Reference
$-CO_2H$	Review (fatty acids)			[58, 280, 281, 301]
	Fatty acids	MeOH, HCl	Me ester	[60]
	Fatty acids	MeOH, H_2SO_4	Me ester	[59]
	Fatty acids	BF_3, MeOH	Me ester	[61, 282, 283]
	Fatty acids	TMH	Me ester	[284]
	Fatty acids	TMAH	Me ester	[285]
	Stearic acid	TMAH or TMH, alkyl iodide	C_1,C_2,C_4 esters	[55]
	Fatty acids	DMF-dialkyl acetals	Me, Et, Pr, Bu and t-Bu esters	[67]
	$C_{10},C_{14},C_{18},C_{22}$ Acids	MSTFA	TMS ester	[286]
	$C_{14}-C_{20}$ Acids	TMTFTI[a]	Me ester	[66]
	Glycerides	Methyl propionate TMH, CH_3I (DMF)	Me ester	[56]
	2-Hydroxytetracosanoic acid	MeOH, HCl BSTFA	Me ester O-TMS ether	[72]

[a]Trimethyl (α,α,α-trifluoro-m-tolyl)ammonium hydroxide.

TABLE 7. Carboxylic Acids (Flame Ionization Detection)

Functional group	Example	Reagent	Derivative	Reference
$-CO_2H$	General	CH_2N_2	Me ester	[53, 54, 267-70]
		MeOH, CDI[e]	Me ester	[68]
		ROH, DCCl[f]	Alkyl or benzyl ester	[271]
		$C_6H_5CHN_2$	Benzyl ester	[70, 272]
	94 Carboxylic acids	BSTFA, TMCS	TMS ester	[220]
	C_2-C_6 Acids[a]	α-OH-Phosphonic acid esters (DCCI)[f]	HPE[c]	[287]
	Caproic acid	MBTFA	Mixed anhydride	[227]
	$C_{10,12,14,16,18}$ Acids	MSA	TMS ester	[210]
	$C_{4,5,6,8,10}$ Acids	PFB-Br $(TBHSO_4)$[b]	PFB ester	[234]
	(2,4-Dichlorophenoxy) acetic acid	BF_3/MeOH vs CH_2N_2	Me ester	[273]
	Acetyl salicylic acid	BSTFA or MSTFA	TMS-ester	[274]
	Dibasic(C_2-C_5) acids	$C_6H_5CHN_2$	Di-benzyl ester	[276]
$>$CH-OH (Alcoholic)	General	BSTFA, TMCS	O-TMS ether TMS ester	[220]
	VMA	BSTFA	O-TMS ether TMS ester	[73]
	Phenyllactic acid	BSTFA	TMS ester O-TMS ether	[74]
	Lactic and 3-OH-butyric	$C_6H_5CHN_2$ or DBI[d]	O-Benzyl ether benzyl ester	[276]
	Vitamin C	MSA	O-TMS ether TMS ester	[231]

	α and β-Hydroxy acids	NBBA	Cyclic butylboronate ester	[176, 275]
	Ecgnine and benzoyl-ecgnine (in cocaine)	BSA	O-TMS ether and ester	[260]
-OH (Phenolic)	o-CH-Phenylacetic acid	BSTFA	O-TMS ether TMS ester	[74]
	VMA, HVA	BSTFA	O-TMS ether TMS ester	[73]
	Dihydroxybenzoic acid	CH_2N_2	O-Me Me ester	[4]
	Dihydroxybenzoic acids	BSA	O-TMS ether and ester	[259]
	Vitamin K_1,K_2 (reduced)	MSA	Di (O-TMS) ether	[277]
=O (Keto)	Phenylpyruvic acid	BSA, HMDS, TMCS NH_2OH	TMS ester =NO-TMS	[278]
	Phenylpyruvic, o-CH-phenylpyruvic	BSTFA	TMS ester O-TMS ether	[74]
	Pyruvic, oxaloacetic, 2-oxoglutaric acids	$C_6H_5CH_2ONH_2$ $C_6H_5CHN_2$	O-Benzyloxime Benzyl ester	[276]
Multifunctional (acid, hy-droxy, enol)	General	BSA (py)	TMS ester TMS ethers (hydroxyl and enol)	[288]
	General	BSTFA, TMCS	TMS	[220]
Other	Indole acids	BSTFA	TMS ester N-TMS	[2]
	Dibasic acids and shikimic acid	BSTFA, TMCS	Di-TMS ester	[3]
Kreb's cycle acids	Phenylacetic, -lactic, -pyruvic acids	TMCS, HMDS NH_2OH	TMS ester, ether, =NO-TMS	[289]

TABLE 7. (Continued)

Functional group	Example	Reagent	Derivative	Reference
	Pyruvic, oxalacetic, 2-oxoglutaric acids	BSTFA CH_3ONH_2	O-TMS ester Methoxime	[290]
	Phenylpyruvic acid	NH_2OH BSTFA, TMCS	=NO-TMS TMS ester	[291]
	Review (urinary acids)	Various		[292]
	Acetate	BDMPH[g]	Benzyl ester	[293]
	Krebs cycle keto acids	CH_2N_2 or MeOH, HCl	Me ester[h]	[294]
	Urinary acids	BSTFA	TMS esters TMS ethers	[279]
	Urinary acids	BSA $C_2H_5ONH_2$	TMS esters O-ethoxime	[75]
$-SO_3H$	Aminobenzene-sulfonic acid	PCl_5 or $POCl_3$	$-SO_2Cl$ $-NHPOCl_2$	[78]
	p-Toluenesulfonic acid	TSIM	$-SO_3TMS$	[77]
Anhy-drides	Glutaric, succinic and phthalic anhydrides	Morpholine BSA	$CO-NC_4H_8O$ O-TMS ester	[295]

[a]Alkaline flame ionization detector.

[b]Extractive alkylation technique.

[c]α-Hydroxy-phosphonate esters.

[d]N,N'-Dicyclohexyl-O-benzylisourea.

[e]N,N'-Carbonyldiimidazole.

[f]N,N'-Dicyclohexylcarbodiimide.

[g]Benzyldimethylphenylammonium hydroxide.

[h]Use of diazomethane leads to complex derivative formation due to reaction with α-keto group.

TABLE 8. Carboxylic Acids (Electron Capture Detection)

Functional group	Example	Reagent	Derivative	Reference
$-CO_2H$	C_1-C_8 Acids	BMDCS(NaI,DEA)	a	[253]
	C_2-C_8 Acids	CCl_3CH_2OH(HFBA)	TCE ester	[71]
	Benzoic, phenylacetic acids	CCl_3CH_2OH(TFAA)	TCE ester	[57]
	Mefenamic acid	$PFPA(ZnCl_2)$	Mixed anhydride	[254]
	Naphthaleneacetic and 2,4-dichlorophenoxy-acetic acid	BMDCS	BMDS ester	[265]
	Probenecid	$C_2F_5CH_2OH$(TFAA)	$C_3F_5H_2$ ester	[255]
	C_2-C_8,$C_{10,12,14,16,18}$ Acids	PFB-Br	PFB-ester	[256]
	Substituted propionic acid	PFB-Br	PFB-ester	[257]
$>$CH-OH (Alcoholic)	HVA	C_3F_6HOH TFAA	C_3F_6 ester O-TFA ester	[258]
-OH (Phenolic)	Phenolic acid metabo-lites of biogenic amines	$C_2F_5CH_2OH$	$C_3F_5H_2$ ester	[25, 261]
	HVA	CH_2N_2 HFBA	Me ester O-HFB	[262]
	5-HIAA[b]	CH_2N_2 HFBI	Me ester N,O-HFB	[263]
Alcoholic and phenolic	Lactic, salicyclic, p-HO phenylacetic acids	HFBA, EtOH (py)	Et ester O-HFB	[76b]
	Bile acids	CMDMCS	CMDMS-ether, ester	[266]

[a] $-O-CH_2Si(CH_3)_2OSi(CH_3)_2CH_2-$.   [b] 5-Hydroxyindol-3-acetic Acid.

TABLE 9. Amines (Flame Ionization Detection)

Functional group	Example	Reagent	Derivative	Reference
Primary and Secondary Amines	Low molecular weight (e.g., n-BuNH$_2$)	Diethyl pyrocarbonate	Urethane	[304]
	General-26 primary and secondary amines	DNPB	DNP	[305]
	Bis(4-aminocyclohexyl) methane, 2,2-bis(4-aminocyclohexyl)propane	DMF-DMA	N-DMAM-[a]	[80]
	n-Octylamine	HMDS, TMCS	N-TMS	[306]
	Cyclooctylamine	DMF-DMA	N-DMAM-[a]	[228]
	m and p-Xylylene diamines (also carbonates and HCl)	TFAA	N,N'-TFA	[307]
	Polyamines (e.g., putrescine, spermidine, spermine)	TFAA	N-TFA	[21]
Biomedicinal Amines	General	BSTFA, TMCS	N,O-TMS	[220]
	PEA, PEOA	PFPAA	N,O-PFP	[94]
	Mono-, di-, tri-ethanolamine	BSA	N,O-TMS	[247]
	PEA	DMF-DMA	N-DMAM[a]	[228]
	Amphetamine, p-OH-amphetamine	MSTFA, TFAAc	O,N-TMS	[222]
	Amphetamine, methamphetamine, diphenylamine	MBTFA or BTFA	N-TFA	[227]
	Ephedrine, norepinephrine	MSTFA MBTFA or BTFA	O-TMS N-TFA	[227, 264]
	Tyramine, ethanolamine	TFAA BSTFA	N,O-TFA[b] phenolic O-TMS	[308]

	Tyramine, DA	CS_2 BSTFA, TMCS	Isocyanate O-TMS	[309]
	Review (DA and related compounds)	Various		[310]
	Dopamine	C_6F_5CHO BSA	$N=CHC_6F_5$ (b) O-TMS	[311]
	Norepinephrine	TSIM BSA, TMCS	O-TMS N-TMS	[1]
	Epinephrine, norepinephrine	BSA TFAA	O-TMS N-TFA	[93]
	Epinephrine, norepinephrine	TSIM HFBI	O-TMS N-HFB	[312]
	NE, E, DA, tyramine tryptamine, histamine	Ac_2O (py)	O,N-Ac	[314]
	MN, NMN, 5-MeO-tryptamine, 3-MeO-tyramine	BSTFA or BSA MBTFA	O-TMS N-TFA	[92]
	Serotonin, 5-MeO- tryptamine	Me_2CO HMDS	Schiff base O-TMS	[303]
	Serotonin	TSIM HFBI	O-TMS N-HFB (indole) N-HFB	[313]
Tertiary	Amitripytyline, promazine	Alkyl halide, Ag_2O (Hofmann degradation)	Olefin deriv.	[81]
Other	Phosphorylethanolamines	BSTFA, TMCS	O,N-TMS P(O)-TMS	[99]
	β-Hydroxyamines (e.g., E, NE, MN, isoprenaline)	NBBA	Butylboronate esters	[275]
	1,2- and 1,3-Hydroxyamines	NBBA	Oxazaborinanes	[176]

[a]N-DMAM; N-Dimethylaminomethylene-.

[b]Mass fragmentography.

DNP: 2,4-Dinitrophenyl

DNPE: DNP-*tert*-butylamine

TABLE 10. Amines (Electron Capture Detection)

Functional group	Example	Reagent	Derivative	Reference
Primary and Secondary Amines	General	HFBA	N-HFB	[28]
	t-Butylamine		N-DNP	[298, 299]
	PEA, N-MePEA	C_6F_5COCl C_6F_5CHO	PF-benzamide Schiff's base	[27]
	Tryptamine and N-Me, N,N-dimethyl	HFBI	N,N-HFB[a]	[32]
	Cyclohexylamine (in aqueous soln.)	DNFB	N-DNP	[300]
	PEA	C_6F_5CHO	Schiff's base	[26]
	PEA, Amphetamine (primary) Methamphetamine, phentermine (secondary)	C_6F_5COCl C_6F_5CHO or HFBA[b]	PF-benzamide PF-benzylidene N-HFB	[87]
	Protriptyline	HFBI	N-HFB	[316]
Hydroxyamines (alcoholic and phenolic)	Phenethylamines (e.g., octopamine)	DNBS BSA	N-DNP O-TMS	[96]
	Phenethylamines (e.g., tyramine)	3,5-Dinitro-4-chlorobenzo-trifluoride BSA	N-DNT O-TMS	[96]
	Propranolol	HFBA	N,O-HFB	[302]
	NE and DA	C_6F_5CHO BSA	PF-benzylimine O-TMS	[97]
	NE, E, DA, NM, serotonin	PFPA	N,O-PFP	[16]

	NE and E	PFPA	O,N-PFP	[303]
	Phenolakylamines	MSTFA MBTFA	O-TMS N-TFA	[90, 91]
	PEF, aminophenols	TFAA, PFPA, HFBA or $(ClCH_2CO)_2O$	N,O-R	[315]
	Catecholamines and indoleamines	PFPA	N,O-PFP (c)	[95]
	Catecholamine and indoleamine acidic metabolites	BCl_3/MeOH PFPA	Me ester N,O-PFP (a)	[95]
	Catecholamines and acidic metabolites[c]	CH_2N_2 PFPA, TFA, HFBA	Me ester N,O-R	[15]
Tertiary amines	Imipramine, amitriptyline, diphenhydramine	Pentafluorobenzyl- chloroformate	PFB-carbamate	[84, 88]
	Trimipramine, 2-chloroimipramine	Pentafluorobenzyl- chloroformate	PFB carbamate	[89]
	N,N-Dimethylaniline	TFAA	C-TFA (C-Acylation)	[82]
	Imipramine, amitryptyline	Methyl chloro- formate HFBA	1. Demethylation 2. N-HFB Acylation	[87]
	DTM[d]	Ethyl chlorocarbonate	Chloroethylene deriv.	[296]
	Imipramine	Ethyl chloroformate HFBA	1. Demethylation 2. N-HFB Acylation	[85]
Quaternary amines	Acetylcholine	Sodium benzothiolate	Demethylation	[297]
Other	Nitrosamines	HFBA	N-HFB	[98]
	Imicazoles	$(Ac)_2O$	N-Ac	[202]
	3-Methylindole	TFAA (TMA)	N-TFA	[201]

TABLE 10. (Continued)

[a]Indole nucleus nitrogen is also acylated.

[b]Comparative study.

[c]GC/MS detection.

[d]N,N-Dimethyldibenzo [b,f] thiepin-10-methylamine.

TABLE 11. Amino Acids (Flame Ionization Detection)

Functional group	Example	Reagent	Derivative	Reference
$-NH_2$; $-CO_2H$	Review	Various		[101, 209]
	18 Protein amino acids	TFAA MeOH, HCl	N-TFA Me ester	[210-212]
	20 Protein amino acids	TFAA n-BuOH, HCl	N-TFA n-Bu ester	[102, 117, 118, 122]
	16 Protein amino acids	TFAA n-BuOH, HCl	N-TFA n-Bu ester	[121]
	19 Protein amino acids (except cystine)	TFAA n-BuOH, HCl	N-TFA n-Bu ester	[213]
	Protein amino acids, incl. asparagine, glutamine, pyroli-dine-carboxylic acid	TFAA n-BuOH, HCl	N-TFA n-Bu ester	[214]
	18 Nonprotein amino acids	TFAA n-BuOH, HCl	TFA n-Bu ester	[120]
	Hydroxyproline	TFAA n-BuOH, HCl	TFA n-Bu ester	[215]
	Histidine	TFAA ethoxyformic anhydride n-BuOH, HCl	N^{α}-TFA N^{im}-carbethoxy n-Bu ester	[123]
	General-23 amino acids	TFAA MeOH, HCl	N-TFA Me ester	[216]
	17 Protein amino acids	Ac_2O n-PrOH, HCl	N-Ac n-Pr ester	[217]
	17 Protein amino acids	TFAA n-BuOH	N-TFA n-Bu ester	[218]
	35 Protein amino acids	TFAA n-BuOH	N-TFA n-Bu ester	[219]

TABLE 11. (Continued)

Functional group	Example	Reagent	Derivative	Reference
	19-20 Protein amino acids, incl. HOPr and HOLys	HFBA n-PrOH, HCl	N-HFB n-Pr ester	[116, 221]
	Rare methylated amino acids[a]	TFAA n-BuOH, HCl	N-TFA n-Bu ester	[311]
	Asparagine, asparagine acid and histidine	MSTFA, TFAc	Tri-TMS	[222]
	Protein amino acids (except cystine)	HFBA n-PrOH, HCl	N-HFB n-Pr ester	[115]
	14 Protein amino acids	HFBA or PFPA n-BuOH	N-HFB or N-PFP n-Bu ester	[223]
	General	BSTFA, TMCS	N,O-TMS	[220]
	18 Protein amino acids	BSTFA	N,O-TMS	[224]
	20 Protein amino acids	BSTFA	N,O-TMS	[103, 104, 106, 225]
	20 Protein amino acids	BSTFA n-BuOH, HCl	N-TMS n-Bu ester	[110]
	17 Protein amino acids	MSA	N,O-TMS	[226]
	Protein amino acids (except Arg)	BSA	N,O-TMS	[105]
	Phe, Tyr, Lys	TMSDEA	N,O-TMS	[108]
	Tyr, Leu, Me ester	MBTFA	N,O-TMS and N-TMS	[227]
	Val, Lys, Phe, Glu, Leu	DMF-DMA[b]	N-DMAM Me ester	[228]
	Review	DNFB	DNP	[111]

	12-α-Methyl substituted alanine derivatives	Benzoyl-Cl (DCCI)	2-Phenyl-4-methyl-oxazolin-5-one	[229]
Sulfur containing	9 Sulfur amino acids	BSTFA	NH-TMS, S-TMS SO_3-TMS, CO_2TMS	[230]
Iodine containing	Tri- and tetraiodo-thyronine	MeOH, HCl	Me ester	[124]
	Tri-iodothyronine and thyroxine	Pivalic anhydride	N,O-Dipivaloyl	[125]

[a]Mass fragmentography detection.

[b]The corresponcing butyl ester may be formed using DMF-di-n-butyl acetal.

TABLE 12. Amino Acids (Electron Capture Detection)

Met, Cys	HFBA n-BuOH, HCl	N-HFB Bu ester	[122]
Ala, Ser, Met, Cys Pro, HOPro, Phe, Tyr	TFAA n-BuOH, HCl	N-TFA Bu ester	[122]
L-Pyrrolidone carboxylic acid	PFPA, $C_2F_5CH_2OH$	$O-CH_2C_2F_5$ ester	[126]
Iodotyrosines	$(CF_2Cl)_2CO^a$	Oxazolidinones	[113, 114]

[a]Tyrosine hydroxyl group is blocked with a variety of groups, e.g., HFB and TMS.

TABLE 13. Barbiturates and Other Amides and Imides (Flame Ionization Detection)

Functional group	Example	Reagent	Derivative	Reference
Amide	Review	Various		[358]
	General (Barbiturates)	Various		[133, 134, 360, 366]
	General	$(CH_3)_2SO_4$	N-Me	[359]
	General	TFAA	N-TFA	[29]
	Barbiturates	BSA	N-TMS	[133, 360]
	DPH and metabolite (HPPH)	BSA (DMF)	N-TMS	[362]
	General (barbiturates and DPH)	1. CH_2N_2 2. BSA and TMCS	$N-CH_3$ N-TMS	[363]
	Methsuximide	BSA	N-TMS	[364]
	DPH (Dilantin)	TMH	1,3-Di(N-Me)	[365]
	Review	TMAH	N-Me	[132]
	General	TMAH	N-Me	[128]
	Various barbiturates	TMAH	N-Me	[368]
	Phenobarbital	TMAH	N-Me	[137]
	Dilantin (DPH)	TMAH	1,3-Di(N-Me)	[131, 367]
	DPH and various barbiturates	TEAH[a]	N-Et	[361, 370]
	Barbiturates and xanthines	TBAH[b]	N-Bu	[372]
	Methyl-phenylbarbitone	TBAH[b]	N-Bu	[135]
	Theophylline	TBAH	N-Bu	[140]
	Phenobarbital and DPH	C_1-C_7 Tetra-alkylammonium hydroxides	N-R	[371]

TABLE 13. (Continued)

Functional group	Compound	Reagent	Derivative	Reference
	General (barbituric acids)	CH_3I, K_2CO_3	N-Me	[136, 138]
	General (barbituric acids)	Various alkylating agents and K_2CO_3	N-R	[369]
	Phenobarbital	Alkyl iodide, TMAH or TMH	N-R	[30]
	General (barbituric acids)	DMF-DMA	Acetal	[139]
-CONHR	Methylurea	TFAA	N-TFA	[31]
$-P{\overset{O}{\underset{NH-}{<}}}$	Cyclophosphamide, isophosphamide	TFAA	N-TFA	[144]
	Saccharin	BSA	N-TMS	[143]
	Saccharin	CH_3I	$N-CH_3$	[374]

[a]Tetraethylammonium hydroxide.

[b]Tetrabutylammonium hydroxide.

TABLE 14. Barbiturates and Other Amides and Imides (Electron Capture Detection)

Functional group	Compound	Reagent	Derivative	Reference
	General	PF-benzoyl chloride	N-PFBz	[29]

TABLE 15. Carbohydrates (Flame Ionization Detection)

Functional group	Example	Reagent	Derivative	Reference
	Review (1962)	Various		[376]
	Review (1968)	Various		[146]
	Review (1971)	Various		[377, 378]
	Review (1973)	Various		[147]
	General	HMDS, TMCS (py)	TMS	[148]
	Review (urinary sugars and sugar alcohols)	CH_3ONH BSA, TMCS or BSTFA	Methoxime TMS-ether	[269]
	Aqueous solutions	HMDS, TMCS (py)	TMS	[379]
		HMDS, TFAAc[a]	TMS	[380]
		TSIM (py)	TMS	[381]
Simple sugars	Fructose, glucose (in syrups)	TSIM, TMCS (py)	TMS-ether	[384]
	α- and β-D-glucose	TSIM (py)	TMS-ether	[385]
	Glucose	CH_3ONH_2 BSTFA	Methoxime penta-TMS ether	[159]
	Hexoses	b	Alditol acetate	[208, 382]
	Hexoses (from neutral glycolipids)	b	Alditol acetate	[383]
	Monosaccharides (glycoproteins)	b	Alditol acetate	[375]
	Neutral aldo-pentoses, -hexoses, -heptoses[c]	HMDS, TMCS (py)		[154]
Amino sugars	Neutral and amino sugars	b	Alditol acetate	[157]
	Hexosamines	BSA, TMCS (py)	N,O-TMS	[382]
	2-Amino-2-deoxyhexoses	Various	TMS	[155]

TABLE 15. (Continued)

Functional group	Example	Reagent	Derivative	Reference
	2-Amino-2-deoxyhexoses	Ac_2O,NaOAc	N,O-Ac	[386]
	Glucosamine, methyl glycoside	Ac_2O (py) BSTFA	N-Ac O-TMS	[387]
	Glucosamine and galactosamine	TFAA	N,O-TFA	[388]
	Glucosamine and galactosamine	Ac_2O HMDS, TMCS (py)	N-Ac TMS-ether	[7]
	Glucosamine and glactosamine	b, d	Alditol acetate	[165]
	N-Acetylneuraminic acid	BSTFA	TMS-ester and ether	[164, 382]
	Acylneuraminic acids	TSIM or HMDS, TMCS (py)	O-TMS	[392]
Sugar phosphates	α-D-Glucose-6-PO_4	BSA, TMCS[e]	O-TMS P(O)-(OTMS)$_2$	[162]
	Aldose phosphates	BSTFA, TMCS or TSIM	P(O)-(OTMS)$_2$(f)	[389]
Glucuronides	Glucuronides of aromatic acids and phenols	BSTFA, TMCS	O-TMS ether and ester	[6]
	Glucuronides of 1-naphthol and androsterone	CH_2N_2 TFAA	Me ester O-TFA	[161]
	Glucuronides of 6-bromo-2-naphthol, chloramphenicol, 4-(N,N-dimethyl-sulfamoyl)phenol (g)	CH_2N_2 BSTFA or BSA	per(TMS ether) or Me ester, TMS ether	[391]

| Other | Deoxyaldonic and hexaric acids and lactones | BSTFA, TMCS | O-TMS | [160] |
| | Indole-3-acetic acid esters of myirositol | BSTFA | O-TMS ether N-TMS? | [390] |

[a] BSTFA may be substituted for HMDS with advantage: personal communication.

[b] Principal reagents for alditol acetate derivatization are Ac_2O and sodium borohydride.

[c] Aldoses are sequentially converted to aldonic acid and 1,4-lactone prior to silylation.

[d] Hexosamines are deaminated to neutral 2,5-anhydrohexoses then are quantitatively determined as the corresponding alditol acetate.

[e] The authors subsequently used BSTFA, TMCS: personal communication.

[f] In some cases, the methoxime or ethoximes were prepared.

[g] Mass spectral studies; no chromatography.

TABLE 16. Steroids (Flame Ionization Detection)

Functional group	Example	Reagent	Derivative	Reference
	Reviews	Various		[166, 178, 335-339]
	17-Ketosteroids, pregnanediol and pregnanetriol	BSTFA	O-TMS	[340]
	C-19, C-21-Dihydroxy-steroids	Various		[342, 343]
	Progesterone (inter alia)	CH_3ONH_2 HMDS, TMCS	Methoxine TMS	[169]
	Progesterone	NH_2OH t-Bu$(CH_3)_2$SiCl	=NO(t-BDMS)	[173]
	Hydroxsteroids (e.g., 17β-ol)	CMDMCS	O-CMDMS	[341]
	Hydroxysteroids (e.g., 3,11,17, 20-ols)	CMDMCS	O-CMDMS	[184]
	Estrone	Dimethylamino-dimethylphosphine	O-P(O)Me$_2$[b]	[344]
	Cholestan-3-one	N,N-Dimethyl-hydrazine	Dimethylhydrazone	[337]
	Hydroxy-keto steroids (e.g., androstane-17-one)	CH_3ONH_2 HMDS, TMCS	Methoxime TMS	[49]
Keto steroids	20-one	CH_3ONH_2	Methoxime	[324]
	3,16,17, and 20-Keto	CH_3ONH_2 HMDS or BSA[a]	Methoxime TMS	[170, 172, 325]

	3,15,17, and 20-one	$C_4H_9ONH_2$	Butoxime	[172, 333]
	3,15,17, and 20-one	$C_6H_5CH_2ONH_2$	Benzyloxime	[172, 327]
Hydroxysteroids				
unhindered	3,15,*sec*-17,20, 21-ol	HMDS-TMCS, or BSA, or BSTFA	TMS ether	[327, 328]
moderately hindered	11β and *t*-17α-ol (adjacent to C-2) keto)	BSA, TMCS	TMS ether	[329, 330]
highly hindered	5,14, and *t*-17α-ol (adjacent to C-2)-ol)	BSA, TSIM	TMS ether	[8, 324, 329-332]
Corticosteroids and metabolites		CH_3ONH_2 BSA, TMCS	Methoxime TMS	[311, 330]
Corticosteroids	17,21-diol	NBBA	Cyclic butyl-boronate ester	[275]
Corticosteroids[c]	Biological sample	BSA, TMCS CH_3ONH_2	TMS Methoxime	[311]
Aldosterone[c]	Biological sample	CH_3ONH_2	Methoxime	[311]
Estrogens[c]	Biological sample	BSTFA (py) or HFBI	O-TMS or O-HFB	[311]
Estrogens	Equine origin	CH_3ONH_2 TSIM, BSA, TMCS	Methoxime TMS	[168]

[a]With or without TMCS catalyst.

[b]Alkali detector.

[c]Mass fragmentography detection.

TABLE 17. Steroids (Electron Capture Detection)

Functional group	Example	Reagent	Derivative	Reference
	Hydroxysteroids (saturated ring)	HFBA, TMCS acetone	$-O-C(CH_3)=CHOC_3F_7$ (conjugated heptafluorobutanoyl)	[319]
	Testosterone	HFBA (py)	3,17(OH-HFB) ester	[180, 317]
	Testosterone	HFBA and PF-benzoic anhydride	3-(O-HFB) ester 17β(-O-PFBx) ester	[320]
	Estrone, pregnanediol, 3-oxo-4-enes	HFBA	O-HFB	[177]
	Estrogens	HFBA	O-HFB	[322]
	Sterols (e.g., cholesterol)	$C_6F_5(CH_3)_2SiN(C_2H_5)_2$ and $C_6F_5(CH_3)_2SiCl$	$O-Si(CH_3)_2C_6F_5$	[182, 318]
	Testosterone, estradiol	$H(CF_2)_8COCl$ $H(CF_2)_{10}COCl$	$O-CO(CF_2)_8H$ $O-CO(CF_2)_{10}H$	[321]
	3,17,20-Monoketo-3,17 and 3,20-diketo	Perfluorobenzyl-hydroxylamine	$O-CH_2C_6F_5$ oxime	[181]
	Zooecdysones (e.g., α-ecdysone)	TSIM $HFBI-C_3F_7CO_2H$	Tri(O-TMS) Di(O-HFB)	[323]

TABLE 18. Miscellaneous (Flame Ionization Detection)

Functional group	Example	Reagent	Derivative	Reference
Purine and pyrimidine Bases	General	BSTFA, TMCS	TMS	[220]
	Uracil	BSTFA	O-TMS	[199]
	Uracil	TMH	Di(N-CH$_3$)	[353]
	C,A	MSTFA, TFA	Di-N-TMS	[222]
	U,C,T,G	BSTFA	N,O-TMS	[195]
	U,T,C,A,G	HMDS, TMCS(py)	Di- or tri-TMS	[351, 352]
	5-Fluorouracil	BSTFA, TMCS(py)	Di-TMS	[354]
	N^6-Methyl-adenine and -adenosine	BSA	N,O-TMS	[5]
Nucleosides	General	BSTFA[a]	N,O-TMS	[194]
	General	BSTFA, TMCS		[50, 315]
	Adenosine, guanosine cytidine, thymidine uridine, inosine	HMDS, TMCS(py)	Tri-, tetra-, penta-TMS	[351]
	Thymidine	BSTFA	O-TMS	[199]
	Pseudouridine, etc.	BSTFA	N,O-TMS	[350]
	7-Methylpurine	BSTFA, TMCS	N,O-TMS	[198]
	Methylated (various)	BSTFA	O-TMS	[196]
Nucleotides	5'-AMP, GMP, UMP, TMP, 3'-AMP	BSTFA, TMCS(py)	N,O-TMS	[200]
	c-AMP, 5'-CMP	BSA, TMCS	N,O-TMS	[200]
Heterocycles	General	BSTFA, TMCS	TMS	[220]
	Pyridine, hydroxyl pyridoxine	MSA	Tri(O-TMS)	[233]
	pyridoxal (hemiacetal)	MSA	Di(O-TMS)	[233]

TABLE 18. (Continued)

Functional group	Example	Reagent	Derivative	Reference
	Hydroxytriazine	BSTFA	O-TMS	[355]
	Triazines (Review)	Silylation Alkylation Methoxylation Hydrolysis Methylation		[185]
	Thiamine [4-methyl-5(β-hydroxy-ethyl)thiazole]	BSA	TMS ether	[356]
Enol	Malonic acid, Malonate ester	BSA	Tris-TMS Mono-TMS	[153]
Thiol	Thioglycol	MBTFA	O,S-TFA	[227]
Aminothiols	3-Mercaptoethano-lamine	BSTFA	N,O,S-TMS	[205]
Mercaptans	n-AmylSH	C_6H_5COCl	Benzoate	[206]
Epoxides	Barbiturate metabolites	TMCS	$\overset{\textstyle{}}{\underset{\textstyle Cl\ OH}{>C-C<}}$	[14]
Lactones	Penicillic acid and patulin	BSA, TMCS	O-TMS ether	[33]
Sulfonamides	p-Toluenesulfonamide	DMF-dialkylacetal	N-DMAM[a]	[345]
Sulfonamides	Various drugs	CH_2N_2 $(Ac_2)O$	N^1-Me N^4-Ac	[346, 347]
Ureas	Phenylureas (tri-substituted)	TMAH	N-Me	[188]
Ureas and carbamates	Review	Silylation TFA, PFP, HFB Alkylation Transesterification Hydrolysis		[185, 186]

Organo phosphorus	Insecticides (Review)	Oxidation Hydrolysis Reduction		[185]
	Alkyl (mono- or di-) phosphoric acids	CH_2N_2	Me ester	[203]
	Alkyl and Aminoalkyl- phosphonates	BSTFA, TMCS	$P(O)-(OTMS)_2$ $N-(TMS)_2$	[204]

[a]Optimal conditions for silylation of nucleosides: 225 molar excess BSTFA at 150° for 15 min. in closed vial followed by chromatography on 4% w/w OV-11 on 100/120 Supelcoport.

[b]N-Dimethylaminomethylene

TABLE 19. Miscellaneous (Electron Capture Detection)

Funcetional group	Example	Reagent	Derivative	Reference
Hydroxyl	Cocaine (reduced to 2-hydroxymethyltropine)	PFPA	Di-O-PFP	[348]
Mercaptan	n- and t-BuSh, n-$C_{10}H_{21}$SH	PFB-Br	PFB thioether	[19]
Biguanides	Phenformin, buformin, metformin	ClF_2CCO_2H	a	[253]
Guanidino	Debrisquin, guanethidine, metabolites	Hexafluoroacetyl-acetone		[100]
Sulfonamides	Benzenesulfonamide	PFB-Br(TBHSO$_4$)[b]	Di(N-PFB)	[349]
	N-Et and N-Phenyl-benzenesulfonamide	TFAA or HFBAA (TMA)	N-TFA or N-HFB	[349]
		PFB-Br(TBHSO$_4$)[b]	N-PFP	[349]

[a]2-Substituted 4-monochlorodifluoromethyl-2,6-diamino-1,3,5-S-triazine.

[b]Extractive alkylation using tetrabutylammonium hydrogensulfate.

REFERENCES

1a. M. G. Horning, A. M. Moss, and E. C. Horning, *Biochim. Biophys. Acta*, *148*, 597(1967).

1b. J. Drozd, *J. Chromatogr.*, *113*, 303(1975).

1c. S. Ahuja, *J. Pharm. Sci.* *65*, 163(1976).

1d. T. Nanbara and J. Goto, *Japan Analyst* (Bunseki Kagaku)　, 704(1974).

1e. K. Blau and G. S. King, *Handbook of Derivatives for Chromatography*, Heyden, Bellmawr, N. J., 1977.

2. C. Grunwald and R. G. Lockard, *J. Chromatogr.*, *52*, 491(1970).

3. A. L. Barta and C. A. Osmond, *J. Agr. Food Chem.*, *21*, 316(1973).

4. C. M. Williams, *Anal. Biochem.*, *11*, 224(1965).

5. I. A. Muni, C. H. Altschuler, and J. C. Neicheril, *Anal. Biochem.*, *50*, 354(1972).

6. J. E. Mrochek and W. T. Rainey, *Anal. Biochem.*, *57*, 173(1974).

7. M. B. Perry, Can. *J. Biochem.*, *42*, 451(1964).

8. E. M. Chambaz and E. C. Horning, *Anal. Biochem.*, *30*, 7(1969).

9. N. E. Hoffman and K. A. Peteranetz, *Anal. Lett.*, *5*, 589(1972).

10. H. Gleispach, *J. Chromatogr.*, *91*, 407(1974).

11. G. D. Brittain, in *Recent Advances in Gas Chromatography* (I. Domsky and J. Perry, eds.), Marcel Dekker, New York, 1971, p. 215.

12. G. D. Brittain, J. E. Sullivan, and L. R. Schewe, in *Recent Advances in Gas Chromatography* (I. Domsky and J. Perry, eds.), Marcel Dekker, New York, 1971, p. 223.

13. H. Morita, *J. Chromatogr.*, *101*, 189(1974).

14. D. J. Harvey, D. B. Johnson, and M. G. Horning, *Anal. Lett.*, *5*, 745(1972).

15. E. Anggard and G. Sedvall, *Anal. Chem.*, *41*, 1250(1969).

16. F. Karoum, F. Cattabeni, E. Costa, C. R. J. Ruthven, and M. Sandler, *Anal. Biochem.*, *47*, 550(1972).

17. N. K. McCallum and R. J. Armstrong, *J. Chromatogr.*, *78*, 303(1973).

18. H. Ehrsson, T. Walle, and H. Brotell, *Acta Pharm. Suecica*, *8*, 319(1971).

19. F. K. Kawahara, *Anal. Chem.*, *40*, 1009(1968).

20. H. Brotell, H. Ehrsson, and O. Gyllenhaal, *J. Chromatogr.*, *78*, 293(1973).

21. C. W. Gehrke, K. C. Kuo, R. W. Zumwalt, and T. P. Waalkes, in *Polyamines in Normal and Neoplastic Growth* (D. Russell, ed.), Raven Press, New York, 1973, p. 343.

22. J. A. F. Wickramasinghe, W. Morozowich, W. E. Hamlin, and S. R. Shaw, *J. Pharm. Sci.*, *62*, 1428(1973).

23. L. G. Johnson, *J. Ass. Offic. Anal. Chem.*, *56*, 1503(1973).

24. F. Karoum, C. R. J. Ruthven, and M. Sandler, *Biochem. Med.*, *5*, 505(1971).

25. E. Watson, S. Wilk, and J. Roboz, *Anal. Biochem.*, *59*, 441(1974).

26. A. C. Moffat and E. C. Horning, *Anal. Lett.*, *3*, 205(1970).

27. A. Moffat, E. C. Horning, S. B. Matin, and M. Rowland, *J. Chromatogr.*, *66*, 255(1972).

28. T. Walle and H. Ehrsson, *Acta Pharm. Suecica*, *7*, 389(1970).

29. H. Ehrsson and R. Mellstrom, *Acta Pharm. Suecica*, *9*, 107(1972).

30.   R. H. Greeley, *Clin. Chem.*, *20*, 192(1974).

31.   R. T. Evans, *J. Chromatogr.*, *88*, 398(1974).

32.   F. Bennington, S. T. Christian, and R. D. Morin, *J. Chromatogr.*, *106*, 435(1975).

33.   L. M. Cummins, in *Recent Advances in Gas Chromatography* (I. Domsky and J. Perry, eds.), Marcel Dekker, New York, 1971, p. 313.

34.   D. O. Edlund and J. R. Anfinsen, *J. Ass. Offic. Anal. Chem.*, *53*, 289(1970).

35.   M. P. Rabinowitz, P. Reisberg, and J. I. Bodin, *J. Pharm. Sci.*, *63*, 1601(1974).

36.   C. Braestrup, *Anal. Biochem.*, *55*, 420(1973).

37.   J. N. Seiber, D. G. Crosby, H. Fouda, and C. J. Soderquist, *J. Chromatogr.*, *73*, 89(1972).

38.   M. Margosis, *J. Chromatogr. Sci.*, *12*, 549(1974).

39.   K. Tsuji and J. H. Robertson, *Anal. Chem.*, *43*, 818(1971).

40.   L. W. Brown and P. B. Bowman, *J. Chromatogr. Sci.*, *12*, 373(1974).

41.   P. T. Russell, A. J. Eberle, and H. C. Ching, *Clin. Chem.*, *21*, 657(1975).

42.   F. Vane and M. G. Horning, *Anal. Lett.*, *2*, 357(1969).

43.   C. Pace-Asciak and L. S. Wolfe, *J. Chromatogr.*, *56*, 129(1971).

44.   D. S. Erley, *Anal. Chem.*, *29*, 1564(1957).

45.   R. W. Kelly, *Anal. Chem.*, *45*, 2079(1973).

46.   B. Samuelson, M. Hamberg, and C. C. Sweeley, *Anal. Biochem.*, *38*, 301(1970).

47.   L. J. Papa and L. P. Turner, *J. Chromatogr. Sci.*, *10*, 744(1972).

48.   H. Kallio, R. R. Linko, and J. Kaitaranta, *J. Chromatogr.*, *65*, 355(1972).

49.   H. M. Fales and T. Luukkainen, *Anal. Chem.*, *37*, 955(1965).

50.   W. C. Butts, *J. Chromatogr. Sci.*, *8*, 474(1970).

51.   J. W. Vogh, *Anal. Chem.*, *43*, 1618(1971).

52.   J. Korolczuk, M. Daniewski, and Z. Mielniczuk, *J. Chromatogr.*, *88*, 177(1974).

53.   O. Mlejnek, *J. Chromatogr.*, *70*, 59(1972).

54.   M. J. Levitt, *Anal. Chem.*, *45*, 618(1973).

55.   R. H. Greeley, *J. Chromatogr.*, *88*, 229(1974).

56.   J. C. West, *Anal. Chem.*, *47*, 1708(1975).

57.   R. V. Smith and S. L. Tsai, *J. Chromatogr.*, *61*, 29(1971).

58.   A. J. Sheppard and J. L. Iverson, *J. Chromatogr. Sci.*, *13*, 448(1975).

59.   *Official Methods of Analysis*, 10th Ed., Ass. Offic. Anal. Chem., Washington, D. C., 1965, Sec. 26.052, p. 429.

60.   A. J. Sheppard and L. A. Ford, *J. Ass. Offic. Anal. Chem.*, *46*, 947(1963).

61.   D. Firestone, *J. Ass. Offic. Anal. Chem.*, *52*, 254(1969).

62.   C. J. F. Bottcher, F. P. Woodford, E. Boelsma-Van Haute, and C. M. VanGent, *Rec. Trav. Chim.*, *78*, 794(1959).

63.   S. W. Christopherson and R. L. Glass, *J. Dairy Sci.*, *52*, 1289(1969).

64.   J. L. Iverson, 87th Annual Meeting, Ass. Offic. Anal. Chem., Washington, D. C., October 9-12, 1973, Paper No. 203.

65.   E. A. Moscatelli, *Lipids*, *7*, 268(1972).

66. J. MacGee and K. G. Allen, *J. Chromatogr.*, *100*, 35(1974).

67. J-P. Thenot, E. C. Horning, M. Stafford, and M. G. Horning, *Anal. Lett.*, *5*, 217(1972).

68. H. Ko and M. E. Royer, *J. Chromatogr.*, *88*, 253(1974).

69. U. Hintze, H. Röper, and G. Gercken, *J. Chromatogr.*, *87*, 481(1973).

70. H-P. Klemm, U. Hintze, and G. Gercken, *J. Chromatogr.*, *75*, 19(1973).

71. C. C. Alley, J. B. Brooks, and G. Choudberry, *Anal. Chem.*, *48*, 387(1976).

72. R. A. Laine, N. D. Young, and C. C. Sweeley, *Biomed. Mass Spectra*, *1*, 10(1974).

73. T. J. Sprinkle, A. H. Porter, M. Greer, and C. M. Williams, *Clin. Chem. Acta*, *25*, 409(1969).

74. N. E. Hoffman and K. M. Gooding, *Anal. Biochem.*, *31*, 471(1969).

75. R. A. Chalmers and R. W. E. Watts, *Analyst*, *97*, 958(1972).

76a. S. Wilk, E. Watson, and S. D. Glick, *Eur. J. Pharmacol.*, *30*, 117(1975).

76b. J. B. Brooks, C. C. Alley, and J. A. Liddle, *Anal. Chem.*, *46*, 1930(1974).

77. J. Eagles and M. E. Knowles, *Anal. Chem.* *43*, 1697(1971).

78. J. S. Parsons, *J. Chromatogr. Sci.*, *11*, 659(1973).

79. R. H. De Wolfe (ed.), *Carboxylic Ortho Acid Derivatives*, Academic Press, New York, 1970.

80. M. W. Scoggins, *J. Chromatogr. Sci.*, *13*, 146(1975).

81. M. V. Street, *J. Chromatogr.*, *73*, 73(1972).

82. T. Walle, *J. Chromatogr.*, *111*, 133(1975).

83. P. H. Degen and W. Riess, *J. Chromatogr.*, *85*, 53(1973).

84. P. Hartvig and J. Vessman, *Anal. Lett.*, *7*, 223(1974).

85. J. Vessman, P. Hartvig, and M. Holandor, *Anal. Lett.*, *6*, 699(1973).

86. P. Hartvig and J. Vessman, *Acta Pharm. Suecica*, *11*, 115(1974).

87. S. B. Matin and M. Rowland, *J. Pharm. Sci.*, *61*, 1235(1972).

88. P. Hartvig and J. Vessman, *J. Chromatogr. Sci.*, *12*, 722(1974).

89. P. Hartvig, W. Handl, J. Vessman, and C. M. Smith, *Anal. Chem.*, *48*, 390(1976).

90. M. Donike, *J. Chromatogr.*, *103*, 91(1975).

91. M. Donike, *Chromatographia*, *7*, 651(1974).

92. G. Schwedt and H. H. Bussemas, *J. Chromatogr.*, *106*, 440(1975).

93. H. G. Lovelady and L. L. Foster, *J. Chromatogr.*, *108*, 43(1975).

94. J. Willner, H. F. LeFevre, and E. Costa, *J. Neurochem.*, *23*, 857(1974).

95. E. Gelpi, E. Peralta, and J. Segura, *J. Chromatogr. Sci.*, *12*, 701(1974).

96. D. J. Edwards and K. Blau, *Anal. Biochem.*, *45*, 387(1972).

97. J-C. Lhuguenot and B. F. Maume, *J. Chromatogr. Sci.*, *12*, 411(1974).

98. J. B. Brooks, C. C. Alley, and R. Jones, *Anal. Chem.*, *44*, 1881(1972).

99. K. A. Karlsson, *Biochim. Biophys. Res. Commun.*, *39*, 847(1970).

100. P. Erdtmansky and T. J. Roehl, *Anal. Chem.*, *47*, 750(1975).

101. K. Blau, in *Biomedical Applications of Gas Chromatography* (H. A. Szymanski, ed.), Plenum Press, New York, 1968, p. 1.

102. F. E. Kaiser, C. W. Gehrke, R. W. Zumwalt, and K. G. Kuo, *J. Chromatogr.*, *94*, 113(1974).

103. C. W. Gehrke, H. Nakamoto, and R. W. Zumwalt, *J. Chromatogr.*, *45*, 24(1969).

104. C. W. Gehrke and K. Leimer, *J. Chromatogr.*, *57*, 219(1971).

105. J. F. Klebe, H. Finkbeiner, and D. M. White, *J. Amer. Chem. Soc.*, *88*, 3390(1966).

106. D. L. Stalling, C. W. Gehrke, and C. D. Ruyle, *Biochem. Biophys. Res. Commun.*, *31*, 616(1968).

107. R. W. Zumwalt, Master's Thesis, University of Missouri, June 1968.

108. E. D. Smith and K. L. Shewbart, *J. Chromatogr. Sci.*, *7*, 704(1969).

109. P. W. Albro and L. Fishbein, *J. Chromatogr.*, *55*, 297(1971).

110. J. P. Hardy and S. L. Kerrin, *Anal. Chem.*, *44*, 1497(1972).

111. J. Rosmus and Z. Deyl, *J. Chromatogr.*, *70*, 221(1972).

112. W. H. Lamkin, J. W. Weatherford, N. S. Jones, T. Pan, and D. N. Ward, *Anal. Biochem.*, *58*, 422(1974).

113. P. Husek, *J. Chromatogr.*, *91*, 475(1974).

114. P. Husek, *J. Chromatogr.*, *91*, 483(1974).

115. J. Jönsson, J. Eyem, and J. Sjöquist, *Anal. Biochem.*, *51*, 204(1973).

116. C. W. Moss and M. A. Lambert, *Anal. Biochem.*, *59*, 259(1974).

117. C. W. Gehrke, K. Kuo, and R. W. Zumwalt, *J. Chromatogr.*, *57*, 209(1971).

118. C. W. Gehrke and H. Takeda, *J. Chromatogr.*, *76*, 63(1973).

119. D. J. Casagrande, *J. Chromatogr.*, *49*, 537(1970).

120. F. Raulin, P. Shapshak, and B. N. Khare, *J. Chromatogr.*, *73*, 35(1972).

121. P. Cancalon and J. D. Klingman, *J. Chromatogr. Sci.*, *12*, 349(1974).

122. R. W. Zumwalt, K. Kuo, and C. W. Gehrke, *J. Chromatogr.*, *57*, 193(1971).

123. I. M. Moodie, *J. Chromatogr.*, *99*, 495(1974).

124. P. I. Jaakonmaki and J. E. Stouffer, *J. Chromatogr. Sci.*, *6*, 303(1967).

125. N. N. Nihei, M. C. Gershengorn, T. Mitsuma, L. R. Stringham, A. Gordy, B. Kuchmy, and C. S. Hollander, *Anal. Biochem.*, *43*, 433(1971).

126. S. Wilk and M. Orlowski, *Fed. Eur. Biochem. Soc.*, *33*, 157(1973).

127. S. Wilk and M. Orlowski, *Anal. Biochem.*, *69*, 100(1975).

128. E. Brochmann-Hansen and T. O. Oke, *J. Pharm. Sci.*, *58*, 370(1969).

129. J. MacGee, Methods of Analysis of Anti-Epileptic Drugs, *Internat. Cong. Ser.*, *286*, 111(1972) Exerpta Medica, Amsterdam.

130. E. W. Robb and J. J. Westbrook, *Anal. Chem.*, *35*, 1944(1963).

131. J. Barrett, *Clin. Chem. Newsletter*, *3*, 16(1971).

132. G. Kananen, R. Osiewicz, and I. Sunshine, *J. Chromatogr. Sci.*, *10*, 283(1972).

133. H. V. Street, *Clin. Chim. Acta*, *34*, 357(1971).

134. E. B. Solow, J. M. Metaxas, and T. R. Summers, *J. Chromatogr. Sci.*, *12*, 256(1974).

135. W. D. Hooper, D. K. Dubetz, M. J. Eadie, and J. H. Tyler, *J. Chromatogr.*, *110*, 206(1975).

136. W. Dünges and E. Bergheim-Irps, *Anal. Lett.*, *6*, 185(1973).

137. R. Osiewicz, V. Aggarwal, R. M. Young, and I. Sunshine, *J. Chromatogr.*, *88*, 157(1974).

138. W. Dünges, *Anal. Chem.*, *45*, 963(1973).

139. V. S. Venturella, V. M. Gualario, and R. E. Lang, *J. Pharm. Sci.*, *62*, 662(1973).

140. G. F. Johnson, W. A. Dechtiaruk, and H. M. Solomon, *Clin. Chem.*, *21*, 144(1975).

141. H. Ehrsson, *Anal. Chem.*, *46*, 922(1974).

142. H. L. Rice and G. R. Pettit, *J. Amer. Chem. Soc.*, *76*, 302(1954).

143. R. Gerstl and K. Ranfft, *Z. Anal. Chim.*, *258*, 110(1972).

144. C. Pantarotto, A. Bossi, G. Belvedere, A. Martini, M. G. Donelli, and A. Frigerio, *J. Pharm. Sci.*, *63*, 1554(1974).

145. A. E. Pierce, *Silylation of Organic Compounds*, Pierce Chemical Company, Rockford, Ill., 1968.

146. J. H. Sloneker, in *Biomedical Applications of Gas Chromatography* (H. Szymanski, ed.), Plenum Press, New York, 1968, p. 87.

147. G. G. S. Dutton, in *Advances in Carbohydrate Chemistry and Biochemistry* (R. Tipson and D. Horton, eds.), Academic Press, New York, 1973, p. 11.

148. C. C. Sweeley, R. Bentley, M. Makita, and W. W. Wells, *J. Amer. Chem. Soc.*, *85*, 2497(1963).

149. A. H. Weiss and H. Tambawala, *J. Chromatogr. Sci.*, *10*, 120(1972).

150. S. H. Langer, R. A. Friedel, I. Wender, and A. G. Sharkey, *Anal. Chem.*, *30*, 1353(1958).

151. R. Bentley and N. Botlock, *Anal. Biochem.*, *20*, 312(1967).

152. K. M. Brobst and C. E. Lott, Jr., *Cereal Chem.*, *43*, 35(1966).

153. O. A. Mamer and S. S. Tjoa, *Clin. Chem.*, *19*, 58(1973).

154. I. M. Morrison and M. B. Perry, *Can. J. Biochem.*, *44*, 1115(1966).

155. R. E. Hurst, *Carbohydrate Res.*, *30*, 143(1973).

156. H. Jacin, J. M. Slanski, and R. J. Moshy, *J. Chromatogr.*, *37*, 103(1968).

157. L. J. Griggs, A. Post, E. R. White, J. A. Finkelstein, W. E. Moeckel, K. G. Holden, J. E. Zarembo, and J. E. Weisbach, *Anal. Biochem.*, *43*, 369(1971).

158. J. J. Myher and A. Kukis, *J. Chromatogr. Sci.*, *13*, 138(1975).

159. R. A. Laine and C. C. Sweeley, *Anal. Biochem.*, *43*, 533(1971).

160. J. Szafranek, C. D. Pfaffenberger, and E. C. Horning, *J. Chromatogr.*, *88*, 149(1974).

161. H. Ehrsson, T. Walle, and S. Wikström, *J. Chromatogr.*, *101*, 206(1974).

162. M. Zimbo and W. R. Sherman, *J. Amer. Chem. Soc.*, *92*, 2105(1970).

163. D. J. Harvey and M. G. Horning, *J. Chromatogr.*, *76*, 51(1973).

164. D. A. Craven and C. W. Gehrke, *J. Chromatogr.*, *37*, 414(1968).

165. W. H. Porter, *Anal. Biochem.*, *63*, 27(1975).

166. F. Polvani, M. Surace, and M. Luisi (eds.), *Gas Chromatographic Determination of Hormonal Steroids*, Academic Press, New York, 1967.

167. E. C. Horning and M. G. Horning, in *Recent Advances in Gas Chromatography* (I. Domsky and J. Perry, eds.), Marcel Dekker, New York, 1971, pg. 341.

168. K. M. McErlane, *J. Chromatogr. Sci.*, *12*, 97(1974).

169. M. Axelson, G. Schumacher, and J. Sjovall, *J. Chromatogr. Sci.*, *12*, 535(1974).

170. M. G. Horning, A. M. Moss, and E. C. Horning, *Anal. Biochem.*, *22*, 284(1968).

171. M. Delaforge, B. F. Maume, F. Bournot, M. Prost, and P. Padieu, *J. Chromatogr. Sci.*, *12*, 545(1974).

172a. J-P. Thenot and E. C. Horning, *Anal. Lett.*, *4*, 683(1971).

172b. T. A. Baillie, C. J. W. Brooks, and E. C. Horning, *Anal. Lett.*, *5*, 351(1972).

173. R. W. Kelly and P. L. Taylor, *Anal. Chem.*, *48*, 465(1976).

174. E. M. Chambaz, G. Defaye, and C. Madani, *Anal. Chem.*, *45*, 1090(1973).

175. G. M. Anthony, C. J. W. Brooks, I. Maclean, and I. Sangster, *J. Chromatogr. Sci.*, *7*, 623(1969).

176. C. J. W. Brooks and I. Maclean, *J. Chromatogr. Sci.*, *9*, 18(1971).

177. L. A. Dehennin and R. Scholler, *Steroids*, *13*, 739(1969).

178. H. H. Wotiz and S. J. Clark, in *Methods of Biochemical Analysis* (D. Glick, ed.), Vol. 18, Interscience, New York, 1970, p. 340.

179. S. J. Clark and H. H. Wotiz, *Steroids*, *2*, 535(1963).

180. P. G. Devaux and E. C. Horning, *Anal. Lett.*, *2*, 637(1969).

181. K. T. Koshy, D. G. Kaiser, and A. L. VanDerSlik, *J. Chromatogr. Sci.*, *13*, 97(1975).

182. E. D. Morgan and C. F. Poole, *J. Chromatogr.*, *104*, 351(1975).

183. B. S. Thomas, *J. Chromatogr.*, *56*, 37(1971).

184. J. R. Chapman and E. Bailey, *Anal. Chem.*, *45*, 1636(1973).

185. W. P. Cochrane, *J. Chromatogr. Sci.*, *13*, 246(1975).

186. H. W. Donough and J. H. Thorstenson, *J. Chromatogr. Sci.*, *13*, 212(1975).

187. D. C. Saunders and L. E. Vanatta, *Anal. Chem.*, *46*, 1319(1974).

188. F. S. Tanaka and R. G. Wien, *J. Chromatogr.*, *87*, 85(1973).

189. I. C. Cohen and B. B. Wheals, *J. Chromatogr.*, *43*, 233(1969).

190. I. C. Cohen, J. Norcup, J. H. A. Ruzicka, and B. B. Wheals, *J. Chromatogr.*, *49*, 215(1971).

191. J. H. Caro, D. E. Glotfelty, H. P. Freeman, and A. W. Taylor, *J. Ass. Offic. Anal. Chem.*, *56*, 1319(1973).

192. J. H. Caro, H. P. Freeman, D. E. Glotfelty, B. C. Turner, and W. M. Edwards, *J. Agr. Food Chem.*, *21*, 1010(1973).

193. B. C. Turner and J. H. Caro, *J. Environ. Qual.*, *2*, 245(1973).

194. C. W. Gehrke and A. B. Patel, *J. Chromatogr.*, *130*, 103(1977).

195. D. B. Lakings and C. W. Gehrke, *Clin. Chem.*, *18*, 810(1972).

196. S. Y. Chang, D. B. Lakings, R. W. Zumwalt, C. W. Gehrke, and T. P. Waalkes, *J. Lab. Clin. Med.*, *83*, 816(1974).

197. D. B. Lakings, C. W. Gehrke, and T. P. Waalkes, *J. Chromatogr.*, *116*, 69(1976).

198. D. L. Von Minden, R. N. Stillwell, W. A. Koenig, K. J. Lyman, and J. A. McCloskey, *Anal. Biochem.*, *50*, 110(1972).

199. V. Rehak and V. Pacakova, *Anal. Biochem.*, *61*, 294(1974).

200. A. M. Lawson, R. N. Stillwell, M. M. Tacker, K. Tsuboyama, and J. A. McCloskey, *J. Amer. Chem. Soc.*, *93*, 1014(1971).

201. H. Ehrsson, *Acta Pharm. Suecica,* 9, 419(1972).

202. C. G. Begg and M. R. Grimmett, *J. Chromatogr.,* 73, 238(1972).

203. J. M. H. Daemen and W. Dankelman, *J. Chromatogr.,* 78, 281(1973).

204. D. J. Harvey and M. G. Horning, *J. Chromatogr.,* 79, 65(1973).

205. R. T. Lofberg, *Anal. Lett.,* 4, 77(1971).

206. J. Korolczuk, M. Daniewski, and Z. Mielniczuk, *J. Chromatogr.,* 100, 165(1974).

207. W. J. A. Vandenheuvel and V. F. Gruber, *J. Chromatogr.,* 112, 513(1975).

208. J. S. Swardeker, J. Sloneker, and A. Jeanes, *Anal. Chem.,* 37, 1602(1965).

209. J. R. Coulter and C. S. Hann, in *New Techniques in Amino Acid, Peptide and Protein Analysis* (A. Niederweiser and G. Pataki, eds.), Ann Arbor Science, Ann Arbor, 1971.

210. L. Birkofer and M. Donike, *J. Chromatogr.,* 26, 270(1967).

211. A. Islam and A. Darbre, *J. Chromatogr.,* 43, 11(1969).

212. A. Islam and A. Darbre, *J. Chromatogr.,* 71, 223(1972).

213. J. Metz, W. Ebert and H. Weicker, *Chromatographia,* 4, 259(1971).

214. H. Hediger, R. L. Stevens, H. Brandenberger and K. Schmid, *Biochem. J.,* 133, 551(1973).

215. J. M. L. Mee, *J. Chromatogr.,* 87, 155(1973).

216. A. J. Cliffe, N. J. Berridge, and D. R. Westgarth, *J. Chromatogr.,* 78, 333(1973).

217. J. Graff, J. P. Wein, and M. Winitz, *Fed. Proc.,* 22, 244(1963).

218. W. J. McBride and J. D. Klingman, *Anal. Biochem.,* 25, 109(1968).

219. D. J. Casagrande, *J. Chromatogr.,* 49, 534(1970).

220. W. C. Butts, *Anal. Biochem.,* 46, 187(1972).

221. C. W. Moss, M. A. Lambert, and F. J. Diaz, *J. Chromatogr.,* 60, 137(1971).

222. M. Donike, *J. Chromatogr.,* 85, 1(1973).

223. G. E. Pollock, *Anal. Chem.,* 39, 1194(1967).

224. K. Bergstrom and J. Guntler, *Acta Chem. Scand.,* 25, 175(1971).

225. C. W. Gehrke and K. Leimer, *J. Chromatogr.,* 53, 201(1970).

226. R. Birkofer and M. Donike, *J. Chromatogr.,* 26, 270(1967).

227. M. Donike, *J. Chromatogr.,* 78, 273(1973).

228. J. P. Thenot and E. C. Horning, *Anal. Lett.,* 5, 519(1972).

229. O. Grahl-Nielsen and E. Solheim, *Anal. Chem.,* 47, 333(1975).

230. F. Shahrokhi and C. W. Gehrke, *J. Chromatogr.,* 36, 31(1968).

231. M. Vecchi and A. Kaiser, *J. Chromatogr.,* 26, 22(1967).

232. M. Vecchi, W. Vetter, and W. Walther, *Helv.,* 50, 1243(1967).

233. K. Richter, M. Becchi, W. Vetter, and W. Walther, *Helv. Chim. Acta,* 50, 364(1967).

234. H. Ehrsson, *Acta Pharm. Suecica,* 8, 113(1971).

235. I. E. Smiley and E. D. Schall, *J. Ass. Offic. Anal. Chem.,* 52, 107(1969).

236. R. V. Smith and A. W. Stocklinski, *Anal. Chem.,* 47, 1321(1975).

237. G. R. Nakamura and E. L. Way, *Anal. Chem.,* 47, 775(1975).

238. N. E. Hoffman and R. K. Yang, *Anal. Lett.,* 5, 7(1972).

239. D. J. Harvey and W. D. M. Paton, *J. Chromatogr.*, *109*, 73(1975).

240. T. Katagi, A. Horii, Y. Oomura, H. Miyakawa, T. Kyu, Y. Ikeda, and K. Isoi, *J. Chromatogr.*, *79*, 45(1973).

241. C. J. W. Brooks and J. Watson, *Chem. Commun.*, 952(1967).

242. K. Samuelson, *Scand. J. Clin. Lab. Invest.*, *27*, 371(1971).

243. J. B. Brooks, J. A. Liddle, and C. C. Alley, *Anal. Chem.*, *47*, 1960(1975).

244. I. C. Cohen, J. Norcup, J. H. A. Ruzicka, and B. B. Wheals, *J. Chromatogr.*, *44*, 251(1969).

245. M. P. Heenan and N. K. McCallum, *J. Chromatogr. Sci.*, *12*, 89(1974).

246. D. C. Fenimore, R. R. Freeman, and P. R. Loy, *Anal. Chem.*, *45*, 2331(1973).

247. R. Piekos, K. Kobylczyk, and J. Grzybowski, *Anal. Chem.*, *47*, 1157(1975).

248. S. Nicosia and G. Galli, *Anal. Biochem.*, *61*, 192(1974).

249. L. Baczynskj, D. J. Duchamp, J. F. Zierserl, Jr., and U. Axen, *Anal. Chem.*, *45*, 479(1973).

250. E. R. Atkinson and S. I. Calouche, *Anal. Chem.*, *43*, 460(1971).

251. S. Suchaec, *Anal. Chem.*, *37*, 1361(1965).

252. M. Margosis, *J. Chromatogr.*, *37*, 46(1968).

253. S. B. Matin, J. H. Karam, and P. H. Forsham, *Anal. Chem.*, *47*, 545(1975).

254. S. A. Bland, J. W. Blake, and R. S. Ray, *J. Chromatogr. Sci.*, *14*, 201(1976).

255. E. Watson and S. Wilk, *J. Neurochem.*, *21*, 1569(1973).

256. F. Kawahara, *Anal. Chem.*, *40*, 2073(1968).

257. D. G. Kaiser, S. R. Shaw, and G. J. Vangiessen, *J. Pharm. Sci.*, *63*, 567(1974).

258. S. W. Dziedzic, L. M. Bertani, D. D. Clarke, and S. E. Gitlow, *Anal. Biochem.*, *47*, 592(1972).

259. H. Morita, *J. Chromatogr.*, *71*, 149(1972).

260. J. M. Moore, *J. Chromatogr.*, *101*, 215(1974).

261. E. Watson, B. Travis, and S. Wilk, *Life Sci.*, *15*, 2167(1974).

262. B. Sjoquist and E. Anggard, *Anal. Chem.*, *44*, 2297(1972).

263. L. Bertilson, A. J. Atkinson, Jr., J. R. Althaus, A. Harfast, J. E. Lindgren, and B. Holmstedt, *Anal. Chem.*, *44*, 1434(1972).

264. M. Donike, *J. Chromatogr.*, *115*, 591(1975).

265. C. A. Bache, L. E. St. John, Jr., and D. L. Lisk, *Anal. Chem.*, *40*, 1241(1968).

266. W. J. A. Vandenheuvel and K. L. Braly, *J. Chromatogr.*, *31*, 9(1967).

267. H. Schlenk and J. L. Gellerman, *Anal. Chem.*, *32*, 1412(1960).

268. M. G. Horning, K. L. Knox, C. E. Dalgliesh, and E. Horning, *Anal. Biochem.*, *17*, 244(1966).

269. E. C. Horning and M. G. Horning, *J. Chromatogr. Sci.*, *9*, 129(1971).

270. D. P. Schwartz and R. S. Bright, *Anal. Biochem.*, *61*, 271(1974).

271. E. Felder, U. Tiepolo, and A. Mengassini, *J. Chromatogr.*, *82*, 291(1973).

272. D. L. Corina, *J. Chromatogr.*, *87*, 254(1973).

273. J. Horner, S. S. QueHee, and R. G. Sutherland, *Anal. Chem.*, *46*, 110(1974).

274. S. L. Ali, *Chromatographia*, *8*, 33(1975).

275. G. M. Anthony, C. J. W. Brooks, I. Maclean, and I. Sangster, *J. Chromatogr. Sci.*, *7*, 623(1969).

276. J. Oehlenschlager, U. Hintze, and G. Gercken, *J. Chromatogr.*, *110*, 53(1975).

277. W. Vetter, M. Vecchi, M. Guttmann, R. Ruegg, W. Walther, and P. Meyer, *Helv. Chim. Acta*, *50*, 1866(1967).

278. R. M. Thompson, B. G. Belanger, R. S. Wappner, and I. K. Brandt, *Clin. Chem. Acta*, *61*, 367(1975).

279. J. A. Thomson and S. P. Markey, *Anal. Chem.*, *47*, 1313(1975).

280. L. D. Metcalfe, *J. Chromatogr. Sci.*, *13*, 516(1975).

281. R. G. Ackman, in *Methods in Enzymology* (J. M. Lowenstein, ed.), Vol. 14, Academic Press, New York, 1969, p. 354.

282a. W. R. Morrison and L. M. Smith, *J. Lipid Res.*, *5*, 600(1964).

282b. *Official and Tentative Methods of the American Oil Chemists Society*, Vols 1 and 2, 3rd Ed., A.O.C.S., Champaign, Ill., 1973.

283. H. L. Solomon, W. D. Hubbard, A. R. Prosser, and A. J. Sheppard, *J. Ass. Offic. Anal. Chem.*, *51*, 424(1974).

284. M. Suzuki and J. Hirano, *Yukagaku*, *19*, 468(1970).

285. B. S. Middleditch and D. M. Desiderio, *Anal. Lett.*, *5*, 605(1972).

286. M. Donike, *Chromatographia*, *6* 190(1973).

287. P. Schulz and R. Vilceanu, *J. Chromatogr.*, *111*, 105(1975).

288. N. E. Hoffman, A. Milling, and D. Parmalee, *Anal. Biochem.*, *32*, 386(1969).

289. J. M. Vavich and R. R. Howell, *J. Lab. Clin. Med.*, *77*, 159(1971).

290. I. Andersson, B. Norkrans, and G. Odham, *Anal. Biochem.*, *53*, 629(1973).

291. Y. H. Loo, J. Scotto, and M. G. Horning, *Personal Communication*.

292. M. G. Horning, in *Biomedical Applications of Gas Chromatography* (H. A. Szymanski, ed.), Plenum, New York, 1964, p. 53.

293. R. Richards, C. Mendenhall, and J. MacGee, *J. Lipid Res.*, *16*, 395(1975).

294. P. G. Simmonds, B. C. Pettit, and A. Zlatkis, *Anal. Chem.*, *39*, 163(1967).

295. L. W. Haas and R. J. DuBois, *Anal. Chem.*, *48*, 385(1976).

296. P. H. Degen and W. Riess, *J. Chromatogr.*, *85*, 53(1973).

297a. D. J. Jenden, R. Booth, and M. A. Roch, *Anal. Chem.*, *44*, 1879(1972).

297b. D. J. Jenden and L. B. Campbell, in *Methods of Biochemical Analysis* (D. Glick, ed.), Vol. 19, Interscience, New York, 1971, p. 183.

297c. D. J. Jenden and L. B. Campbell, in *Analysis of Biogenic Amines and Their Enzymes* (D. Glick, ed.), Interscience, New York, 1970.

298. E. W. Day, T. Golab, and J. R. Koons, *Anal. Chem.*, *38*, 1053(1966).

299. S. Baba, I. Hashimoto, and Y. Ishitoya, *J. Chromatogr.*, *88*, 373(1974).

300. M. D. Solomon, W. E. Pereira, and A. M. Duffield, *Anal. Lett.*, *4*, 301(1971).

301. A. Kukis, *Chromatogr. Rev.*, *8*, 192(1966).

302. E. DiSalle, K. M. Parker, and G. R. Dareggi, W. D. Watkins, C. A. Chidsey, A. Frigerio, and P. L. Morselli, *J. Chromatogr.*, *84*, 347(1973).

303. E. C. Horning, M. G. Horning, W. J. A. Vandenheuval, K. L. Knox, B. Holmstedt, and C. J. W. Brooks, *Anal. Chem.*, *36*, 1546(1964).

304. T. Gejvall, *J. Chromatogr.*, *90*, 157(1974).

305. S. Baba, *J. Chromatogr.*, *88*, 373(1974).

306. P. S. Mason and E. D. Smith, *J. Gas Chromatogr.*, *4*, 398(1966).

307. S. Mori, M. Furusawa and T. Takuchi, *J. Chromatogr. Sci.*, *8*, 477(1970).

308. P. Cancalon and J. D. Klingman, *J. Chromatogr. Sci.*, *10*, 253(1972).

309. N. Narasimhachari and P. Vouros, *Anal. Biochem.*, *45*, 154(1972).

310. D. D. Clarke, S. Wilk, and S. E. Gitlow, in *Biomedical Applications of Gas Chromatography* (H. A. Szymanski, ed.), Plenum Press, New York, 1964, p. 53.

311. B. Maume, P. Bournot, J. C. Lhuguenot, and C. Baron; F. Barbier, G. Maume, M. Prost, and P. Padieu, *Anal. Chem.*, *45*, 1073(1973).

312. M. G. Horning, A. M. Moss, E. A. Boucher, E. C. Horning, *Anal. Lett.*, *1*, 311-321(1968).

313. J. Vessman, A. M. Moss, M. G. Horning, and E. C. Horning, *Anal. Lett.*, *2*, 81(1969).

314. C. J. W. Brooks and E. C. Horning, *Anal. Chem.*, *36*, 1540(1964).

315. D. D. Clarke, S. Wilk, and S. E. Gitlow, *J. Gas Chromatogr.*, *4*, 310(1966).

316. S. Sisenwine, *Anal. Lett.*, *2*, 315(1969).

317. D. Exley, *Biochem. J.*, *107*, 285(1968).

318. E. D. Morgan and C. F. Poole, *J. Chromatogr.*, *89*, 225(1974).

319. L. A. Dehennin and R. Scholler, *J. Chromatogr.*, *111*, 238(1975).

320. L. A. Dehennin, A. Reifsteck, and R. Scholler, *J. Chromatogr. Sci.*, *10*, 224(1972).

321. M. A. Kirschner and J. P. Taylor, *Anal. Biochem.*, *30*, 346(1969).

322. F. L. Rigby, H. J. Karavoles, D. W. Norgard, and R. C. Wolfe, *Steroids*, *16*, 703(1970).

323. H. Miyazaki, M. Ishibashi, C. Mori, and N. Ikekawa, *Anal. Chem.*, *45*, 1164(1973).

324. N. Sakauchi and E. C. Horning, *Anal. Lett.*, *4*, 41(1971).

325. W. L. Gardiner and E. C. Horning, *Biochem. Biophys. Acta*, *115*, 524(1966).

326. T. A. Baillie, C. J. W. Brooks, and B. S. Middleditch, *Anal. Chem.*, *44*, 30(1972).

327. P. G. Devaux, M. G. Horning, and E. C. Horning, *Anal. Lett.*, *4*, 151(1971).

328. M. Makita and W. W. Wells, *Anal Biochem.*, *5*, 523(1963).

329. E. M. Chambaz and E. C. Horning, *Anal. Lett.*, *1*, 201(1967).

330. E. M. Chambaz, B. F. Maume, G. Maume, and E. C. Horning, *Anal Letters*, *1*, 749(1968).

331. J. P. Thenot and E. C. Horning, *Anal. Lett.*, *5*, 21(1972).

332. P. Vouros and D. J. Harvey, *Anal. Chem.*, *45*, 7(1973).

333. T. A. Baillie, C. J. W. Brooks, and E. C. Horning, *Anal. Lett.*, *5*, 351(1972).

334. T. Suzuki, Y. Fujimoto, Y. Hoshino, and A. Tanaka, *J. Chromatog.*, *105*, 95(1975).

335. A. Kukis, in *Methods of Biochemical Analysis* (D. Glick, ed.), Vol. 14, Academic Press, New York, 1966, p. 325.

336. F. Vandenheuvel, *J. Chromatogr.*, *96*, 47(1974).

337. W. J. A. Vandenheuvel and E. C. Horning, *Biochem. Biophys. Acta*, *74*, 560(1963).

338. E. C. Horning, W. J. A. Vandenheuvel, and B. G. Creech, in *Methods of Biochemical Analysis* (D. Glick, ed.), Vol. II, Interscience, New York, 1963, p. 69.

339. W. J. A. Vandenheuvel, in *Gas Chromatography of Steroids in Biological Fluids* (M. B. Lipsett, ed.), Plenum Press, New York, 1965.

340. W. R. Holub, *Clin. Chem.*, *17*, 1083(1971).

341. C. J. W. Brooks and B. S. Middleditch, *Anal. Lett.*, *5*, 611(1972).

342. F. Berthou, L. Bardou, and H. H. Floch, *J. Chromatogr.*, *93*, 149(1974).

343. F. Berthou, *J. Chromatogr. Sci.*, *12*, 662(1974).

344. W. Vogt, I. Fisher, and M. Knedel, *Z. Anal. Chem.*, *267*, 28(1973).

345. W. J. A. Vandenheuvel and V. F. Gruber, *J. Chromatogr.*, *112*, 513(1975).

346. E. Roder and W. Stuthe, *Z. Anal. Chem.*, *266*, 358(1973).

347. E. Roder and W. Stuthe, *Z. Anal. Chem.*, *271*, 281(1974).

348. J. W. Blake, R. S. Roy, J. S. Noonan, and P. W. Murdick, *Anal. Chem.*, *46*, 288(1974).

349. O. Gyllenhaal and H. Ehrsson, *J. Chromatogr.*, *107*, 327(1975).

350. R. L. Hancock, *J. Chromatogr. Sci.*, *7*, 366(1969).

351. T. Hashizume and Y. Sasaki, *Anal. Biochem.*, *16*, 1(1966).

352. T. Hashizume and Y. Sasaki, *Anal. Biochem.*, *24*, 232(1968).

353. J. MacGee, *Anal. Biochem.*, *14*, 305(1966).

354. J. L. Cohen and P. B. Brennan, *J. Pharm. Sci.*, *62*, 572(1973).

355. G. T. Flint and W. A. Aue, *J. Chromatogr.*, *52*, 487(1970).

356. B. K. Dwivedi and R. G. Arnold, *J. Food Sci.*, *37*, 889(1972).

357. R. J. Soukup, R. J. Scarpellino, and E. Danielczik, *Anal. Chem.*, *36* 2255(1964).

358. N. C. Jain and R. H. Cravey, *J. Chromatogr. Sci.*, *12*, 231(1974).

359. H. F. Martin and J. L. Driscoll, *Anal. Chem.*, *38*, 345(1966).

360. H. V. Street, *J. Chromatogr.*, *41*, 358(1969).

361. J. MacGee, *Clin. Chem.*, *17*, 587(1971).

362. T. Chang and A. J. Glazko, *J. Lab. Clin. Med.*, *75*, 145(1970).

363. M. G. Horning, K. Lertratanangkoon, J. Nowlin, W. G. Stillwell, T. E. Zion, P. Kellaway, and R. M. Hill, *J. Chromatogr. Sci.*, *12*, 630(1974).

364. I. A. Muni, C. H. Altschuler and J. C. Neicheril, *J. Pharm. Sci.*, *62*, 1820(1973).

365. J. MacGee, *Anal. Chem.*, *42*, 421(1970).

366. G. W. Stevenson, *Anal. Chem.*, *38*, 1948(1966).

367. H. J. Kupferberg, *Clin. Chim. Acta*, *29*, 283(1970).

368. M. Novotny and K. D. Bartle, *J. Chem. Educ.*, *51*, 333(1974).

369. W. Dunges, *Chromatographia*, *6*, 196(1973).

370. P. Friel and A. S. Troupin, *Clin. Chem.*, *21*, 751(1975).

371. J. Peci and T. J. Giovanniello, *J. Chromatogr.*, *109*, 163(1975).

372. M. Kowblansky, B. Scheinthal, G. Cravello, and L. Chafetz, *J. Chromatogr.*, *76*, 467(1973).

373. W. Dunges, H. Heinemann, and K. J. Netter, in *Chromatography 72* (S. G. Perry, ed.), The Institute of Petroleum, London, 1973, p. 99.

374.   R. Daun, *J. Ass. Offic. Anal. Chem.*, *54*, 1140(1971).

375.   R. G. Spiro, in *Methods in Enzymology* (V. Ginsburg, ed.), Academic Press, New York, 1972.

376.   C. T. Bishop, in *Methods of Biochemical Analysis* (D. Glick, ed.), Vol. X, Interscience, New York, 1962, p. 1.

377.   P. M. Holligan and E. A. Drew, *New Phytologist*, *70*, 239(1971).

378.   J. R. Clamp. T. Bhatti, and R. E. Chambers, in *Methods of Biochemical Analysis* (D. Glick, ed.), Vol. 19, Interscience, New York, 1971, p. 229.

379.   A. H. Weiss and H. Tambawala, *J. Chromatogr. Sci.*, *10*, 120(1972).

380.   J. B. Beadle, *J. Agr. Food Chem.*, *17*, 904(1969).

381.   G. D. Brittain, J. E. Sullivan, and L. R. Schewe, in *Recent Advances in Gas Chromatography* (I. Domsky and J. A. Perry, eds.), Marcel Dekker, New York, 1971, p. 223.

382.   D. B. Weinstein, J. B. Marsh, M. C. Glick, and L. Warren, *J. Biol. Chem.*, *245*, 3928(1970).

383.   R. Kannam, P. N. Seng, and H. Debuch, *J. Chromatogr.*, *92*, 95(1974).

384.   L. Sennello, *J. Chromatogr.*, *56*, 121(1971).

385.   B. Coxon and R. Schaffer, *Anal. Chem.*, *43*, 1567(1971).

386.   M. B. Perry and A. C. Webb, *Can. J. Biochem.*, *46*, 1163(1968).

387.   J. R. Etchison and J. J. Holland, *Anal. Biochem.*, *66*, 87(1975).

388.   J. M. L. Mee, *J. Chromatogr.*, *94*, 298(1974).

389.   D. J. Harvey and M. G. Horning, *J. Chromatogr.*, *76*, 51(1973).

390.   A. Ehrman and R. S. Bandurski, *J. Chromatogr.*, *72*, 61 (1972).

391.   S. Billets, P. S. Lietman, and C. Fenselau, *J. Med. Chem.*, *16*, 30(1973).

392.   J. Casals-Stenzel, H. P. Buscher, and R. Schauer, *Anal. Biochem.* *65*, 507(1975).

Chapter 5

DERIVATIZATION TECHNIQUES IN HPLC

Walter Morozowich

Moo J. Cho

The Upjohn Company
Kalamazoo, Michigan

I. INTRODUCTION

Derivatization is rapidly becoming commonplace in HPLC. The main reasons for employing derivatives are

(a) To permit detection of non-uv absorbing drugs.

(b) To achieve increased sensitivity through fluorescence monitoring.

(c) To improve resolution of closely related compounds.

Even though a number of detectors have become available, uv is still one of the most common and reliable modes of detection with good sensitivity. The universal detectors, such as refractive index or flame ionization, are generally less sensitive and somewhat more troublesome than uv detectors. Many non-uv absorbing drugs contain reactive functional groups suitable for attachment to uv absorbing moieties, and therefore, derivatization presents a logical approach to explore before seeking

alternate modes of detection. With the modern high-efficiency HPLC columns, it is
not uncommon to achieve limits of detection in the low-nanogram range with uv absorbing compounds with typical ε values between 10,000 and 20,000. This range of sensitivity is sufficient for determining the blood levels of many drugs and HPLC is now
becoming widely used in therapeutic monitoring of drugs.

The recent availability of fluorescence monitors for HPLC has opened the door to
the determination of exceptionally low levels of fluorescent compounds. Most drugs
are not highly fluorescent and suitable derivatives must be prepared for detection by
this mode. It is anticipated that with the development of suitable methodology for
derivatization, detection of subnanogram quantities of many drugs may become routine
through fluorescence monitoring.

Derivatives are widely employed in the GLC analysis of drugs, but as yet only
few drugs have been analyzed by HPLC after derivatization. A wealth of information
on potentially useful derivatives is available from the areas of qualitative organic
analysis [1-5] and protective group synthesis [6]. In choosing a derivative for HPLC,
ideally the reaction should be specific, quantitative, free from side reactions, and
complete in a relatively short time period under mild conditions. Much of this information is not available in the literature, and therefore derivatization studies can
be a rather time consuming venture. The design or choice of a derivatizing agent,
furthermore, is critical since the rate of reaction with the drug is often strongly
dependent on the substituents present in the derivatizing agent as well as the solvent, the stochiometry, and the concentration employed.

This chapter will review the derivatives currently employed in HPLC followed by
a description of the essential theory necessary for the optimization of the conditions
for derivatization.

II. DERIVATIVES EMPLOYED IN HPLC

This survey will describe the derivatives used in HPLC according to functional group,
including carboxyl, carbonyl, hydroxyl, and amino derivatives. The method of derivatization will be described in detail to give the reader sufficient information for
evaluation of the various derivatives reported.

A. Derivatives to Improve Detection

1. Carboxyl Derivatives

Of the carboxyl group derivatives reported thus far for HPLC (Table 1), the *p*-nitro-
phenacyl ester appears to be the one most rapidly formed at room temperature.
Morozowich and Douglas found that the carboxyl group of prostaglandins is esterified
quantitatively in less than 6 min at room temperature under the following conditions
[7]. Up to 4.5 mg (12.9 μmole) of PGE_2 was mixed with 9.6 mg (39.3 μmole) of

TABLE 1. Carboxyl Derivatives

Compound or class	Reagent	Derivative	Reference
Prostaglandins	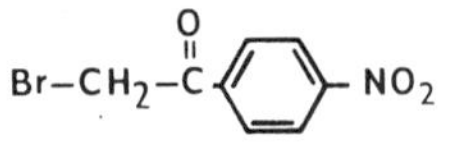 $Br-CH_2-C(=O)-\text{Ar}-NO_2$ (p-Nitrophenacyl Bromide)	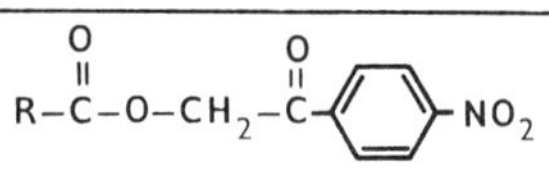 $R-C(=O)-O-CH_2-C(=O)-\text{Ar}-NO_2$	7
Fatty Acids	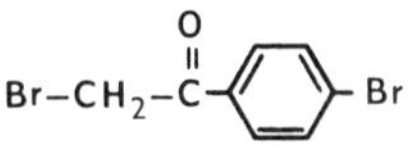 $Br-CH_2-C(=O)-\text{Ar}-Br$ (p-Bromophenacyl Bromide)	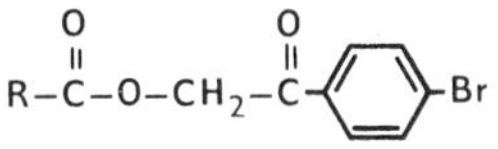 $R-C(=O)-O-CH_2-C(=O)-\text{Ar}-Br$	8
Unspecified	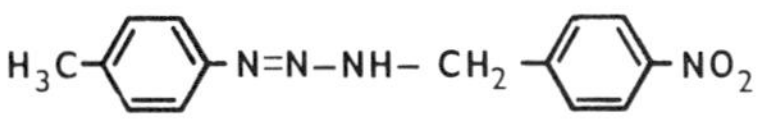 $H_3C-\text{Ar}-N=N-NH-CH_2-\text{Ar}-NO_2$ (1-p-Nitrobenzyl-3-p-tolyltriazene)	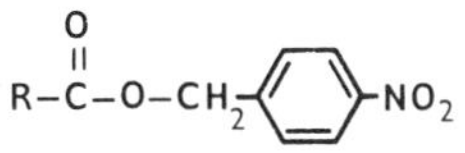 $R-C(=O)-O-CH_2-\text{Ar}-NO_2$	10
Fatty Acids	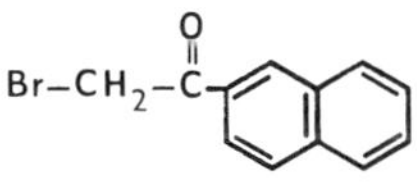 $Br-CH_2-C(=O)-\text{Naphthyl}$ (2-Naphthacyl Bromide)	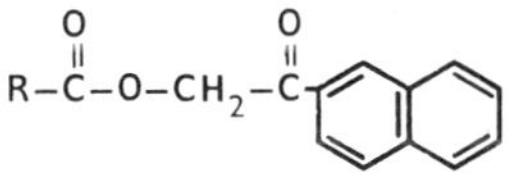 $R-C(=O)-O-CH_2-C(=O)-\text{Naphthyl}$	9
Prostaglandins	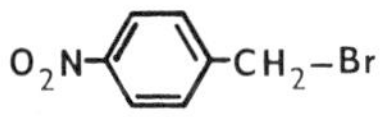 $O_2N-\text{Ar}-CH_2-Br$ (p-Nitrobenzyl Bromide)	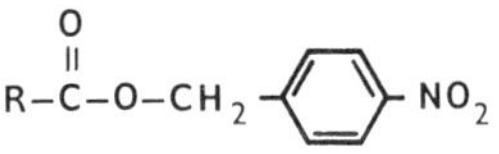 $R-C(=O)-O-CH_2-\text{Ar}-NO_2$	12
Fatty Acids	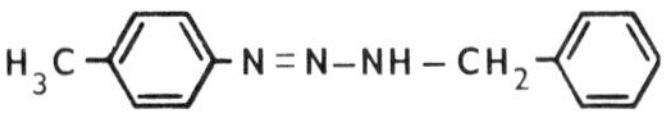 $H_3C-\text{Ar}-N=N-NH-CH_2-\text{Ar}$ (1-Benzyl-3-p-tolyltriazene)	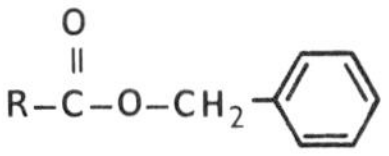 $R-C(=O)-O-CH_2-\text{Ar}$	13

p-nitrophenacyl bromide and 2.9 µl (16.9 µmole) of N,N-diisopropylethylamine in 1 ml of acetonitrile at room temperature. Excellent resolution of closely related prostaglandin *p*-nitrophenacyl esters was achieved by HPLC on Zorbax-Sil with limits of detection of about 1 ng. The *p*-nitrophenacyl esters have a molar absorptivity (ϵ) of 12,777 at 254 nm.

Durst et al. converted the potassium salts of fatty acids to *p*-bromophenacyl esters by reaction with *p*-bromophenacyl bromide [8]. In this procedure the acid (0.5-20 mmole) was neutralized with KOH using phenolphthalein as an indicator. The solvent was removed and the residue was co-distilled with benzene. An excess of a mixture of *p*-bromophenacyl bromide and a crown ether (18-crown-6 or dicyclohexyl-18-crown-6) in a molar ratio of 20:1 was added and the resulting solution was heated at

80° for 15 min. The derivatization procedure was applied to a variety of aliphatic acids using Corasil II for HPLC analysis. The *p*-bromophenacyl esters have an ε of 18,700 at 254 nm.

Cooper and Anders synthesized 2-naphthacyl esters of a variety of long-chain fatty acids and employed Corasil-C_{18} in HPLC analysis [9]. The derivatives were synthesized by reaction of the acid (10 µmole), 2-naphthacyl bromide (20 µmole), and N,N-diisopropylethylamine (40 µmole) in 1 ml of N,N-dimethylformamide. Reaction was complete at 60° in 10 min. The 2-naphthacyl esters have an ε of about 13,000 at 254 nm.

Recently, the Regis Chemical Company introduced 1-*p*-nitrobenzyl-3-*p*-tolyltriazene (PNBTT) for synthesis of *p*-nitrobenzyl esters of carboxylic acids [10]. For derivatization, 2-3 mg of the acid was dissolved in 10 ml of ethanol containing 50 mg of PNBTT. The solution was heated at 65° for 1 hr to achieve esterification. The solution may be chromatographed directly or following a cleanup procedure in which the solvent is removed by evaporation and the residue is dissolved in ether followed by extraction with dilute aqueous HCl. Detailed chromatograms have not been published as yet. The compounds should be strongly uv absorbing with λ_{max} values near that of *p*-nitrobenzene (λ_{max} 268.5 nm; ε = 7800) [11].

Prostaglandins have been converted to *p*-nitrobenzyl esters by Morozowich [12] using *p*-nitrobenzyl bromide and N,N-diisopropylethylamine in acetonitrile. Although excellent chromatography on Zipax-ANH is observed, formation of this derivative requires 4 hr at 60° and the more rapidly formed *p*-nitrophenacyl derivative is preferred for prostaglandins [7].

Politzer et al. employed benzyl esters of fatty acids using Corasil II for HPLC analysis [13]. The esters were prepared by reaction of 2 mmole of the acid with 10 mmole of 1-benzyl-3-*p*-tolyltriazene in 15 ml of ether. The reaction was complete after heating at 36° for 3 hr. The disadvantage of unsubstituted benzyl esters is their low uv absorbance at 254 nm which presumably would be similar to toluene with λ_{max} values at 206.5 nm (ε = 7,000) and 261 (ε = 225) [11]. In addition, interfering impurities and/or by-products accompanying the triazene benzylation technique were reported as troublesome by the authors.

2. Carbonyl Derivatives

Non-uv absorbing carbonyl compounds have been converted to strongly UV absorbing derivatives by means of hydrazone or oxime formation (Table 2). Henry et al. employed 2,4-dinitrophenylhydrazone derivatives of non-uv-absorbing keto steroids as a means of obtaining uv detection at 254 nm [14]. Derivatization was achieved by the method of Siggia [4], in which a slight excess of 2,4-dinitrophenylhydrazine in one ml of methanol containing one drop of concentrated HCl was mixed with the keto

TABLE 2. Carbonyl Derivatives

Compound or class	Reagent	Derivative	Reference
Steroids, Aliphatic Carbonyls	H_2N-NH—(ring)—NO_2, NO_2 (2,4-Dinitrophenylhydrazine)	$\begin{smallmatrix}R'\\R\end{smallmatrix}C=N-NH$—(ring)—$NO_2$, NO_2	14-19
Aldehydes	$H_2N-NH-\overset{O}{\overset{\|}{C}}-NH_2$ (Semicarbazide)	$R-CH=N-NH-\overset{O}{\overset{\|}{C}}-NH_2$	21
Unspecified	$H_2N-O-CH_2$—(ring)—NO_2 (p-Nitrobenzyloxyamine)	$\begin{smallmatrix}R'\\R\end{smallmatrix}C=N-O-CH_2$—(ring)—$NO_2$	10
Aliphatic Carbonyls	Dansyl Hydrazone: CH_3, CH_3 N, naphthalene, SO_2NH-NH_2	CH_3, CH_3 N, naphthalene, $SO_2NH-N=C\begin{smallmatrix}R'\\R\end{smallmatrix}$	23

steroids (0.1-1%). The solution was heated at 50° for a few minutes to achieve reaction. Without work-up, the solution was chromatographed and symmetrical peaks were
obtained on BOP-Zipax.

Other non-uv absorbing keto-steroids were converted to 2,4-dinitrophenylhydra-
zones by Fitzpatrick and Siggia [15], who used similar methods for derivatization.
HPLC was conducted on BOP-Zipax and on Corasil C_{18} using pure keto steroids, as well
as steroids extracted from plasma and urine. The λ_{max} of 2,4-dinitrophenylhydrazones
of saturated carbonyl compounds is typically at 360 nm with an ε of 20,000 [11]. At
254 nm, ε values of 10,000 are usually observed [14]. Keto acids were derivatized
with 2,4-dinitrophenylhydrazine by related methods and separation was achieved on
Sephadex LH-20 by Winchester [16]. A novel post-column procedure for conversion of
cyclohexanone to its 2,4-dinitrophenylhydrazone was developed by Deedler and Hendricks
[17]. The presence of the acid labile cyclohexanone oxime in admixture with cyclohexanone precluded a reaction with 2,4-dinitrophenylhydrazine prior to chromatography.

Non-uv absorbing aliphatic carbonyl compounds were similarly converted to 2,4-dinitrophenylhydrazones for HPLC by Carey and Persinger [18] and Pappa and Turner [19].

The simplicity of derivatization and the strong uv absorbance of these derivatives indicates that 2,4-dinitrophenylhydrazine should be broadly applicable in the HPLC detection of carbonyl containing drugs. It remains to be seen if the new high-resolution columns will result in separation of syn and anti isomers as discrete peaks since it is known that some drugs give isomeric 2,4-dinitrophenylhydrazones which are separable by TLC [20].

Semicarbazone derivatives of the sedative drug 1-chloro-3-ethylpent-1-en-4-yn-3-ol were prepared for HPLC by Needham and Kockhar [21] wherein an aldehyde carbonyl group was first generated by treatment of the drug with acid. A heptane solution of the aldehyde was shaken with 5 ml of 0.5 N semicarbazide·HCl buffered to pH 3.5 with sodium acetate for 3 min at room temperature. Aliquots of the aqueous phase were subjected to HPLC analysis on a C_{18} µBondapak column, and single peaks were obtained for the derivative.

The rapidity of formation of semicarbazones offers encouragement for its broad utility; however, it should be noted that semicarbazone formation is a reversible reaction and the formation constant for aldehydes is about 3 orders of magnitude greater than that of ketones [22].

With semicarbazones of saturated aldehydes or ketones, the typical absorption spectra show a λ_{max} at 225-230 nm with an ε of 11,000 [11]. Semicarbazones of conjugated carbonyl groups show enhanced absorption with a λ_{max} at 265 and an ε of 25,000 [11]. Thiosemicarbazones are promising derivatives due to intense absorption, and in the case of saturated carbonyl compounds, two λ_{max} values are seen at 230 and 280 nm with an ε of 7,000 and 20,000 respectively [11].

p-Nitrobenzyloxamine·HCl has been proposed as a reagent for derivatizing carbonyl groups by the Regis Chemical Company [10]. Reaction is achieved by dissolving 1-5 mg of the carbonyl compound and 40 mg of nitrobenzyloxime·HCl in 4 ml of methanol. After heating at 65° for 1 hr, the reaction mixture is directly subjected to HPLC analysis. Detailed chromatograms have not been reported as yet, but it is anticipated that syn and anti isomers will be obtained since the related reagent benzyloxyamine provides a mixture of *syn* and *anti* isomers with carbonyl compounds and the isomers are easily separated by silica gel TLC.

Dansyl hydrazine was used to convert carbonyl compounds to fluorogenic derivatives [23]. Derivatization is achieved by heating the carbonyl compound in 1 ml of methanol containing a twofold molar excess of dansyl hydrazine and two drops of glacial acetic acid at 70° for 15 min. The solvent is removed under vacuum and the residue is dissolved in benzene. HPLC was conducted on Corasil II or BOP-Zipax.

TABLE 3. Hydroxyl Derivatives

Compound or class	Reagent	Derivative	Reference
Steriod	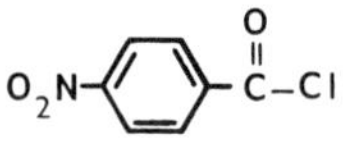 (p-Nitrobenzoyl Chloride)	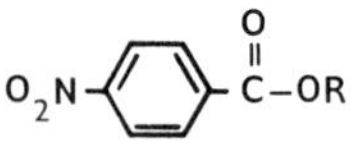	24
Glycols	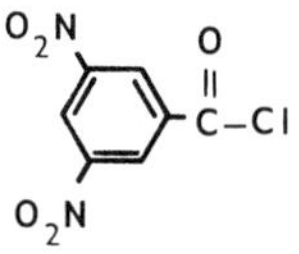 (3,5-Dinitrobenzoyl Chloride)	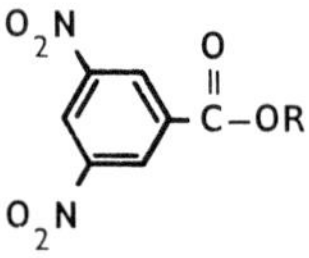	10,18
Steroids, Glycosphingolipids	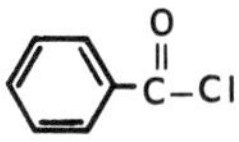 (Benzoyl Chloride)	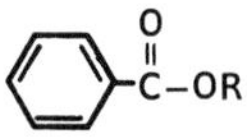	24,28
Pentaerythritol, Hexachlorophene	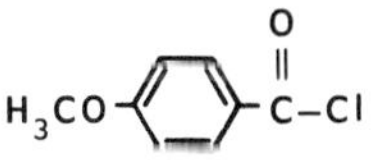 (p-Methoxybenzoyl Chloride)	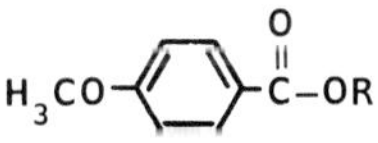	25,26
Aliphatic Alcohols	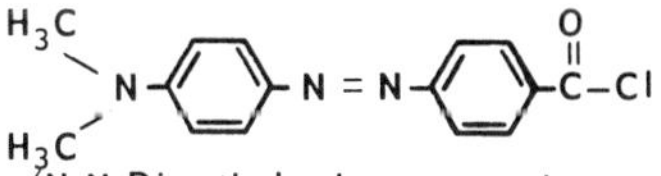 (N,N-Dimethyl-p-benzeneazobenzoyl Chloride)	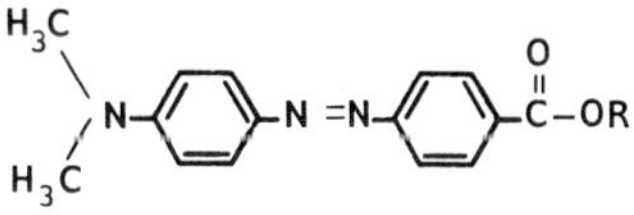	29
Hydroxybiphenyls	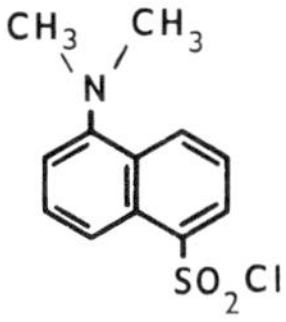 (Dansyl Chloride)	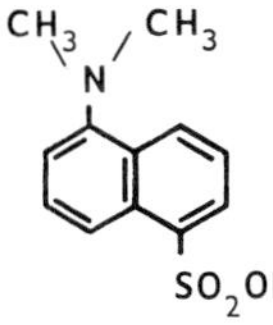	31

3. Hydroxyl Derivatives

UV-absorbing derivatives of alcohols and phenols are readily prepared using acid chlorides or anhydrides (Table 3). Fitzpatrick and Siggia employed benzoate steroids for HPLC on Corasil C_{18} [24]. Derivatization was achieved by reaction of 0.5-50 mg of the steroids in 4 ml of pyridine with a threefold molar excess of the acid chloride at 80° for 15 min. Samples of the reaction mixture could be chromatographed directly or following removal of pyridine by evaporation or extraction with dilute

HCl upon addition of ether. The sensitivity of detection of the *p*-nitrobenzoates was about 10 times better than that of the benzoate esters at 254 nm. The *p*-nitrobenzoates have a λ_{max} at 254 nm with an ε greater than 10,000, whereas the benzoates have a λ_{max} at 230 nm with an ε value of about 13,000. The lower limit of detection of the *p*-nitrobenzoates was estimated to be 1 ng.

Carey and Perisinger converted glycols to 3,5-dinitrobenzoate esters and optimistically maintain that "it appears that a general solution to the problem of detecting and determining low levels of hydroxyl compounds by LC is at hand" [18]. The esters are synthesized by mixing 1 mequiv of polyethylene glycol, or other alcohols, with a solution of 0.5 gm of 3,5-dinitrobenzoyl chloride in 20-30 ml of pyridine. After 15 min at 60°, the solution was acidified to pH 2.5 and extracted with 25 ml of butyl acetate. The organic phase was extracted with 50 ml of 1% sodium carbonate followed by 25 ml of 0.5 N H_2SO_4 and finally by 10 ml of water. Aliquots of the organic phase were subjected to HPLC on Corasil II. As little as 0.05 μg of ethylene glycol could be detected at 254 nm.

The Regis Chemical Company suggested the following procedure for reaction of 3,5-dinitrobenzoyl chloride with alcohols or phenols [10]. A solution of 1-5 mg of the hydroxyl compound in 5 ml of tetrahydrofuran containing 40 mg of 3,5 dinitrobenzoyl chloride is heated at 60° for 1 hr. HPLC analysis can be conducted directly on silicaceous supports according to this report.

p-Methoxybenzoate esters of pentaerythritol were employed for HPLC analysis by Bighley et al. [25]. Pentaerythritol in plasma samples (5 ml) was mixed with 2 ml of 10% trichloracetic acid followed by centrifugation at 2,000*g* for 15 min. The supernatant was treated with 2 ml of 10% sodium hydroxide in water followed by addition of 0.1 ml of *p*-methoxybenzoyl chloride and 4 ml of heptane-chloroform (3:2). After shaking for 60 min at 40°, the organic phase was subjected to HPLC analysis on Corasil II. Pyridine-catalyzed esterification was deemed undesirable since repeated injections resulted in decreased retention times through either reaction with the support hydroxyl groups or by unknown physical adsorption mechanisms. With highly reactive derivatizing agents, reaction of the support may occur and suitable deactivating or cleanup procedures may prove advantageous.

Hexachlorophene is an example of a compound which shows poor uv absorption near 254 nm due to a minimum in the absorption curve in this region. Maxima are observed at about 290 nm (ε = 6000) and 240 nm [26]. To enhance detector sensitivity at 254 nm, Pocaro and Shubiak prepared the bis-*p*-methoxybenzoate ester thereby achieving nanogram level detection by HPLC on Sil-X [27].

Neutral glycosphinogolipids were converted to benzoyl derivatives for HPLC by Evans and McCluer [28]. Benzoylation was achieved by heating 0.01-0.1 μmole of the glycosphingolipid in 0.6 ml of 20% benzoyl chloride in pyridine at 60° for 1 hr. After removal of the solvent, the residue was dissolved in 5 ml of hexane followed by successive extraction with 3 ml each of 95% methanol saturated with $NaHCO_3$, 95% methanol, 0.6 *M* HCl in 95% methanol, and finally, 95% methanol. HPLC on Zipax gave a full-scale recorder response with approximately 1 nmole of glycolipid.

Aliphatic alcohols were converted to N,N-dimethyl-*p*-aminobenzeneazobenzoate esters for HPLC by Churacek and Jandera [29]. The alcohol (2-5 µl) was mixed with 1 mg of N,N-dimethylamino-*p*-benzeneazobenzoyl chloride in one drop of pyridine. Esterification was complete after heating at $100°$ for 5-10 min [30].

Phenols have been converted to dansyl esters for HPLC analysis with uv and fluorescence detection by Cassidy et al. [31]. To achieve reaction, 10 µl of a phenol solution (concentration unspecified) was mixed with 200 µl of a 0.1% solution of dansyl chloride in acetone, Following addition of 30 µl of aqueous 0.1 *M* sodium carbonate, the mixture was heated at $42-45°$ for 15-20 min. Two drops of 1 *N* sodium hydroxide were added and the dansyl derivative was extracted with 500 µl of hexane. HPLC analysis of the hexane extract was conducted on 7- to 18-µm silica. The limit of detection of the *p*-hydroxybiphenyl dansyl ester was 0.1 ng using fluorescence detection.

4. Amine Derivatives

The reagents for conversion of amines to uv-absorbing or fluorescent derivatives are given in Table 4. The Regis Chemical Company indicated that amines may be derivatized with 3,5-dinitrobenzoyl chloride [10]. In their procedure, 1-5 mg of the compound is dissolved in 5 ml of tetrahydrofuran containing 40 mg of 3,5-dinitrobenzoyl chloride. The solution is heated at $60°$ for 1 hr, and samples can be chromatographed directly. Alternatively, triethylamine or pyridine may be added as a proton acceptor. If pyridine is used, it should be removed before chromatography through an acid-extraction step using a water-immiscible solvent.

The amino anesthetic ketamine was converted to a *p*-nitrobenzamide by Needam and Kochhar to provide increased sensitivity for HPLC [32]. Using chloroform extracts of the drug from urine, the solution (0.2 ml) was treated with 10 mg of *p*-nitrobenzoyl chloride and 0.4 ml of 5% NaOH. After 20 min at room temperature, aliquots of the organic layer were subjected to HPLC on C_{18} µBondapak.

Jungawala et al. detected ethanolamine and serine containing phospholipids by HPLC after amide formation with 4-biphenylcarbonyl chloride [33]. In one procedure, 10-100 nmole of the phospholipid was dissolved in 0.1-0.25 ml of 4-biphenylcarbonyl chloride (0.2%) in acetonitrile. After adding 2-5 µl of triethylamine, the mixture was heated at $37°$ for 30 min. HPLC on MicroPak SI-10 allowed detection of as little as 10-13 pmole of phospholipids at 280 nm. The derivatives showed λ_{max} values at 268 nm with ϵ = 20,000.

The diamines putrescine, spermidine and spermine were determined by Sugiura et al. [34] using tosyl derivatives. For derivatization, 1 ml of an aqueous solution of the

TABLE 4. Amine Derivatives

Compound or class	Reagent	Derivative	Reference
Ketamine	(p-Nitrobenzoyl Chloride)		32
Unspecified	(3,5-Dinitrobenzoyl Chloride)		10
Amino Phospholipids	(4-Biphenylcarbonyl Chloride)		33
Aliphatic Amines, Carbamates, Aliphatic Polyamines	(Dansyl Chloride)		23,36,41
Aliphatic Diamines	(Tosyl Chloride)		34
Aliphatic Amines	(4-Chloro-7-nitrobenzo-2,1,3-oxadiazole)		23

TABLE 4. (Continued)

Compound or class	Reagent	Derivative	Reference
Aliphatic Secondary Amines	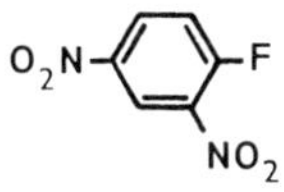 (2,4-Dinitrofluorobenzene)	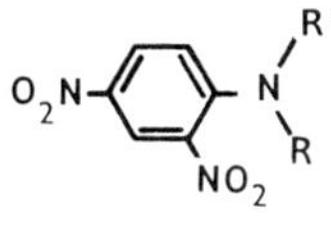	37
Amino Acids	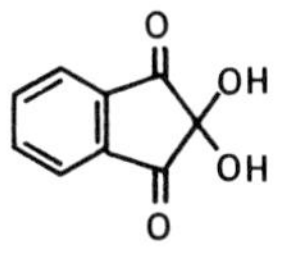 (Ninhydrin)	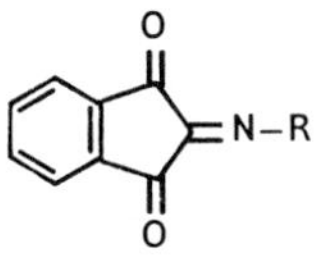	39,40
Amino Acids	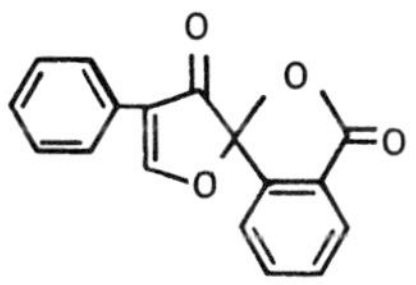 (Fluorescamine)	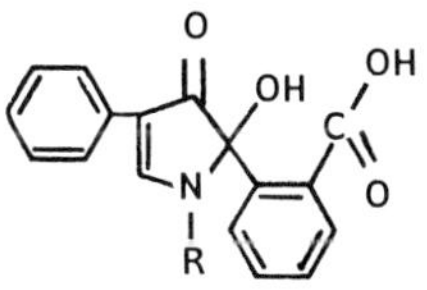	43,45
Amino Acids	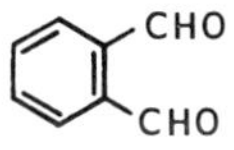 (o-Phthaldehyde)	Unknown	46,48

amine (amount unspecified) was mixed with 1 ml of 0.5 M sodium bicarbonate and 2 ml of acetone containing 1% tosyl chloride. After 1 hr at 70°, 0.1 ml of an internal standard solution (1 mg of tosylated 1,10-diaminodecane in methanol) was added along with 10 ml of 1 N sodium hydroxide. The mixture was washed with four 5-ml volumes of n-hexane and the aqueous layer was mixed with 15 ml of 1 N HCl. After extraction with 10 ml of chloroform, the solvent was removed from the aqueous layer and the residue was dissolved in a few drops of methanol. Aliquots were subjected to HPLC on an ETH Zipax column and quantities down to 1 µg were chromatographed.

Putrescine and related diamines were detected by HPLC using fluorescence monitoring after reaction with fluorescamine by Samejima [35]. Derivatization was achieved by mixing one volume of the amine (1 nmole/µl) with one volume of 0.1 M borate buffer (pH 8.0) and 1 volume of fluorescamine (2 mg/ml) in acetone under rapid agitation. Aliquots of the reaction mixture were chromatographed directly on a Vydac reversed-phase column. The limit of detection was estimated to be 1 pmole.

Dansyl derivatives of polyamines were prepared for HPLC by Abdel-Monem and Ohno [36]. Details for synthesis were deferred to a later paper, however; chromatograms were shown using columns of 10-μm silica gel, 5-μm alumina and Corasil II while monitoring at 280 nm. With this procedure, a variety of polyamines and monoacetyldiamines were detected.

Nitrosamines were reduced electrochemically and the resulting secondary amines were converted to 2,4-dinitrophenyl derivatives for HPLC analysis on carbowax-silica gel or Corasil C_{18} by Cox [37]. Derivatization was achieved by treating a solution of the secondary amine (dimethylamine, diethylamine, or pyrrolidine) in 1% aqueous sodium tetraborate (1.5 ml) with 0.2 ml of 3% 1-fluoro-2,4-dinitrobenzene in dioxane. After heating at 60° for 25 min, 0.2 ml of 2 N sodium hydroxide was added and the solution was further heated for 15 min. The solution was shaken with 1 ml of cyclohexane and then extracted with 0.1 M sodium carbonate (3 x 2 ml). Aliquots of the organic phase were analyzed by HPLC. The derivatives are photolabile and must be stored in the dark or in foil-wrapped flasks. A variable-wavelength monitor was used at 350 nm corresponding to the λ_{max} of the derivatives.

Carbamates were converted to amines by Frei and Lawrence upon hydrolysis with 2 M sodium hydroxide [38]. The solution was treated with two drops of methylisobutyl ketone and two drops of dansyl chloride solution (0.1% in acetone) and heated at 65° for 30 min. The solution was acidified, extracted with 0.3 ml of benzene and subjected to HPLC. The limit of detection was 1-10 ng using fluorescence detection. In another method, carbamates were converted to amines which were reacted with 4-chloro-7-nitrobenzo-2,1,3-oxadiazole (NBD-Cl). In this method [38], the carbamate was mixed with 0.5 ml of 0.1 M NaHCO$_3$ and 0.5 ml of 1% NBD-Cl in methylisobutyl ketone. After 30 min at 80°, aliquots of the upper phase were subjected to HPLC using fluorescence detection.

The same authors [23] prepared dansyl amino acids by heating the amino acid in one ml of acetone containing 0.1% dansyl chloride with addition of 2-3 drops of 0.5 M sodium carbonate. After 1 hr at 65°, the solvent was evaporated and the derivative was dissolved in benzene and analyzed by HPLC.

Ninhydrin has been widely used to detect amino acids [39] and other amines [40] using post-column derivatization wherein the reagent is added to the column effluent prior to going through the HPLC detector. Other reagents useful for detection of amino acids by fluorescence monitoring include dansyl chloride [41], fluorescamine [43], and o-phthaldehyde [44]. The latter three also employ post-column derivatization methods. Reaction of primary amines with fluorescamine proceeds with a "halftime" of a fraction of a second in aqueous media. The excess reagent is destroyed by reaction with water in a competitive reaction with a "half-time" of 5-10 sec [43]. The water reaction product is nonfluorescent whereas the amino derivative is strongly fluorescent. Detection of amino acids in the lower picomole range has been reported using fluorescamine [43-45].

o-Phthaldehyde was used for derivatization of amino acids by Roth [46,47]. The reaction conditions were improved by Benson and Hare [48] and the following advantages are claimed over fluorescamine: (a) the *o*-phthaldehyde derivatives exhibit 5-10 times greater sensitivity of detection; (b) the reagent is soluble and stable in aqueous media; (c) baseline stability is improved; and (d) the reagent is less expensive. Derivatization was achieved as follows: A solution consisting of 800 mg of *o*-phthalaldehyde in 10 ml of ethanol was diluted to 11 ml with 0.40 M borate buffer (pH 9.7), containing 2 ml of 2-mercaptoethanol and 1 g of Brij per 100 ml. This solution is metered into the column effluent in an equal volume ratio and the mixture is passed directly into a fluorometer. The amino acids were chromatographed on the cation exchange resin DC-4A using sodium formate and sodium formate-borate buffers in sequence. Routine analyses were conducted using 10-100 pmole of each amino acid. The authors concluded that fmole quantities (10^{-15}) of amino acids could be detected if the buffers employed were free of trace amino contaminants.

B. Derivatives for Improved Resolution

The highly efficient small-particle supports have eliminated many problems with resolution in the drug area. Two types of compounds that still provide problems of separation are enantiomeric pairs and isomers or other closely related structures. The separation of enantiomers is easily achieved by gas chromatography after formation of diastereomers with chiral derivatizing reagents [48-53]. Using HPLC, the R and S isomers of α-phenylethylamine were separated by Helmchen and Strubert [55] after conversion to diastereomeric (S)-O-methylmandelyl amides. Reaction was achieved by treating 50 μl of α-phenylethylamine with a 20% excess of (S)-O-methylmandelyl chloride in dioxane containing 70 μl of triethylamine at room temperature (time unspecified). An ether-benzene solution of the crude amides was extracted with 1 N mineral acid followed by extraction with 1 N base to remove the amine and excess derivatizing reagent. Separation of the resulting diastereomeric N-(1-phenylethyl)-O-methyl-mandelamides was easily achieved on Merckosorb SI 60 (5 μm) using isooctane-ethyl acetate (4:1).

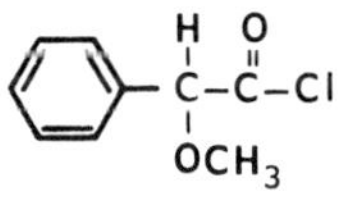

$$\text{C}_6\text{H}_5 - \overset{\displaystyle \text{H}}{\underset{\displaystyle \text{OCH}_3}{\text{C}}} - \overset{\displaystyle \text{O}}{\text{C}} - \text{Cl}$$

Koreeda et al. [56] reported the separation of the dl isomers of the abscisic acid precursor. Reaction of the C_3 hydroxyl group with (+)-α-methoxy-α-trifluromethylphenylacetyl chloride was conducted under conditions similar to those used by others [57, 58]. Five recycles on a 3/8-in. o.d. x 9 ft Porasil T column resulted in separation of about 100 mg of the isomeric mixture [59].

Abscisic Acid Precursor

(+)-α-Methoxy-trifluoro-methylphenylacetyl Chloride

Little information is available on the use of derivatives in HPLC to improve resolution of compounds other than the optical isomers described above. Fitzpatrick and
Siggia provided an example in which two non-uv-absorbing steroids were poorly resolved
as the 2,4-dinitrophenylhydrazone derivatives whereas the benzoate ester derivatives
showed markedly improved resolution [24]. Improved resolution may be due to the decreased polarity of the benzoate ester. In adsorption chromatography, retention time
is determined primarily by the group contribution energies of the various polar groups
present [60]. Hydrocarbons have negligible interaction with the absorptive sites on
silica gel. In principle, it should be possible to improve resolution of closely related compounds by forming less polar derivatives of those highly polar functional
groups common to the compounds in question. Thus, in the case of a compound mixture
consisting of the hydroxyl isomers of a fatty acid, conversion of the carboxyl group
to the less polar carboxyl ester function may result in improved separation of the
hydroxyl isomers.

III. KINETIC CONSIDERATIONS IN DERIVATIZATION

The ideal derivative is one that is rapidly formed under mild conditions, preferably
at room temperature without significant side-product formation. Not all reactions
utilized in characterizing organic compounds and in protective group synthesis meet
these requirements, and modifications are usually necessary to further optimize the
reaction conditions without losing the detection sensitivity in HPLC. The factors
influencing reaction rates are discussed in this section as an aid in developing the
reaction conditions as well as optimizing the reactivity of the reagent through substituent effects.

A. Stoichiometry and Concentration

For an irreversible reaction of the type A + B → X, the concentration of the product
(x) at a given time (t) is related to the rate constant (k) by the following equation:

$$\frac{2.3}{a - b} \log \frac{b(a - x)}{a(b - x)} = kt \tag{1}$$

where a and b are the initial concentrations of the drug and the reagent, respectively [61]. The time required for 90 or 99% of a reaction (t_{90} or t_{99}) can be calculated under the given conditions by solving Eq. (1) with respect to t using x = 0.90a or 0.99a.

First, as shown in Eq. (1), t_{99} decreases proportionally as k increases for a given set of a and b. Thus, proper manipulation of the factors determining the magnitude of k can reduce the time required for derivatization remarkably. In subsequent sections, we will discuss the effect of temperature, solvent, and substituent on k.

The second point to be noted is that there exists a nonlinear relationship between the stoichiometric ratio of b to a and t_{99}. As shown in Fig. 1 where t_{90} and t_{99} are plotted as a function of the stoichiometric ratio (b/a) using $k = 100 \ min^{-1}M^{-1}$ and $a = 10^{-3}M$, the time required for reaction decreases markedly as the stoichiometric ratio is increased only from 1:1 to 1:3 (990 min at 1:1 versus 20 min at 1:3). In contrast, an increase in the ratio beyond 5 does not affect t_{99} significantly. When a large excess of reagent is used, the reaction becomes pseudo first order with respect to the reagent and, hence, t_{99} decreases only linearly as b/a becomes larger. From this analysis, it can be concluded that for practical purposes a ratio greater than 3 does not give much advantage in terms of t_{99}. A large excess of reagent can cause resolution problems in HPLC analysis and, for this reason, a large excess is undesirable.

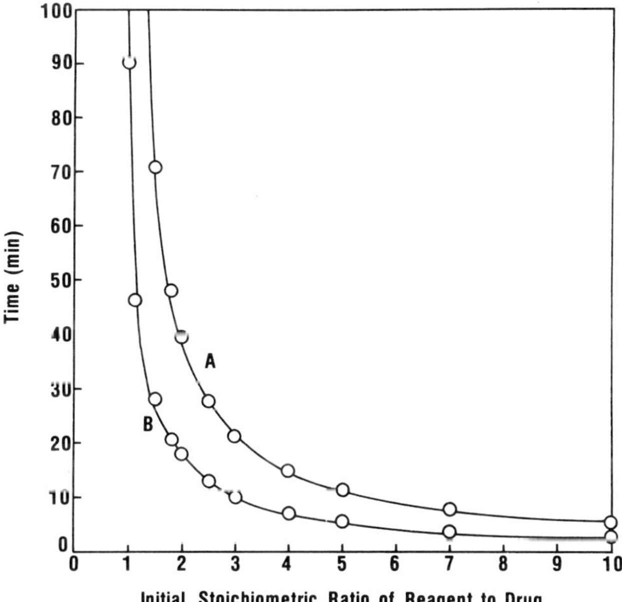

Fig. 1. Time required for 99% (curve A) and 90% (curve B) derivative formation as a function of the initial stoichiometric ratio of reagent to drug. Curves were generated under conditions where $k = 100 \ min^{-1}M^{-1}$ and drug concentration is $10^{-3}M$.

Another important consideration in derivatization is the effect of the absolute initial concentration of a drug and a reagent. As indicated by the first term in Eq. (1), t_{99} decreases as a and b decrease. Thus, derivatization conducted on a µg/ml scale will be 1000 slower than that conducted on a mg/ml scale. For this reason, it is wise to concentrate a dilute sample prior to derivatization, whenever possible.

B. Temperature

The rate of many reactions increases about 2- to 3-fold with each $10^{\circ}C$ rise in temperature. Thus, the rate may increase by a factor of 30-240 in going from 25° to $75^{\circ}C$. The advantage of this effect, however, may give rise to complications for the following reasons. First, the compound to be derivatized should be stable at the elevated temperature at which derivatization is carried out. This aspect is particularly important in pharmaceutical analysis where the thermal degradation products formed during the derivatization process could be the same as those degradation products which one wants to quantitate in studies on solution stability of the drug. Second, if the activation energy (E_a) for the side-product formation is considerably larger than that for the desired reaction, the side reaction will become more significant as temperature increases. For instance, if the values of E_a are 10 and 15 kcal/mole and the frequency factors in the Arrhenius equation are 10^7 and 10^{10} $sec^{-1}M^{-1}$ for the desired and side reactions, respectively, the extent of the side reaction will increase from 18% to 42% in going from 25° to $75^{\circ}C$. An analogous temperature effect on a pair of parallel reactions was treated by Eriksen and H. Stelmach [62].

In practice, the extent of thermal degradation of a drug and side-product formation during a derivatization reaction should be determined along with the extent of the desired reaction, either by HPLC using an internal standard or by simple TLC monitoring of the derivatization reaction.

C. Solvent

Solvent molecules actively participate in the important microscopic reaction steps resulting in a large difference in the overall reaction rate when different solvents are employed. The influence of solvent can be generalized as follows. If the solvent stabilizes the transition state through association or solvation more effectively than the ground state of the reactant, the rate of reaction increases. Thus, for a reaction where the transition state is more polar than the ground state, a polar solvent would facilitate the reaction, or vice versa.

In bimolecular substitution reactions where an anionic nucleophile attacks a neutral molecule, the rate of reaction is generally slower in protic solvents (H_2O or alcohols) than in aprotic solvents due to better ground-state solvation of the anionic nucleophile in the protic solvents [63]. For these reasons, esterification of ionized carboxylic acids with alkyl halides such as benzyl halides or phenacyl halides are more rapid in polar aprotic solvents such as acetonitrile or dimethylformamide than in protic solvents such as alcohols [7-9, 12]. Esterification of alcohols with acid chlorides proceeds more rapidly in polar solvents such as pyridine [18, 24] or dioxane [64]. Many *p*-substituted benzoyl halides react with alcohols and amines very rapidly and therefore various polar or nonpolar solvents can be employed. However, strict exclusion of moisture is required to maintain the acid halide. In many cases, the amount of water present in the solvent or head space can be greater than that of the drug, and a significant fraction of the acid halide can be present as the anhydride unless a large excess of the acid chloride is employed.

Formation of phenylhydrazones, oximes, or semicarbazones from carbonyl compounds is generally carried out in aqueous alcohol containing an acid [14, 15, 18, 19]. In the absence of acid, dehydration from the carbinolamine produced by addition of nitrogen nuclophiles becomes rate determining.

D. Substituent Effects

When a drug is an alcohol or an amine, in most derivatizations it behaves as a nucleophile. If the drug is a carboxylic acid, the reaction conditions usually are modified to generate a carboxylate anion which is a much better nucleophile than the free acid. Thus, the derivatizing reagent should be very suseptible to nucleophilic attack. Reagents with unsaturation at the carbon *beta* to a leaving group invariably undergo nucleophilic substitution readily because the interaction between the attacking nucleophile and the π-electron system present in the reagent lowers E_a leading to the transition state. For this reason, phenacyl, benzyl, and allylic compounds are frequently employed in derivatization.

If a drug is a carbonyl compound, most derivatization reactions proceed with initial attack of a nucleophilic reagent followed by dehydration of the intermediate adduct. In the presence of acid, the dehydration step becomes fast and the nucleophilic attack controls the reaction rate. The carbonyl carbon bears a partial positive charge, and addition of the nucleophile to the carbonyl group resembles protonation of the nucleophile. Therefore, basicity can be a good measure of nucleophilicity. A reagent with a negative charge is invariably a powerful nucleophile and, among neutral compounds, the reactivity is in the order of N > O > S nucleophiles. For this reason, hydrazines are most commonly employed for derivatization of carbonyl compounds.

The reactivity of the reagent can be altered by appropriate substituents as predicted by linear free-energy relationships. The Hammett equation [Eq. (2)] is obeyed with many derivatizing reagents and

$$\log \frac{k}{k_o} = \sigma\rho \tag{2}$$

this can serve as a guide for predicting the influence of substituents on rate of reaction. For instance, the ρ values for esterification of alcohols and amide formation from amines with substituted benzoyl chlorides are in the range of 1.2 to 1.5 [65], and therefore substituents with large σ values will react with a drug more rapidly. *p*-Nitrobenzoyl chloride is therefore expected to react with alcohols or amines approximately 18-37 times faster than *p*-methoxylbenzoyl chloride (σ values of $-NO_2$ and $-OCH_3$ are 0.778 and -0.268, respectively).

Similarly, phenacylation of the carboxylic acid prostaglandin $F_2\alpha$ with seven *p*-substituted phenacyl bromides ($-OCH_3$, $-CH_3$, $-C_6H_5$, $-H$, $-Cl$, $-Br$, and $-NO_2$) in acetonitrile containing diisopropylethylamine was found to yield a ρ value of 0.78 at 25° [66]. Sensitivity of this series of reactions to the electrical effects from *para* substituents are not fully transmitted to the reaction site presumably because it is two carbons away from the benzene ring. Consequently, the ρ value is rather low. Nevertheless, as shown in Fig. 2, there is a substantial difference in the rate between p-NO_2 and p-OCH_3 substituted reagents: 9.98 and 1.27 $min^{-1}M^{-1}$, respectively. *p*-Nitrophenacyl bromide is obviously the most reactive *para*-substituted phenacyl bromide.

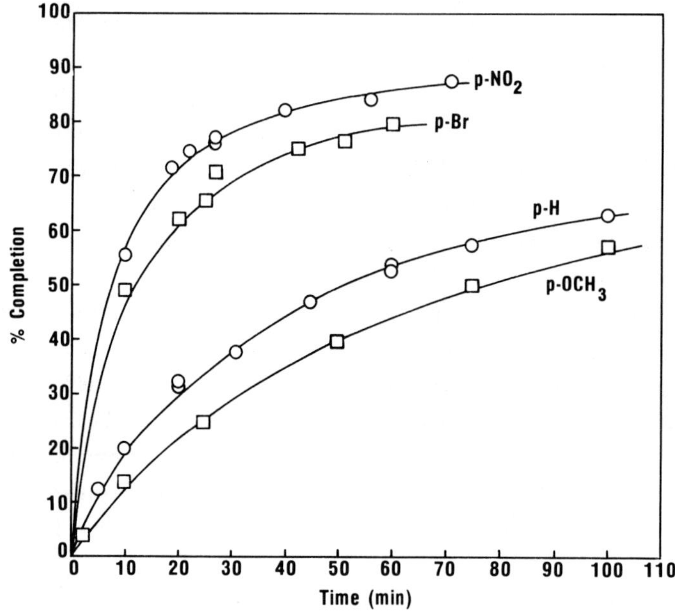

FIG. 2. Phenacylation of prostaglandin $F_2\alpha$ with *p*-substituted phenacyl bromide in acetonitrile at 25° in the presence of diisopropylethylamine. The initial concentrations of the drug and the reagent were both $10^{-2}M$ and that of the amine was $1.2 \times 10^{-2}M$. Phenacylation was monitored by HPLC using hydrocortisone as an internal standard [66].

In many cases, formation of hydrazones, oximes, and semicarbazones do not follow a simple Hammett relationship [67]. However, for a series of reactions where no change in rate-determining step takes place and homogeneous steric effects are prevailing, there exists a linear free energy relationship. In the condensation of benzaldehyde with substituted anilines a ρ value of -2.00 is observed and, therefore, in this series of reactions, reagents with electron-donating substituents (negative σ values) react more rapidly than those with electron-withdrawing substituents (positive σ values) [68].

It is obvious that the design of the reactivity of a reagent requires an understanding of the mechanism of reaction and the prevailing linear free energy relationships. A more detailed consideration of this subject is beyond the scope of this review. It is anticipated that many existing derivatizing reagents can be optimized for their use in HPLC to give ultimately a selective group of reagents suitable for rapid and quantitative derivatization in HPLC.

IV. OTHER PROMISING REAGENTS

Table 5 lists several reagents that may be suitable for derivatization in HPLC. Undoubtedly many more could be added to this list. Most of the resulting derivatives should be strongly uv-absorbing derivatives and, in some cases, (2-diphenylacetyl-1,3-indane-dione-1-hydrazone) fluorescent derivatives should result.

RECENT DEVELOPMENTS

Recent reviews also worthy of mention include a book, "Chemical Derivatization in Liquid Chromatography" by Lawrence and Frei [79], and excellent articles by Dixon et al. [80], Drozd [81], and Ahuja [82]. The latter two deal with chemical derivatization for gas chromatography, but many of the reagents can be employed in liquid chromatography as well.

In addition, new reagents for derivatization include the following. A new reagent for synthesis of *p*-nitrobenzyl esters, namely, O-*p*-nitrobenzyl-N,N-diisopropylurea, offers the advantage of rapid esterification of carboxylic acids in methylene choloride [83]. Carboxylic acids have been converted to fluorescent esters by reaction with 4-bromomethyl-7-methoxy-coumarin [84]. Pyruvoyl chloride, 2,6 dinitro phenyl hydrazone, and *p*-iodobenzenesulfonyl chloride have been used to esterify alcoholic steroids [85]. Enantiomeric amino acids have been separated by HPLC upon reaction of the amino group with ( ) α methoxy-α-methyl-1-naphthaleneacetic acid [86]. Enantiomeric carboxylic acids have been resolved upon amide formation with (R)-(+)-α-methyl-*p*-nitrobenzylamine [87]. *p*-Nitrobenzyloxime and pentafluorobenzyloxime have been used for conversion of carbonyl groups to UV-absorbing oximes for HPLC detection [88]. Selective acylation of amines has been reported using the novel reagent N-succinimidyl-*p*-nitrophenylacetate [89].

TABLE 5. Potential Reagents for Use in HPLC

Compound or class	Reagent	Derivative	Reference
A. $R-\overset{O}{\underset{\|}{C}}-OH$			
Aliphatic and Aromatic Acids	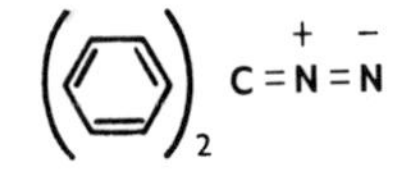 (Diphenyldiazomethane)	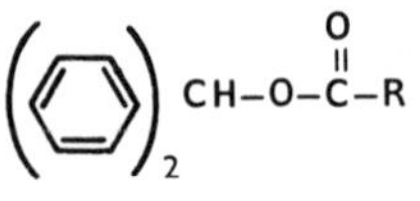	69
Prostaglandins and Fatty Acids	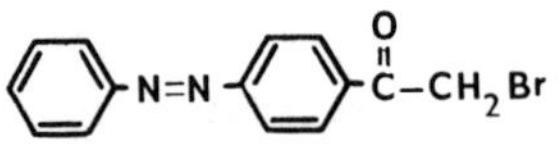 (p-Phenylazophenacyl Bromide)	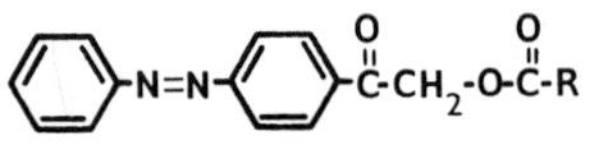	7-9
B. $R-\overset{O}{\underset{\|}{C}}-R'$			
Aldehydes and Ketones	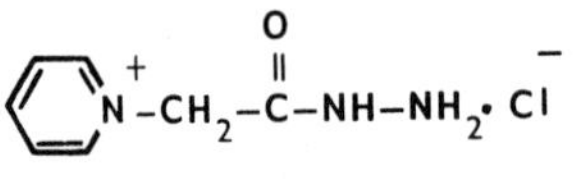 (Girard's Reagent P)	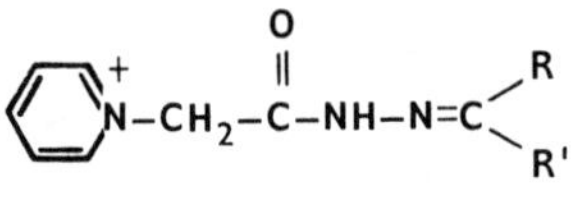	70
Aldehydes	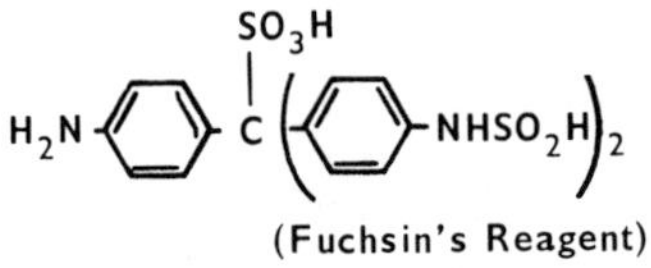 (Fuchsin's Reagent)	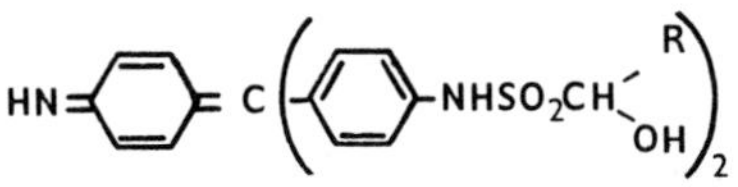	71
Aromatic Aldehydes	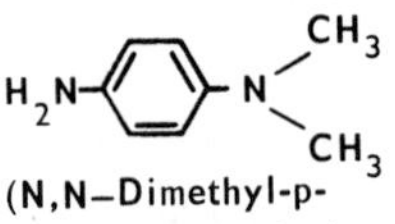 (N,N-Dimethyl-p-phenylenediamine)	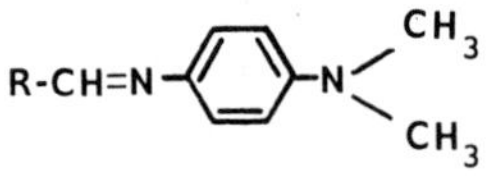	69
Alehydes and Ketones	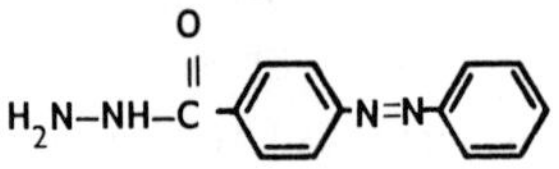 (Azobenzene–4–carboxylic Acid Hydrazide)	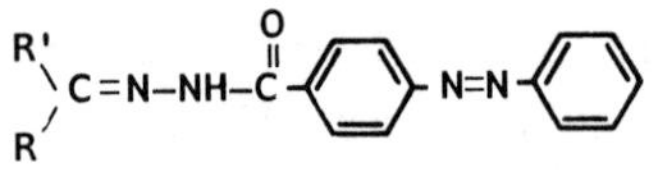	69
Aliphatic Aldehydes	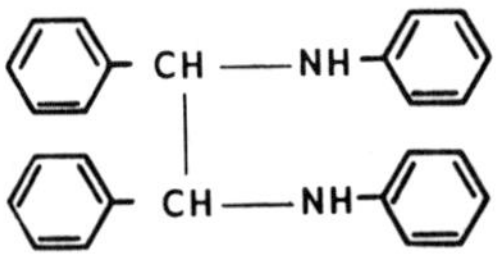 (dl–Dianilino–1,2–diphenylethane)	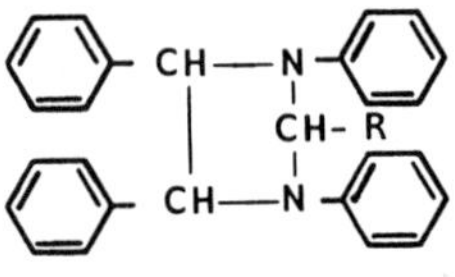	69

TABLE 5. (Continued)

Compound or class	Reagent	Derivative	Reference
B. R–C(=O)–R' (Cont.)			
Formaldehyde	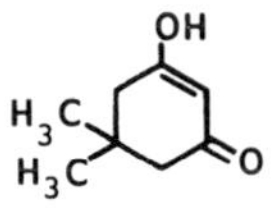 (Dimedone)	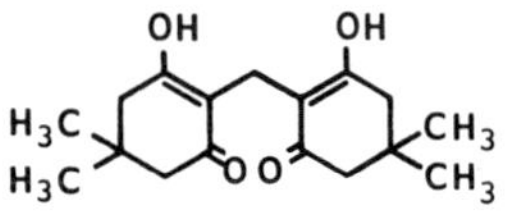	69
Aldehydes and Ketones	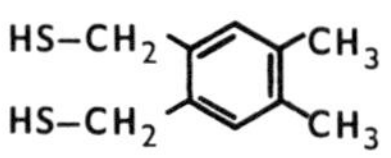 (1,2-Dimethyl-4,5-di(mercapto-methyl) benzene)	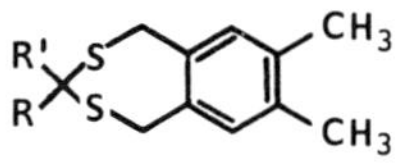	69
Aldehydes and Ketones	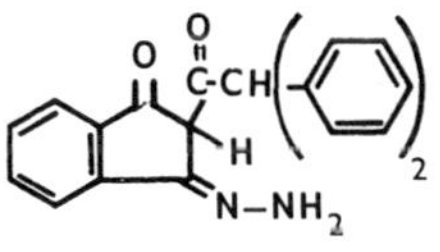 (2-Diphenylacetyl-1,3-indanedione-1-hydrazone)	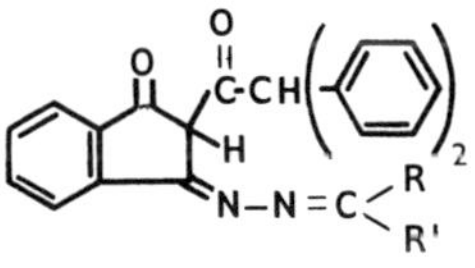	69,71
C. R – OH			
Alcohols and Phenols	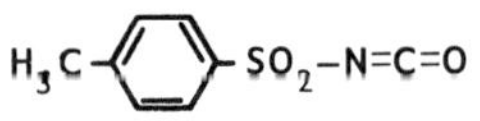 (p-Toluenesulfonylisocyanate)	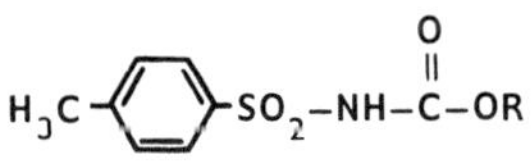	73
Alcohols and Phenols	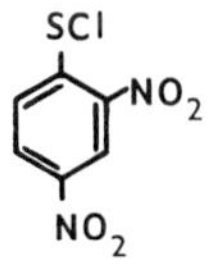 (2,4 Dinitrobenzenesulfonyl Chloride)	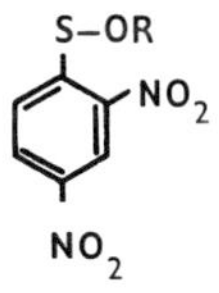	69
Alcohols and Phenols	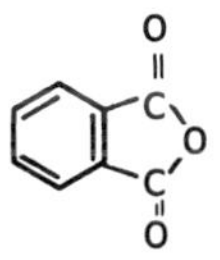 (Phthalic Anhydride)	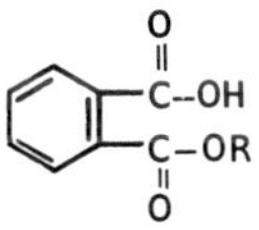	74,75

TABLE 5.   (Continued)

Compound or class	Reagent	Derivative	Reference
C. R—OH (Cont.)			
Alcohols and and Phenols	(Phenylisocyanate) ϕ—N=C=O	ϕ—NH—C(=O)—OR	76
Alcohols and Phenols	(2,4—Dinitrofluorobenzene)	O_2N—⟨⟩(—NO_2)—OR	77
Alcohols and Thiols	(4'-Nitroazobenzene-4-carboxylic Acid Chloride) O_2N—⟨⟩—N=N—⟨⟩—C(=O)—Cl	O_2N—⟨⟩—N=N—⟨⟩—C(=O)—OR	69
Allylic Alcohols	(p-Phenylsulfonylbenzoyl Chloride) ϕ—SO_2—⟨⟩—C(=O)—Cl	ϕ—SO_2—⟨⟩—C(=O)—OR	69
Diols	(Phenylboronic Acid) HO—B(—OH)—ϕ	—CH—O / $(CH_2)_n$ / —CH—O \ B—ϕ	69
D. R—NH₂			
Primary and Secondary Amines	(2,4-Dinitrobenzenesulfenyl Chloride) S—Cl, NO_2, NO_2	S—N⟨R,R'⟩, NO_2, NO_2	69
Proteins	(2-Naphthylisothiocyanate) naphthyl—N=C=S	naphthyl—N—C(=S)—N⟨R,R'⟩	69

TABLE 5. (Continued)

Compound or class	Reagent	Derivative	Reference
D. R—NH₂ (Cont.)			
Primary and Secondary Amines	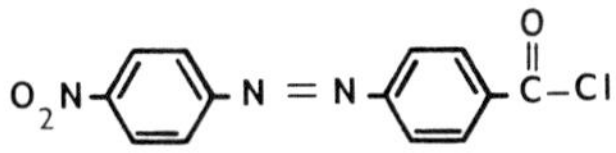 (4'-Nitroazobenzene-4-carboxylic Acid Chloride)	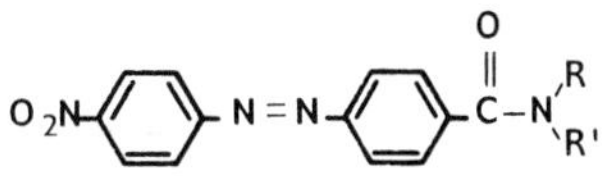	69
Primary Amines	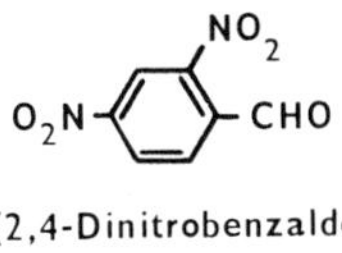 (2,4-Dinitrobenzaldehyde)	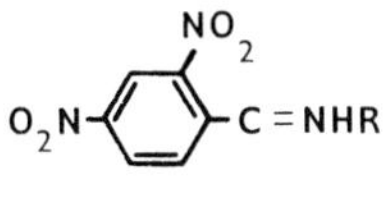	69
Peptides	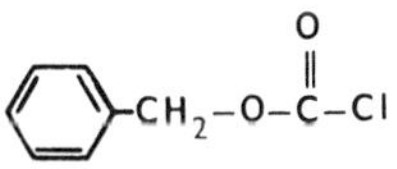 (Carbobenzoxy Chloride)	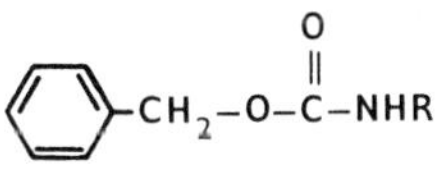	69
Amines	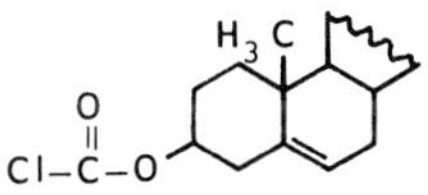 (Cholesteryl Chloroformate)	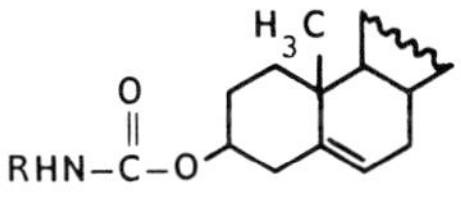	69
Amines	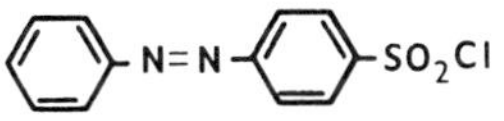 (p-Phenylazobenzenesulfonyl Chloride	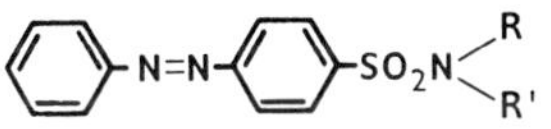	69
Amides, Imides, Urea, and Barbiturates	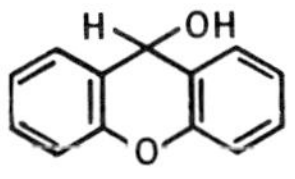 (Xanthydrol)	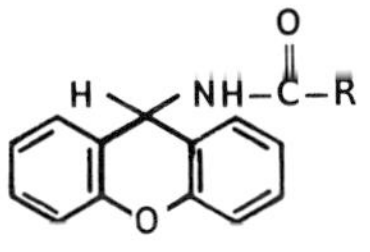	78

REFERENCES

1. N. D. Cheronis, J. B. Entrikin, and E. M. Hodnett, *Semimicro Quantitative Organic Analysis*, Interscience, New York, 1965.

2. N. D. Cheronis and T. S. Ma, *Organic Functional Group Analysis by Micro and Semimicro Methods*, Interscience, New York, 1964.

3. F. Wild, *Characterization of Organic Compounds*, Cambridge University Press, London, 1960.

4. S. Siggia, *Quantitative Organic Analysis*, Wiley, New York, 1963.

5. R. L. Shriner, R. C. Fuson, D. Y. Curton, *The Systematic Identification of Organic Compounds*, Wiley, New York, 1956.

6. J. F. W. McOmie, *Protective Gropus in Organic Chemistry*, Plenum, New York, 1973.

7. W. Morozowich and S. Douglas, *Prostaglandins*, *10*, 19(1975).

8. H. Durst, M. Milano, D. J. Kikta, S. A. Connely, and E. Gruska, *Anal. Chem.*, *47*, 1799(1975).

9. M. J. Cooper and M. W. Anders, *Anal. Chem.*, *46*, 1849(1974).

10. Regis Lab Notes, No. 17 (1974).

11. C. N. R. Rao, *Ultraviolet and Visible Spectroscopy, Chemical Applications*, Butterworths, London, 1961.

12. W. Morozowich, *Abst. 119th Meeting A. Ph. A. Academy of Pharmaceutical Sciences*, April 1972.

13. I. R. Politzer, G. W. Griffin, B. J. Dowty, and J. L. Laseter, *Anal. Lett.*, *6*, 539(1973).

14. R. A. Henry, J. A. Schmidt, and J. F. Dieckman, *J. Chromatogr. Sci.*, *9*, 513(1971).

15. F. A. Fitzpatrick and S. Siggia, *Anal. Chem.*, *44*, 2211(1972).

16. R. V. Winchester, *J. Chromatogr.*, *78*, 429(1973).

17. R. S. Deelder and P. J. H. Hendricks, *J. Chromatogr.*, *83*, 343(1973).

18. M. A. Carey and H. E. Persinger, *J. Chromatogr. Sci.*, *10*, 537(1972).

19. L. J. Pappa and L. P. Turner, *J. Chromatogr. Sci.*, *10*, 747(1972).

20. R. H. King, L. T. Grady, and J. T. Reamer, *J. Pharm. Sci.*, *63*, 1591(1974).

21. L. L. Needham and M. N. Kockhar, *J. Chromatogr.*, *111*, 422(1975).

22. J. S. Fritz and G. Hammond, *Quantitative Organic Analysis*, Wiley, New York, 1957, p. 9.

23. R. W. Frei and J. F. Lawrence, *J. Chromatogr.*, *83*, 321(1973).

24. F. A. Fitzpatrick and S. Siggia, *Anal. Chem.*, *45*, 2310(1973).

25. L. D. Bighley, D. E. Wurster, and D. Cruden-Loeb, *J. Chromatogr.*, *110*, 375(1975).

26. A. L. Hayden, O. R. Sammul, G. B. Selzer, and J. Carol, *J. Ass. Offic. Agr. Chem.*, *45*, 797(1962).

27. P. J. Pocaro and P. Shubiak, *Anal. Chem.*, *44*, 1865(1972).

28. J. E. Evans and R. H. McCluer, *Biochim. Biophys. Acta*, *370*, 565(1972).

29. J. Churaček and P. Jandera, *J. Chromatogr.*, *53*, 69(1970).

30. J. Churaček, M. Husková, H. Pechova, and J. Řiha, *J. Chromatogr.*, *41*, 511(1970).

31. R. M. Cassidy, D. S. LeGay, and R. W. Frei, *J. Chromatogr. Sci.*, *12*, 85(1974).

32. L. L. Needham and M. M. Kochar, *J. Chromatogr.*, *14*, 220(1975).

33. F. B. Jungalowala, R. J. Turel, J. E. Evans, and R. H. McCluer, *Biochem. J.*, *145*, 517(1975).

34. T. Sugiura, T. Hayashi, S. Kawai, and T. Ohno, *J. Chromatogr.*, *110*, 385(1975).

35. K. Samejima, *J. Chromatogr.*, *96*, 250(1974).

36. M. M. Abdel-Monem and K. Ohno, *J. Chromatogr.*, *107*, 416(1975).

37. G. B. Cox, *J. Chromatogr.*, *83*, 471(1973).

38. J. F. Lawrence and R. W. Frei, *Anal. Chem.*, *44*, 2046(1972).

39. D. H. Spackman, W. H. Stein, and S. Moore, *Anal. Chem.*, *30*, 1190(1958).

40. C. W. Gehrke, K. C. Kuo, R. W. Zumwalt, and T. P. Waalkes, *J. Chromatogr.*, *89*, 231(1974).

41. V. A. Spivak, V. A. Fedoseev, V. M. Orlov, and J. A. M. Varshavsky, *Anal. Biochem.*, *44*, 12(1971).

42. N. Lustenberger, H. Lanbe, and K. Hempel, Angew. *Chem. Int. Ed.*, *11*, 227(1972).

43. S. Udenfriend, S. Stein, P. Böhlen, W. Dairman, W. Leimgruber, and M. Weigele, *Science*, *178*, 871(1972).

44. P. Böhlen, S. Stein, W. Dairman, and S. Udenfriend, *Arch. Biochem. Biophys.*, *155*, 213(1973).

45. W. Voelter and K. Zech, *J. Chromatogr.*, *112*, 643(1975).

46. M. Roth, *Anal. Chem.*, *43*, 880(1971).

47. M. Roth and A. Hampoi, *J. Chromatogr.*, *83*, 353(1973).

48. J. R. Benson and P. E. Hare, *Proc. Nat. Acad. Sci.*, *72*, 619(1975).

49. B. Halpern and J. W. Estley, *Biochem. Biophys. Res. Commun.*, *19*, 361(1965).

50. G. E. Pollock, V. I. Oyama, and R. D. Johnson, *J. Gas Chromatogr.*, *3*, 174(1965).

51. S. B. Martin, M. Rowland, and N. Castagnoli, *J. Pharm. Sci.*, *62*, 831(1973).

52. H. Iwase and A. Murai, *Chem. Pharm. Bull.*, *22*, 8(1974).

53. T. Nambara, J. Goto, K. Taguchi, and T. Iwata, *J. Chromatogr.*, *100*, 180(1974).

54. C. J. W. Brooks, M. T. Gilbert, and J. D. Gilbert, *Anal. Chem.*, *45*, 896(1973).

55. G. Helmchen and W. Strubert, *Chromatographia*, *7*, 713(1974).

56. M. Koreeda, G. Weiss, and K. NaKanishi, *J. Am. Chem. Soc.*, *95*, 125(1973).

57. J. A. Dale, D. L. Dull, and H. S. Mosher, *J. Org. Chem.*, *34*, 2543(1969).

58. K. Nakanishi, D. A. Schooley, M. Koreeda, and J. Dillon, *Chem. Commun.*, 1235(1971).

59. G. J. Fallick, in *Advances in Chromatography* (J. C. Giddings, E. Gruska, R. A. Keller, and J. Cazes, eds.), Vol. 12, Marcel Dekker, New York, 1975, p. 93.

60. L. R. Snyder, *Principles of Adsorption Chromatography*, Marcel Dekker, New York, 1968, pp. 295-333.

61. A. A. Frost and R. G. Pearson, *Kinetics and Mechanism*, 2nd Ed., John Wiley and Sons, New York, 1961, p. 17.

62. S. P. Eriksen and H. Stelmach, *J. Pharm. Sci.*, *54*, 1029(1965).

63. A. J. Parker, *Chem. Rev.*, *69*, 1(1969).

64. N. O. V. Sonntag, *Chem. Rev.*, *52*, 237(1953).

65. H. H. Jaffé, *Chem. Rev.*, *53*, 191(1953).

66. M. J. Cho, unpublished results (1975).

67. Y. Ogata and A. Kawasaki, in *The Chemistry of the Carbonyl Group* (J. Zabicky, ed.), Vol. 2, Interscience, New York, 1970, pp. 55-58 (and references therein).

68. E. F. Pratt and M. J. Kammlet, *J. Org. Chem.*, *26*, 4029(1961).

69. L. F. Fieser and M. Fieser, *Reagents for Organic Synthesis*, John Wiley and Sons, New York, 1967.

70. O. H. Weeler, *Chem. Rev.*, *62*, 205(1962).

71. J. G. Hanna, *The Chemistry of the Carbonyl Group* (S. Patai, ed.), Interscience, New York, 1966, pp. 375-420.

72. D. J. Pietrzyk and E. P. Chan, *Anal. Chem.*, *42*, 37(1970).

73. J. W. McFarland and J. B. Howard, *J. Org. Chem.*, *30*, 957(1965); J. W. McFarland, D. E. Lenz, and D. J. Grosse, *J. Org. Chem.*, *31*, 3798(1966).

74. V. C. Mehlenbacher, *Organic Analysis* (J. Mitchell, Jr., I. M. Kolthoff, E. S. Proskauer, and A. Weissberger, eds.), Interscience, New York, 1953, pp. 1-65.

75. K. G. Stone, *Determination of Organic Compounds*, McGraw-Hill, New York, 1956, p. 42.

76. L. F. Fieser and M. Fieser, *Reagents for Organic Synthesis*, Vol. 2, John Wiley and Sons, New York, 1969.

77. J. D. Reinheimer, J. P. Douglass, H. Leister, and M. B. Voelkel, *J. Org. Chem.*, *22*, 1743(1975).

78. D. J. Pasto and C. R. Johnson, *Organic Structure Determination*, Prentice-Hall, Englewood Cliffs, N. J., 1969, p. 435.

79. J. F. Lawrence and R. W. Frei, *Chemical Derivatization in Liquid Chromatography*, Elsevier, Amsterdam, 1976.

80. P. F. Dixon, M. S. Stoll, and C. K. Lim, *Ann. Clin. Biochem.*, *13*, 409(1976).

81. J. Drozd, *J. Chromatogr.*, *113*, 303(1975).

82. S. Ahuja, *J. Pharm. Sci.*, *65*, 163(1976).

83. C. W. Knapp and M. A. Krueger, *Anal. Letters*, *8*, 603(1975).

84. W. Dunges, *Anal. Chem.*, *49*, 442(1977).

85. R. W. Roos, *J. Chromatogr. Sci.*, *14*, 505(1976).

86. J. Goto, M. Hasegawa, S. Nakamara, K. Shimada, T. Nambara, *Chem. Pharm. Bull.*, *25*, 847(1977).

87. D. Valentine, K. K. Chan, C. G. Scott, K. K. Johnson, K. Toth, and G. Saucy, *J. Org. Chem.*, *41*, 62(1976).

88. F. A. Fitzpatrick, M. A. Wynalda, and D. G. Kaiser, *Anal. Chem.*, *49*, 1032(1977).

89. Regis Lab Notes, No. 18, 1975.

Chapter 6

THE MASS SPECTROMETER AS A DETECTOR FOR
GAS-LIQUID CHROMATOGRAPHY

Richard M. Milberg

J. Carter Cook, Jr.

School of Chemical Sciences
University of Illinois
Urbana, Illinois

I. INTRODUCTION

In the 20 years since a gas chromatograph was first coupled to a mass spectrometer
[1], the technology of the combined gas chromatography-mass spectrometer (GC-MS) has

advanced so that today the method is used extensively, as is reflected in the growth of the body of GC-MS literature. A recent book by McFadden [2] is devoted to the techniques, theory and practice of combined GC-MS and numerous books on mass spectrometric instrumentation [3-7], organic spectra interpretation [8-13] and current developments in mass spectrometry [14-16] are available. Several review articles [17-20] as well as a section in the biennial "Fundamental Reviews Issue" of *Analytical Chemistry* [21] deal with mass spectrometry. Two journals of mass spectrometry [22, 23] and one covering biomedical applications of mass spectrometry [24] are published. The annual conference on mass spectrometry and allied topics sponsored by the American Society for Mass Spectrometry and ASTM Committee E-14 on Mass Spectrometry provide an opportunity for the presentation of papers dealing with the current developments in mass spectrometry. Reference 25 is devoted entirely to the biochemical and medical applications of mass spectrometry.

This review, because of its length limitations, can only touch briefly upon the theory, instrumentation, techniques, and methodology of GC-MS. We will try to approach the subject on a practical, operational point of view.

II. ADVANTAGES OF A MASS SPECTROMETER AS A DETECTOR FOR GAS CHROMATOGRAPHY

As a detector for gas chromatography, the mass spectrometer has several advantages over the conventional flame ionization detector (FID). The combined GC-MS allows the separation and identification of a complex mixture of 100 or more organic compounds, which would be almost impossible by any other method. The GC-MS can provide confirmation of structure or identify an unknown compound. The GC-MS may also be operated in the selected ion detection (SID) mode (see the following discussion), monitoring one or more characteristic ions of a compound rather than scanning an entire mass spectrum, thereby increasing the detection limit [26, 27]. The FID is capable of *detecting* as little as 5×10^{-10} g of material, and modern electron capture detectors (ECD) operated in the pulsed mode can *detect* as little as 2×10^{-13} g of some compounds. The GC-MS is capable of identifying as little as 10^{-10} g of material, and when operated in the SID mode can *detect* and provide *positive identification* of as little 10^{-12} g of material.

III. DIFFERENT TYPES OF MASS SPECTROMETERS

We will describe here the two mass spectrometers most commonly coupled to gas chromatographs in GC-MS systems, the magnetic-scanning sector mass spectrometer and the quadrupole mass spectrometer.

A. Magnetic Sector Mass Spectrometer

The magnetic-scanning sector mass spectrometer [28] is the instrument used most frequently not only in GC-MS but in mass spectrometry in general. The simple single-

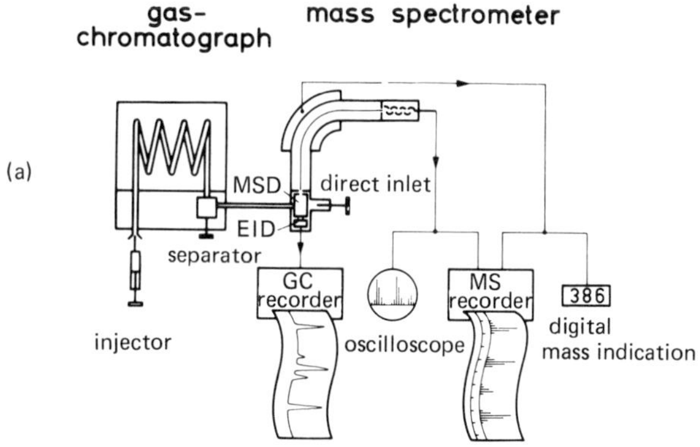

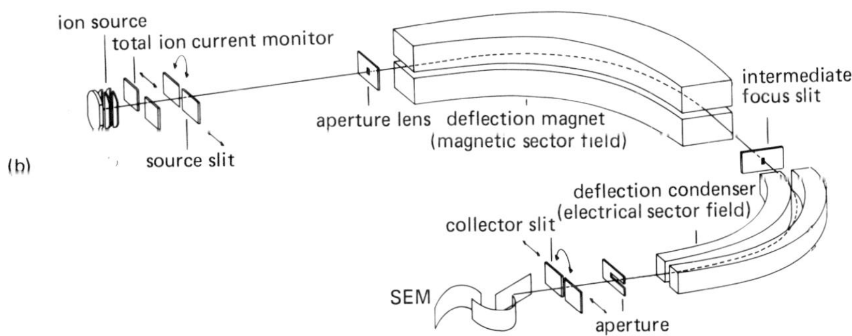

FIG. 1. (a) Varian-MAT 111 GC-MS (MSD and EID are Varian MS components), (b) Varian-MAT 311A, reverse Neir-Johnson geometry double-focusing mass spectrometer.

focusing type, which offers low to medium resolution (300-5000) in a fairly compact form, is the most popular magnetic scanning MS. Resolution is selected by adjustment of source and collector slits and ions are accelerated through a magnetic field by high voltage (0.8-8 kV depending on the instrument). Figure 1(a) is a diagram of a Varian MAT 111 single focusing magnetic sector GC-MS which incorporates a Brunee slit separator and two separate ion sources (see the following discussion).

Normally a mass scan is obtained by varying the magnetic field, resulting in uniform transmission of all masses. If the accelerating potential is scanned, higher masses have lower energy and hence lower efficiency of transmission. Modern electronic circuitry is available [29] which provides extremely reproducible magnetic scans at high speeds for long periods of time to eliminate problems previously associated with continuous cyclic scanning and nonlinear scans.

Since magnetic-scanning instruments have long flight paths, operating pressures must be low to prevent excessive scattering of the ions by residual gas molecules, but this is no problem with modern vacuum techniques. Instruments should be equipped

with differential pumping, and the connection between the ion source and the analyzer
should be an opening only large enough to allow the ion beam to pass through. The
source and analyzer have their own pumping systems so the analyzer can operate at low
pressure and the source can receive a higher total flow from the chromatograph system.

The resolution of a magnetic scanning instrument is determined by the radius of
the ion path in the magnet and the source and detector slit widths [30]. The mass
dispersion or distance between two adjacent m/e value ions is proportional to the
magnet-path radius and to the beam width (determined primarily by the slit width).
Because ions are not perfectly focused by the magnetic field and not all ions are
formed with identical energy and initial direction, the resolution of a single-
focusing instrument is limited by these so-called aberrations. In order to obtain
higher resolution, a radial electrostatic field is added, either before the magnetic
field (the common Nier-Johnson [31] or Mattauch-Herzog geometry [32]) or after the
magnetic field (the reverse Nier-Johnson geometry). Figure 1(b) is a diagram of a
Varian-MAT 311A MS, a double-focusing reverse Nier-Johnson geometry instrument.

With properly designed geometry, a mass spectrometer containing an electric and
a magnetic sector can correct for both direction and energy aberrations to give the
so-called double-focusing effect. The Mattauch-Herzog geometry gives double focusing
in a plane and can be used with photographic detection on sensitized glass plates.
Smaller, medium-resolution double-focusing instruments such as the Varian MAT-311A
and the double-beam AE1 MS-30 offer resolution up to 25,000 as well as very good
performances at low resolution. Higher resolution is needed to separate doublets
of the same normal integer mass but of different atomic masses. This is achieved by
comparison of the mass of the unknown ion to that of a standard such as mixtures of
perfluoroalkanes and serves to separate mass doublets and the reference compound peak
from the sample peaks. High-resolution research instruments such as the Varian-MAT
731 and 711 and the AEI MS-50 are capable of resolution in excess of 100,000. When
a particular GC-MS problem requires resolution in excess of 20,000, photographic recording on photoplates in an instrument with Mattauch-Herzog geometry is almost always required because few ions are collected at any one mass during the scan of an
entire spectrum. High-resolution GC-MS is currently a very specialized technique,
but with limitations set upon resolution requirements (8-10,000) and increased sample size, it is possible to obtain reasonable mass measurement accuracy utilizing
many of the modern-day medium-resolution double-focusing mass spectrometers commercially available.

B. Quadrupole Mass Spectrometer

The quadrupole mass spectrometer [33] is often chosen for its comparatively low cost
and simplicity of operation and maintenance. It is a mass "filter" consisting of
four parallel metal rods held rigidly at the corners of a square. Opposed rods are

connected together to dc-voltage and radio-frequency (RF) supplies. Ions extracted
from an ion source are directed along the longitudinal axis of the quadrupole field
where they undergo oscillation due to the dc and RF fields. Normally the RF is held
constant and the dc voltage and RF voltage are scanned with a fixed dc-voltage-to-RF-
voltage ratio. At a given value of this ratio only ions of a particular mass to
charge ratio will have stable oscillation through the field and will reach the
detector.

The quadrupole MS can usually be operated at higher pressures than the sector
MS because it has a shorter flight path. Since the ion energy required for injection
into the quadrupole is of the order of 5-15 V, the quadrupole does not have high voltage (1 kV or greater) on its source. This means there are fewer inherent problems
in source design and fewer discharge problems in a high-pressure chemical ionization
source (see the following discussion). The quadrupole MS is capable of scanning as
fast as 1 msec per mass and is ideally suited to selective ion detection by computer
automation. The mass scale is linear and all masses are equally spaced.

The quadrupole MS does, however, have disadvantages. By the nature of the physical laws governing the instrument, all quadrupole mass spectrometers have a mass dis-
crimination above m/e 200, a serious drawback for those who must work in this region.
Resolution adjustment is accomplished not by slits but by adjusting the parameters of
the dc and RF fields, and the ion transmission is lower at high mass than at low mass
as compared to a magnetic-scanning sector instrument. Most GC-MS quadrupole systems
operate at unit resolution across their mass range with reasonable sensitivity, although some quadrupole systems are offered with resolutions exceeding 5000 [34].
Though the quadrupole MS is capable of scanning as fast as 1 msec per mass, it is
rarely used at that rapid a rate for high-sensitivity work due to the small number
of ions collected at each mass.

IV. IONIZATION SOURCES

The ionization source is one of the most critical parts of a GC-MS system (along with
the coupling system between the GC and the MS and the pumping system) and several
types are commercially available. Diffusion pumps, provided they are not exposed to
large amounts of air, and mechanical vacuum pumps can work for months or even years
with very little maintenance. Quadrupole rods and electric sector field plates perform well for long periods of time provided they do not become coated with vacuum
pump fluid. However, sample analysis done with an excessive amount of material can
contaminate an ion source quickly. Sample quantity should be no larger than necessary
to give good mass spectra. Modern GC-MS instruments in *routine operation* should give
complete spectra of organic molecules with molecular weights of 500-600 with as little
as 10 ng of material injected onto the GC column. Much of the sensitivity is dependent on good source design by instrument manufacturers.

A. Electron Impact Source

The electron impact (EI) source, which is most commonly used in GC-MS, is relatively
simple in construction, reliable in operation, and produces a beam of ions with
narrow kinetic energy spread. Perhaps more importantly, the EI source gives reproducible and characteristic mass spectra of many classes of chemical compounds, not
only from day to day but also from different makes and designs. A tungsten or rhen-
ium cathode is heated by an electric current, causing it to emit electrons which are
then accelerated towards a collector electrode. The energy of the electrons is controlled by the potential between the cathode and the collector and usually varies
between 5 and 100 electron volts (eV). Although the ionization potential (IP) of
most organic molecules is 10-20 eV, EI sources, by convention, are most often operated at 70 eV. This provides sufficient energy to ionize the molecules and cause
them to undergo their characteristic fragmentation. The positive ions produced are
extracted from the source by a positive potential and accelerated towards the analyzer. Normally the source or analyzer has a monitor electrode which collects a fraction of the total ion beam to monitor the progress of the chromatographic run. Because many helium ions are produced when helium is used as the carrier gas, it is
desirable in a GC-MS to have a means of "bucking out" the ion current due to helium,
thus enabling the source to operate at 70 eV full time. Some systems operate at
20 eV (4 eV below the IP of helium, but sufficient to ionize organic molecules) during monitoring and switch to 70 eV during a scan. This partially eliminates the
high helium background current. Another method of eliminating the helium background
is to have a second source which operates at 20 eV as part of the main ion-producing
source. This is used on the Varian-MAT 311A, 111, and 112 GC-MS systems. The device
labeled EID in Fig. 1(a) is the second source which receives a portion of the GC effluent and operates independently of the MSD main source.

It is generally desirable to observe the molecular ion of a compound (which gives
the molecular weight) as well as its characteristic fragmentation pattern, but many
classes of compounds do not give molecular ions by EI because 70 eV is sufficient to
fragment completely any molecular ions produced. In some cases the IP can be lowered,
causing the relative abundance of the molecular ion to increase with respect to that
of the fragment ions.

B. Chemical Ionization Source

The need for molecular ion formation led to the development of the second major ionization method used in GC-MS, the chemical ionization (CI) source [35, 36]. The CI
source operates with the transfer of less energy and hence produces less fragmentation. It utilizes the principle of ion-molecule reactions between molecules of the
sample gas (10^{-5} torr or less) and a high-pressure (0.2-2 torr) plasma of reagent gas

(generally methane or isobutane). The reagent gas can be substituted as the chromatographic carrier gas in GC-MS and can be introduced with the sample directly into the source with little effect on chromatographic resolution, especially if it is methane or isobutane. Because of the high pressure of reagent gas, a much higher electron ionization potential (200-500 eV) is needed and the reagent gas undergoes many direct ionizations and ion molecule reactions to produce predominantly CH_3^+ and $C_2H_5^+$ in methane [35]. The reagent-gas ions react with the sample molecules as either Bronsted acids or hydride abstractors, depending on whether the sample molecule is a good proton donor or acceptor. For a molecule MH, species such as MH_2^+ and M^+ may be produced, the so-called *quasi-molecular* ions. It should be noted that these quasi-molecular ions are even-electron species which are more stable than the odd-electron molecular ions produced in an EI process. The CI source combined with GC-MS has proven to be useful in the analysis of therapeutic agents and their metabolites in body fluids [37-39].

Diffusion pumps, with pumping speeds of 600-1200 liters per second, or efficient turbomolecular pumps are required to handle the large quantities of reagent gas (1-10 ml/min) in a CI system. The CI source has an overall efficiency approaching that of an EI source because of the increased abundance of the quasi-molecular ion and has a high ionization efficiency because of the large number of collisions between the excess reagent gas ions and the sample molecules. It requires careful tuning and optimization of the reagent gas pressure to maximize the sample ion production and extraction into the analyzer, but once set up it is fairly stable in operation. If a combination EI-CI source is used, with some form of mechanical switching to "tighten up" the source configuration, the system must be reoptimized with each change from EI to CI unless focusing controls are provided for each operation independent of each other.

Figure 2 demonstrates the usefulness of CI-MS. Figure 2(a) is the 70 eV EI mass spectrum of Ortal (hexathal) in which the molecular ion is very small. Figure 2(b) is the methane CI mass spectrum of Ortal, in which the quasi-molecular ion $(M + H)^+$ is now the base peak and there is much less fragmentation. Figure 2(c) is the isobutane CI mass spectrum of Ortal. Again the $(M + H)^+$ ion is the base peak and there is almost no fragmentation. If some structural information is desired from CI spectra, then methane is the regent gas of choice. Isobutane is generally used for high sensitivity work and for quantitative measurements in the molecular ion region as it produces less fragmentation. Other reagent gases have been used for selective ionization and for detection of specific functional groups [40].

C. Field Ionization Source

A third ionization method in use is the field ionization (FI) source [41], which produces ionization by quantum mechanical tunneling with reduced fragmentation. A sharp

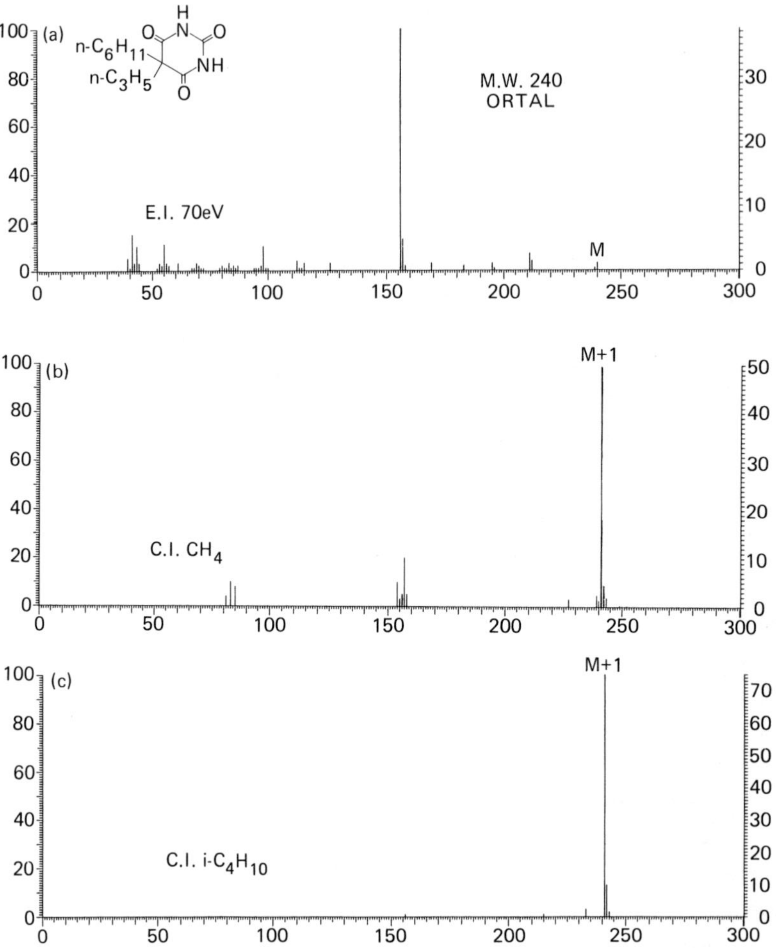

FIG. 2.   (a) Electron impact mass spectrum of Ortal (hexathal), 70 eV; (b) methane chemical ionization mass spectrum of Ortal; (c) isobutane chemical ionization mass spectrum of Ortal.

piece of razor blade or an activated fine wire (10 μ) held at a high potential (8-14 kV) produces positive ions which are then extracted from the chamber and accelerated towards the analyzer.   The FI source suffers from sensitivity problems which make it 5-10 times less sensitive than a comparable EI source, though for compounds which give no molecular ion by EI the FI mode represents an increase in sensitivity for identifying the molecular ion.   In our laboratory GC-FI-MS is a routine analytical technique providing molecular ion determination of GC effluents as an adjunct to EI.

With all three types of sources the compound to be studied must be sufficiently volatile to give a pressure of 10^{-7} to 10^{-5} torr in the source.   If the molecule is thermally labile or very polar, than no EI, CI, or FI source will give a true characterization mass spectrum or, for that matter, any mass spectrum. An additional use of the FI source is in the field desorption (FD) mode [41] in which the FI emitter surface is dipped into or has deposited on it [42] the sample solution. Excellent sensitivity is possible and even when the other ionization sources fail, good molecular ions often result by FD. The method is not amenable to direct GC-MS since the sample is deposited directly on the emitter but is an excellent method for the study and quantitation of mixtures as primarily only molecular ions result.

It is good practice, if possible, to run an initial off-line GC with a flame ionization detector (FID) for each sample. Unless other information is at hand, the first attempt at GC-MS should usually be made with EI because of its simplicity of operation and high sensitivity. If EI results indicate the absence of molecular ions or excessive fragmentation, then thought should be given to GC-CI-MS or GC-FI-MS. Most routine operations will be done in the EI mode but the capability for occasional CI or FI operation is extremely valuable.

D. Atmospheric-Pressure Ionization Source

An interesting new source development is the atmospheric-pressure ionization (API) source developed by Horning's group [43, 44]. Ion-molecule reactions occur in a nitrogen plasma at 1 atm pressure in a source external to the mass spectrometer vacuum system. Ions are produced by electrons from a nickel-63 foil or by a high-voltage corona discharge. A pin hole allows molecules to be drawn into the vacuum region containing a quadrupole mass spectrometer. This is similar in theory to a CI source, but the reagent gas is nitrogen and samples are introduced by direct injection into the source in a solvent such as chloroform or benzene, through a liquid chromatograph (see Chap. 7), or through a GC capillary column coupled directly to the source [45]. The source offers ultrahigh sensitivity [30 fg (10^{-15} g) or less detected by single ion detection] because of the high efficiency of the ionization procedure. The method is currently undergoing major developmental work [46] and may prove *the* method of the future for trace analysis mass spectrometry. The source is ideally suited for direct coupling to a GC capillary column since it operates at nitrogen flow rates of 1-6 ml/min and make-up gas flow is easily provided. No splitter or restrictor is required since the source operates at one atmosphere pressure.

V. COUPLING TO MASS SPECTROMETERS

The coupling of the GC to the MS imposes many problems. The outlet of the GC column is at one atmosphere pressure and the source of the MS operates at 10^{-5} torr or less. The quantity of carrier gas that the vacuum system of the MS can handle is generally much lower than that at which packed columns operate.

There are three methods of coupling the chromatograph to the MS: (1) direct coupling,
(2) split connection, and (3) use of a molecular separator (enrichment device).

A. Direct Coupling

In the direct coupling method the total effluent of the GC column is directed into the
MS source giving 100% sample ulitization. This is only possible for a highly efficient, differentially pumped MS directly coupled to a capillary column at 1-6 ml/min.

B. Split Connection

In the split connection the effluent from the GC column is split into two portions,
one portion going to the atmosphere or to an auxiliary GC detector such as an FID.
The other portion flows through a capillary restriction to the mass spectrometer.
If the MS is not differentially pumped, then only a few percent of the GC column
flow can be split to the MS. For a differentially pumped system much more effluent
can be handled. The split ratio allowable is a function of the capacity of the pumping system and the conductance of the associated hardware.

C. Molecular Separator

Sample utilization in the split connection coupling method is rather inefficient,
which is acceptable if sample size is not limited, but where a wide dynamic range is
required, as in trace analyses, a better use of sample is desirable. A molecular
separator is an enrichment device which increases the ratio of sample to carrier gas
beyond the GC column but before the MS. All enrichment devices are described by two
performance parameters [47], the enrichment factor N and the yield or efficiency Y.
N is the ratio of the concentration of sample in the carrier gas entering the MS from
the separator to the concentration of sample in the carrier gas leaving the GC, and
may vary from 1 to 1000 depending on the type of separator and vacuum system, the
carrier gas flow, and the molecular weight of the sample. Y is the ratio of the amount
of sample entering the MS from the separator to the amount entering the separator from
the GC, usually given in percent.

Molecular separators remove more low molecular-weight carrier-gas molecules than
high molecular-weight sample molecules by one of three main principles: (1) preferential diffusion carrier gas in an expanding jet stream, (2) effusion of carrier gas
through fine pores or slits, and (3) selective solubility of sample or carrier gas
through a semipermeable membrane.

1. Diffusion Separator

The diffusion separator, commonly called the jet separator (48,49), Fig. 3(a), passes
the GC effluent through a restriction under pressure, to a nozzle with a small diameter

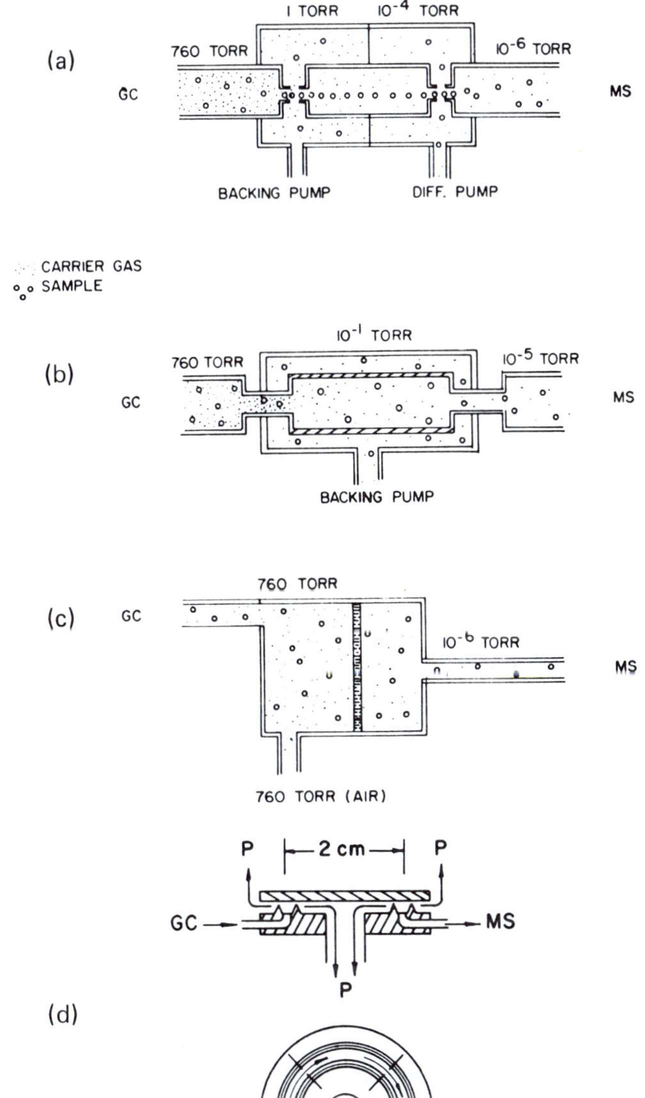

FIG. 3. (a) Ryhage two-stage jet separator, (b) Watson-Biemann separator, (c) Llewellyn-Littlejohn membrane separator, and (d) Brunee slit separator.

from which it flows as an expanding jet stream into region which is evacuated by a vacuum pump. The gas expands under viscous flow conditions in the central core of the beam into the region between the orifice and a second orifice. The carrier gas

(He), with a much higher diffusion rate (inversely proportional to its molecular
weight), is pumped away while the center of the jet is enriched in the much heavier
sample molecules, which then pass into the MS source.

Jet separators are available from several sources as well as GC-MS instrument
manufacturers in single-stage and two-stage (with a second jet in series) models of
all-glass or metal construction, and can be operated to over 300° to prevent sample
condensation. Their major deficiency is the ease of clogging of the orifices by
dirt, column packing, and tiny pieces of glass. They have minimal surface area and
both glass and metal separators can be silylated very easily.

Single- and dual-stage separators operate over a wide range of flow rates, up
to 30-40 cm^3 atm/min. Single-stage separators can give N of 12-20 and Y to 80%,
while dual-stage separators can give N to 100 and Y to 60% [2, 20]. Jet separators
are difficult to construct in the laboratory. All-glass models require considerable
design and glassblowing skill and all-metal models require precise design and machining. An all-glass single-stage jet separator with adjustable distance between the
two jets is commercially available and allows optimization for different carrier-gas
flows and sample molecular weights [50]. The single-stage glass-jet separator provided with the Finnigan GC-MS (51) has been proven in several hundred instruments
and, except for occasional clogging, gives excellent performance for a wide variety
of compounds. If the separator has straight leads into it, clogs can be dislodged
by the careful use of drawn tungsten wires and liquids forced through the pump-out
lead into the jets.

2. Effusive Separators

Effusive separators use the principle of differential effusion of molecules of different molecular weights during molecular flow through a porous material or through
slits. There are several designs which can be fabricated easily and ruggedly from
glass, ceramic or metal and are not prone to clogging. They do have large surface
areas and work best when silylated, which can be done in place with an attached GC
column. The Watson-Biemann (W-B) separator [52, 53] [Fig. 3(b)], is constructed of
a fritted glass tube housed in a chamber evacuated by a mechanical vacuum pump, with
a pore size of the order of 1 μm. The GC effluent enters the separator through a
capillary restriction which limits the pressure to the molecular flow region. The
lighter helium carrier gas effuses preferentially through the pores, so the remaining
carrier gas becomes enriched with the higher molecular-weight sample molecules.

Single-stage separators operate at flow rates of up to 20 ml/min, dual-stage
separators up to 50 ml/min. Flow is optimized at a particular fixed rate since the
pressure inside the separator is determined by the inlet restriction and lower flow
rates cause decreased overall efficiency. Effusive separators achieve N to 100 with
Y to 60% [20, 54].

Several modifications of the porous tube separator exist and porous ceramic tubing with more uniform pore size and tube thickness is now generally used. McCloskey [55] devised a rugged separator, easy to fabricate and readily disassembled for cleaning, which must be silylated because it is all stainless steel.

An interesting effusion separator is the variable-conductance slit separator, Fig. 3(d), designed by Brunee et al. [56, 57] and made by Varian-MAT. It uses the same principle as the porous frit separator, but effusion occurs between a flat quartz plate and two annular rings. The device is enclosed in an evacuated heated housing with the interior metal parts gold plated and is adjustable by a screw mechanism. The flow rate can be varied continuously over a wide range by adjusting the spacing between the slits and the plate, from passing all the GC effluent to the MS (slits closed) to bypassing all the GC effluent to the roughing pump (slits open). When a sample is injected into the GC column, the separator is readjusted to a predetermined MS source pressure corresponding to a desired GC carrier flow. This rapid procedure, requiring only a few seconds to reset the separator, is a great advantage. Because of the wide range of flows permitted, the separator can be coupled with glass capillary columns as well as packed columns.

3. Membrane Separator

The third major class of enrichment device is the semipermeable membrane separator, which uses the principle of selective solubility of the carrier gas or sample molecules through a membrane. Most often used is the Llewellyn-Littlejohn membrane separator [58] [Fig. 3(c)] which employs a thin silicon-rubber membrane. GC effluent flows in on one side of the membrane and the MS inlet is connected to the other. The helium carrier gas is not soluble in the membrane, but most organic compounds will pass through. Excess carrier and sample flow out to the atmosphere or an auxiliary FID or ECD. This provides an excellent method for monitoring a GC-MS run or using an auxiliary selective detector such as an ECD. The single-stage model requires no pumping. In a dual-stage model a second membrane is added and the middle chamber is pumped with a mechanical vacuum pump. The device operates over a wide flow range (10-50 ml/min) and is capable of yields to 70% and enrichment factor of up to 1000 [20].

The separator does have several drawbacks. It depends on a solubility principle and, to optimize solubility, the temperature must be carefully controlled, usually by placing the separator in the GC oven or in a separate oven with programmable temperature control. The membrane will rupture if the temperature reaches 230°-250°. For GC analyses below 200°, where a wide range of column flows is required, the carefully used membrane separator is an interesting device.

Unfortunately, most manufacturers offer only one or two types of separators. If a high GC-column flow rate (40-60 ml/min), wide temperature range, good efficiency

and high yield are needed, then a two-stage W-B porous ceramic separator is the first
choice. If an extreme range of flow rates (1-60 ml/min) is required for coupling to
both packed and capillary columns, then the Brunee separator is desirable, if it is
available.

VI. MASS SPECTROMETER AS A DETECTOR FOR HIGH-RESOLUTION CAPILLARY GC

Recent advances in the preparation, technology, and methodology of high-resolution
surface (or wall) coated open tubular (WCOT) glass capillary columns [59-62] have
been applied to several problems, including the study of therapeutic agents by GC-MS.
The capillary column is ideal for coupling to an MS [63-68] due to its low carrier
flow rates (1-5 ml/min). All or some of the effluent, depending on pumping capacity,
can be handled by a differential pumping system on the MS. The transition from packed
to capillary columns is not easy. Great care must be exercised to eliminate all potential dead volumes, and to insure uniform temperatures and leak-free couplings.
Although Teflon shrink tubing can be used to couple glass to metal and glass to glass,
above 170° the Teflon tubing becomes increasingly permeable to air, increasing the
source pressure of the MS. Special devices such as surrounding the joint with glass
tubing and a continuous helium flow [68] have been used to eliminate this problem.
Because a vacuum-tight seal is required between the end of the column and the mass
spectrometer and it is difficult to seal small glass capillaries to most metals,
platinum capillary tubing has been used [64]. This can either be fused to the glass
directly or Teflon shrink tubing (with the previous precautions observed) can be used
for coupling glass to metal. An excellent review of capillary column literature [69]
is available. Future developments in WCOT columns may make high-resolution GC-column
technology suitable for most laboratories.

VII. TECHNIQUES OF SCANNING AND DATA GATHERING

The huge amount of data that may result from a single GC-MS run can be overwhelming.
There are several methods of scanning as well as data gathering and reduction. Almost all GC-MS operations require the use of a computerized data system for gathering
and reduction of data, background subtraction, and output of data. An excellent review [70] of MS data systems is available for those who wish further information.
One GC-MS run with 20 GC peaks might require 30 min to acquire the data by a light
spot recorder, and a week's labor to mass mark, count peaks, measure abundances, determine relative abundances, tabulate, subtract background, retabulate, and plot.
Twenty GC peaks with three scans per peak might also yield a small mountain of oscillographic paper, but all GC-MS systems should have an oscillographic recorder for
troubleshooting when the very highest sensitivity or the widest dynamic range of
recording is required. An attempt to do GC-MS without a good data system is very
foolish and will cost more in operator man-labor than would be saved by not obtaining

a data system. A good data system is based on a minicomputer with a disk (or disks) for operating software (programs) and data storage, a keyboard-display scope terminal for communications and graphic display of spectra and data, and a high-speed method of outputting data such as an electrostatic printer/plotter or a hard copy unit for the display scope.

There are several methods of scanning the mass spectrometer. The normal method involves triggering the scan operation at a desired point to scan an entire mass spectrum. In cyclic scanning, the MS is continuously scanned across some mass region and all scans are stored. The cyclic scan produces large amounts of empty record where no peaks occur, but when the sum of all the ions in each scan is plotted vs scan number, it gives an excellent reproduction of what an FID trace would look like (the so-called reconstructed gas chromatogram or RGC). The cyclic scan requires a very reproducible scan such as is provided by a quadrupole MS or a field-regulated magnet-sector instrument.

Another method of operation is to scan over a small mass range. By devoting more time to a smaller region, an increase in sensitivity can be achieved.

The most elegant method for increasing the detection limit and specificity of the MS as a GC detector is selective-ion detection (SID) [26, 27], also known as multiple-ion detection (MID) or mass fragmentography. One mass (or a few) is measured as a function of time and the signal at each mass can be integrated to give a marked increase in sensitivity. Levels down to 30 fg have been reported using a direct injection atmospheric pressure ionization source MS [43, 44]. The SID is now routinely used for detection and quantitation of therapeutic agents and metabolites in body fluids, tissues, and other samples [71-73]. The method is ideally suited to the incorporation of stable isotope-labeled internal standards, providing for greater accuracy and reliability [37].

Magnetic-scanning instruments can vary either the magnetic field or the source-ion accelerating potential for SID. It is more difficult to vary the magnetic field because the mass scale is not linear with applied current and because of hysteresis effects. The accelerating potential SID method generally used [26] is limited to a mass range of 10-30%. Quadrupole instruments perform SID by varying the dc ramp voltage.

VIII. APPLICATIONS

It is beyond the scope of a chapter this size to give a complete review of the applications of GC-MS to biological applications. GC-MS has been used to identify unknown drugs and their metabolites in body fluids and tissues and to verify the structure of synthesized and/or isolated agents. The method is also used for quantitation of drugs and metabolites in body fluids and tissues. A small number of recent uses of GC-MS for analyzing therapeutic agents will be presented. The detection of inborn errors of

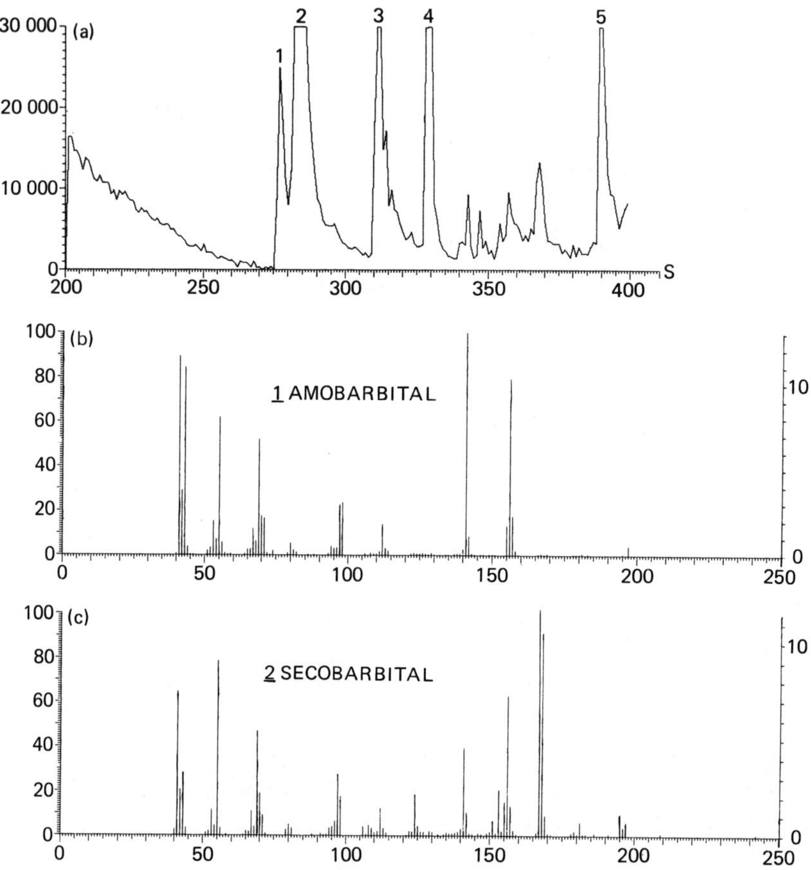

30 000 (a)
20 000
10 000
0
200 250 300 350 400 S
1 2 3 4 5

100 (b)
80
60
40
20
0
10
0
1 AMOBARBITAL
0 50 100 150 200 250

100 (c)
80
60
40
20
0
10
0
2 SECOBARBITAL
0 50 100 150 200 250

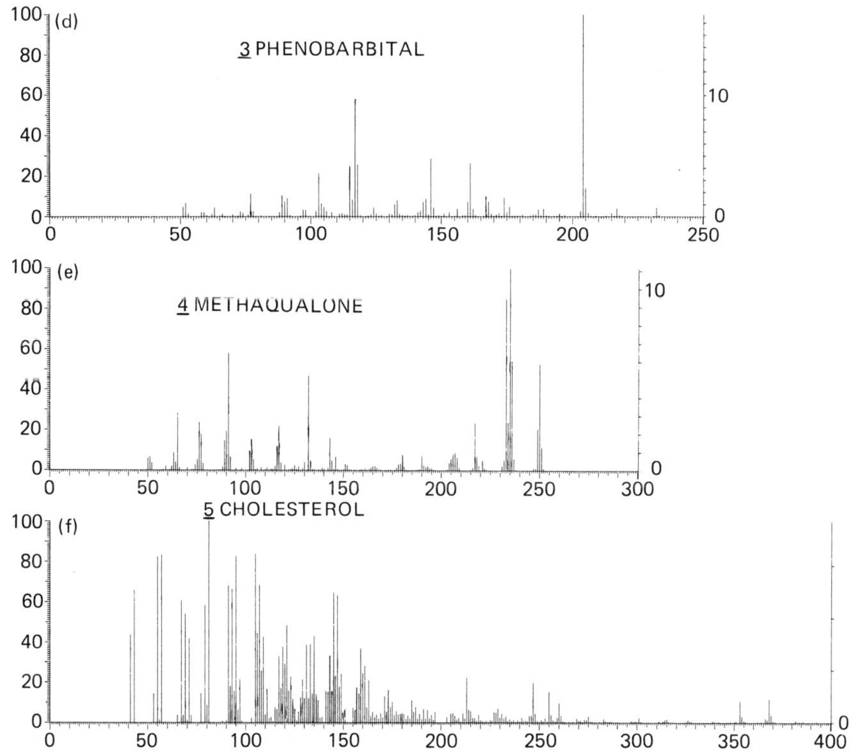

FIG. 4. (a) Reconstructed gas chromatogram of GC-MS from plasma extract of drug-overdose victim. Numbered peaks correspond to those in (b) (f): (b) peak 1, amobarbital; (c) peak 2, secobarbital; (d) peak 3, phenobarbital; (e) peak 4, methaqualone; and (f) peak 5, cholesterol.

metabolism and the study of body natural product metabolism will not be covered. For those interested in the latter, a recent review article by Lawson [19] gives an excellent overview of clinical applications of mass spectrometry.

Several recent papers have dealt with the isolation, derivatization and identification of the glucuronides of several different classes of drugs [74-77]. The work done with the glucuronides indicates the great power of GC-MS to deal with molecules of high molecular weight and polarity. The proper derivatization method combined with GC-MS makes possible the identification of intact glucuronides without prior hydrolysis. The routine use of GC-MS to identify drugs and metabolites in the body fluids of drug overdose patients was pioneered by Biemann and co-workers [78] and is a very important development in identifying abused therapeutic agents.

Our mass spectrometry laboratory has an agreement with local hospitals to identify drugs present in the body fluids of suspected drug overdose victims using GC-MS. A chloroform extract of the particular body fluid is concentrated by evaporation and run on our Varian-MAT 311A GC-MS and SS-100 data system. Figure 4 is an example of a particular sample submitted for analysis. Figure 4(a) is the reconstructed gas chromatogram (RGC) of a plasma sample run on the MS. The numbered peaks correspond to those in Fig. 4(b)-(f). Peaks were identified by computer comparison to a file of known drug mass spectra stored on a disk-based library. The following peaks were identified: peak 1, amobarbital; peak 2, secobarbital; peak 3, phenobarbital; peak 4, methaqualone; and peak 5, cholesterol. The cholesterol is, of course, naturally present in the plasma. The whole procedure from injection of the sample to identification of the data requires less than one hour to perform. Figure 5 is a specific mass RGC of Fig. 4. One or more specific ions of each compound are plotted vs scan number and as can be seen each peak is sharply differentiated from the background. This method is very useful for selective output of data and for very small amounts of material which give very small peaks in the total RGC.

The accuracy of GC-MS as a quantitation tool in measuring low levels of drugs and metabolites has been increased by the use of stable isotope-labeled internal standards and selective ion detection developed by M. G. Horning and co-workers [37]. As little as 50-200 μl of human plasma is needed for quantitation of various drugs. M. G. Horning et al. [79] have also shown the utility of GC-MS and SID for studies in prenatal pharmacology, pharmacokinetics and pediatric toxicology. Sullivan et al. [80] have applied GC-MS to the quantitative determination of methadone in human heroin addict plasma. This use is important in the new area of drug screening and maintenance programs. Wong and Biemann [81] and Lin et al. [82] have simultaneously published articles dealing with the identification and quantitation of phencyclidine and its metabolites in human blood and urine. Milberg et al. [83] have used GC-MS to quantitate methylphenidate and ritalinic acid in human plasma and urine. Schulman and Abramson [84] have published a GC-MS analysis of plasma amino acids in infants being fed by hyperalimentation.

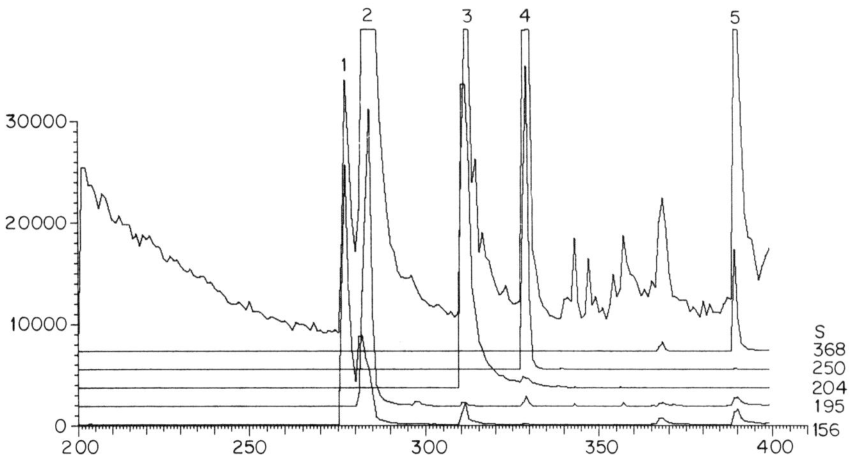

FIG. 5. Selected-ion-mass reconstructed gas chromatogram of Fig. 4(a), with m/e 156, 195, 204, 250, and 368 plotted vs scan number. S is the total ion trace from Fig. 4(a).

Sullivan et al. [85] used GC-MS to identify the metabolites of α-1-*trans*-4-dimethylaminotetrahydro-3-furyl-cyclohexanephenylglycolate and to identify species differences in metabolism. Belvidere et al. [86] used GC-MS to determine plasma levels of imipramine and desipramine in chronically treated patients. Nakagawa et al. [87] used GC-MS to identify chloramphenicol, thiamphenicol and their metabolites in human urine. Walle et al. [88] determined plasma levels of propanolol and 4-hydroxypropanolol using GC-MS. Belvedere et al. [89] used GC-MS for determination of trazodone in rat plasma. Marcucci et al. [90] identified the C_3-hydroxylated glucuronide metabolites of benzodiazepine in human urine. Essien et al. [91] used GC-MS to identify N-oxygenated metabolites of phenothiazines in body fluids. Nash et al. [92] used GC-MS to confirm the structure of propoxyphene and its major metabolites in heroin addict plasma after dosage of propoxyphene napsylate. Bayne et al. [93] employed GC-MS to quantitate scopolamine in plasma and urine with a detector limit of 50 pg/ml for 1 ml of fluid using a deuterated internal standard. Hintze et al. [94] studied the metabolism of cyproheptadine in rats, mice and humans and identified a stable epoxide metabolite using GC-MS. Murphy et al. [95] performed a study of the metabolism of propionyl erythromycin lauryl sulfate in rats aided by GC-MS. Porter et al. [96] studied the metabolism of cyproheptadine in humans using GC-MS. Stillwell et al. [97] discovered an epoxide-diol pathway of metabolism for methaqualone in man and the rat using GC-MS. Allen et al. [98] studied the metabolism of debrisoquine sulfate in the rat and man using GC-MS. Alton et al. [99] examined the biotransformation of a 1,5-benzodiazepine triflubazam in man with the aid of GC-MS. Tocco et al. [100] performed a study of the metabolism of timolol in man and laboratory animals using GC-MS. Cone and Gorodetzky [101] published a comment on the choice of analytical controls in drug metabolism where GC-MS is employed. DeSagher et al. [102] identified *p*-chlorophenoxyacetamide as a metabolite of iproclozide in humans studied its urinary excretion using GC-MS. Alton et al. [103] studied the comparative biotransformation of triflubazam in rats, dogs and monkeys using GC-MS. Tocco et al. [104] examined the metabolism of S-(2',4'-difluorophenyl)-salicyclic acid in the rat, dog and man by GC-MS. Karim et al. [105] studied the metabolism of spironolactone in man using GC-MS. Tang et al. [106] identified N-hydroxyamobarbital as a metabolite of amobarbital in man by GC-MS.

Sano et al. [107] developed a new technique for the detection of stable isotope-labeled drugs by GC-pyrolysis-MS. Karashima et al. [108] identified the metabolism products of halothane with the aid of GC-MS. Ford et al. [109] measured diphenoxylic acid levels in plasma following the administration of diphenoxylate using GC-CI-MS. Miyazaki et al. [110] characterized various estrogen glucuronides by GC-MS. Baty et al. [111] developed a method for the estimation of acetanalide, paracetamol, and phenacetin in plasma and urine using GC-MS. Coutts et al. [112] studied the *in vitro* metabolism of 1-phenyl-2-propanone oxime, a metabolite of amphetamine, in rat liver using GC-MS. Rash and Lynch [113] studied the metabolism of a new anticoccidial, a 6-azauricil derivative, using GC-MS. Knapp et al. [114] examined the qualitative metabolic fate of phenoxybenzamine in the rat, dog, and man using GC-CI-MS with an ^{15}N-labeled phenoxybenzamine standard and using a twin ion-labeling technique. Tyler et al. [115] identified 5-hydroxybenzimidiazole as a major metabolite of benzimidiazole in the rat with the aid of GC-MS. Coutts et al. [116] identified several major phenolic metabolites of *n*-alkyl-substituted amphetamines in the rat using GC-MS.

IX. CONCLUSIONS

Combined GC-MS has proven to be of great value in the analysis of therapeutic agents. With the new developments in electronics and vacuum systems modern GC-MS equipment can be operated by anyone who is willing to learn how to care for and maintain the instrument.

ACKNOWLEDGMENTS

Part of our work was supported by U.S. EPA contract 68-01-3234 and equipment provided by NIH grants CA-11388 and GM 16864.

Our thanks to Mrs. Lois Shield for editing and proofreading this chapter and to Mrs. Annetta Crist for an excellent job of typing.

REFERENCES

1. J. C. Holmes and F. A. Morrell, *Appl. Spectrosc.*, *11*, 86(1957).

2. W. H. McFadden, *Techniques of Combined Gas Chromatography-Mass Spectrometry*, Wiley-Interscience, New York, 1973.

3. H. C. Hill, *Introduction to Mass Spectrometry*, Heyden and Son, London, 1966.

4. J. Roboz, *Introduction to Mass Spectrometry*, Wiley-Interscience, New York, 1968.

5. C. A. McDowell (ed.), *Mass Spectrometry*, McGraw-Hill, New York, 1963.

6. R. W. Kiser, *Introduction to Mass Spectrometry and Its Applications*, Prentice-Hall, Englewood Cliffs, N. J., 1965.

7. J. H. Beynon, *Mass Spectrometry and Its Application to Organic Chemistry*, Elsevier, Amsterdam, 1960.

8. K. Biemann, *Mass Spectrometry: Organic Chemistry Applications*, McGraw-Hill, New York, 1962.

9. F. W. McLafferty (ed.), *Mass Spectrometry of Organic Ions*, Academic Press, New York, 1963.

10. F. W. McLafferty, *Interpretation of Mass Spectra*, 2nd Ed., W. H. Benjamin, Reading, Mass., 1973.

11. H. Budzikiewicz, C. Kjerassi, and D. H. Williams, *Mass Spectra of Organic Compounds*, Holden-Day, San Francisco, 1967.

12. J. H. Beynon, R. A. Saunders, and A. E. Williams, *The Mass Spectra of Organic Molecules*, Elsevier, Amsterdam, 1968.

13. G. R. Waller (ed.), *Biomedical Applications of Mass Spectrometry*, Wiley-Interscience, New York, 1972.

14. A. L. Burlingame (ed.), *Topics in Organic Mass Spectrometry*, Wiley-Interscience, New York, 1970.

15. D. H. Williams (ed.), *Mass Spectrometry*, Vols. 1, 2, and 3, The Chemical Society, London, 1971, 1973, 1975.

16. G. W. A. Milne (ed.), *Mass Spectrometry*, Wiley-Interscience, New York, 1971.

17. G. W. A. Milne and M. J. Lacey, *in CRC Critical Reviews in Analytical Chemistry*, (L. Meites and B. H. Campbell, eds.), Vol. 4, CRC Press, Cleveland, 1974, pp. 45-104.

18. C. F. Simpson, *in CRC Critical Reviews in Analytical Chemistry* (L. Meites, ed.), Vol. 3, CRC Press, Cleveland, 1972, pp. 1-40.

19. A. M. Lawson, *Clin. Chem.*, *21*, 803(1975).

20. G. A. Junk, *Int. J. Mass Spectrosc. Ion Phys.*, *8*, 1(1972).

21. *Anal. Chem.*, Fundamental Reviews Issue.

22. *Organic Mass Spectrometry*, Heyden and Son, London.

23. *International Journal of Mass Spectrometry and Ion Physics*, Elsevier, Amsterdam.

24. *Biomedical Mass Spectrometry*, Heyden and Son, London.

25. A. Frigerio and N. Castignoli (eds.), *Mass Spectrometry in Biochemistry and Medicine*, Raven Press, New York, 1975.

26. C. C. Sweely, W. H. Elliot, I. Fries, and R. Ryhage, *Anal. Chem.*, *38*, 1549(1966).

27. J. F. Holland, C. C. Sweeley, R. E. Thrush, R. E. Teets, and M. A. Bieber, *Anal. Chem.*, *45*, 308(1973).

28. A. O. Nier, *Rev. Sci. Instrum.*, *18*, 398(1947).

29. Varian-MAT, Magnetic Field Regulator.

30. J. Roboz, *Introduction to Mass Spectrometry*, Wiley-Interscience, New York, 1968, pp. 55-59.

31. E. G. Johnson and A. O. Nier, *Phys. Rev.*, *91*, 10(1953).

32. J. Mattauch, *Phys. Rev.*, *50*, 627(1936).

33. W. Paul, H. P. Reinhard, and U. von Zahn, *Z. Phys.*, *152*, 143(1958).

34. Extranuclear Laboratories, Pittsburgh, Pennsylvania.

35. F. H. Field, *Acc. Chem. Res.*, *1*, 42(1968).

36. G. P. Arsenault, in *Biomedical Applications of Mass Spectrometry* (G. R. Waller, ed.), Wiley-Interscience, New York, 1972, pp. 817-832.

37. M. G. Horning, J. Nowlin, K. Lertratanangkoon, R. N. Stillwell, W. G. Stillwell, and R. M. Hill, *Clin. Chem.*, *19*, 845(1973).

38. M. G. Horning, P. Gregory, J. Nowlin, M. Stafford, K. Lertratanangkoon, C. Butler, W. G. Stillwell, and R. M. Hill, *Clin. Chem.*, *20*, 282(1974).

39. E. C. Horning and M. G. Horning, *Clin. Chem.*, *17*, 802(1971).

40. D. F. Hunt, *Finnigan Spectra*, *6*, 1(1976), Finnigan Corporation, Sunnyvale, California.

41. H. D. Beckey, in *Biomedical Applications of Mass Spectrometry* (G. R. Waller, ed.), Wiley-Interscience, New York, 1972, pp. 795-816.

42. K. L. Olson, J. Carter Cook, Jr., and K. L. Rinehart, Jr., *Biomed. Mass Spec.*, *1*, 358(1974).

43. E. C. Horning, M. G. Horning, D. I. Carroll, I. Dzidic, and R. N. Stillwell, *Anal. Chem.*, *45*, 936(1973).

44. D. I. Carroll, I. Dzidic, R. N. Stillwell, M. G. Horning, and E. C. Horning, *Anal. Chem.*, *46*, 706(1974).

45. D. I. Carroll, C. Oestli, I. Dzidic, R. N. Stillwell, and E. C. Horning, presented at the 23rd Annual Conference on Mass Spectrometry and Allied Topics, Houston, Texas, 1975.

46a. E. C. Horning, M. G. Horning, D. I. Carroll, and I. Dzidic, Institute for Lipid Research, Baylor University School of Medicine, Houston, Texas.

46b. Extranuclear Corp., Pittsburgh, Pennsylvania.

47. M. A. Grayson and C. J. Wolf, *Anal. Chem.*, *39*, 1438(1967).

48. R. Ryhage, *Anal. Chem.*, *36*, 759(1964).

49. R. Ryhage, S. Wikstrom and G. R. Waller, *Anal. Chem.*, *37*, 435(1965).

50. R. H. Allen Company, Boulder, Colorado.

51. Finnigan Corporation, Sunnyvale, California.

52. J. T. Watson and K. Biemann, *Anal. Chem.*, *36*, 1135(1964).

53. J. T. Watson and K. Biemann, *Anal. Chem.*, *37*, 844(1965).

54. J. T. Watson, in *Biomedical Applications of Mass Spectrometry* (G. R. Waller, ed.), Wiley-Interscience, New York, 1972, pp. 23-49.

55. P. M. Krueger and J. A. McCloskey, *Anal. Chem.*, *41*, 1930(1969).

56. C. Brunnee, H. J. Bulteman, and G. Kappus, presented at the 17th Annual Conference on Mass Spectrometry and Allied Topics, Dallas, Texas, 1969.

57. C. Brunnee, L. Delgmann, K. Habfast, and S. Meier, presented at the 18th Annual Conference on Mass Spectrometry and Allied Topics, San Francisco, California, 1970.

58. P. M. Llewellyn and D. P. Littlejohn, presented at the Pittsburgh Conference on Analytical Chemistry and Applied Spectroscopy, Pittsburgh, 1966.

59. K. Grob, *Chromatographia, 7,* 94(1974).

60. K. Grob, *Chromatographia, 8,* 423(1975).

61. A. L. German, C. D. Pfaffenberger, J-P. Thenot, M. G. Horning, and E. C. Horning, *Anal. Chem., 45,* 930(1973).

62. M. Novotny and A. Zlatkis, *J. Chromatogr. Sci., 8,* 346(1970).

63. R. Fluckiger, *Chromatographia, 8,* 434(1975).

64. D. Henneberg, U. Henrichs and G. Schomberg, *J. Chromatogr., 112,* 343(1975).

65. J. Eyem, *Chromatographia, 8,* 456(1975).

66. P. Christiansen, *Chromatographia, 8,* 463(1975).

67. U. Rapp, U. Schroder, S. Meier, and H. Elmenhorst, *Chromatographia, 8,* 474(1975).

68. K. Gorb and H. Jaeggi, *Anal. Chem., 45,* 1788(1973).

69. M. Novotny and A. Zlatkis, *Chromatogr. Rev., 14,* 1(1971).

70. G. R. Waller (ed.), *Biomedical Applications of Mass Spectrometry,* Wiley-Interscience, New York, 1972, pp. 51-132.

71. C. C. Sweeley, N. D. Young, J. F. Holland, and S. C. Gates, *J. Chromatogr., 99,* 507(1974).

72. R. W. Kelley, *J. Chromatogr., 54,* 345(1971).

73. B. Samuelsson, M. Hamberg, and C. C. Sweeley, *Anal. Biochem., 38,* 301(1970).

74. R. M. Thompson, N. Gerber, R. A. Seibert, and D. M. Desiderio, *Drug Metab. Disp., 1,* 489(1973).

75. R. Bonnichsen, Y. Marde, and R. Ryhage, *Clin. Chem., 20,* 230(1974).

76. H. Ehrsson, T. Walle, and S. Wikstrom, *J. Chromatogr., 101,* 206(1974).

77. J. F. Mrocheck and W. T. Rainey, Jr., *Anal. Biochem., 57,* 173(1974).

78. C. E. Costello, H. S. Hertz, T. Sakai, and K. Biemann, *Clin. Chem., 20,* 256(1974).

79. M. G. Horning, J. Nowlin, C. M. Butler, K. Lertratanangkoon, K. Sommer, and R. M. Hill, *Clin. Chem., 21,* 1282(1974).

80. H. R. Sullivan, F. J. Marshall, R. E. McMahon, E. Anggard, L. M. Gunne, and J. H. Holmstrand, *Biomed. Mass Spec., 2,* 179(1975).

81. L. K. Wong and K. Biemann, *Biomed. Mass Spec., 2,* 204(1975).

82. D. C. K. Lin, A. F. Fentiman, Jr., R. L. Foltz, R. D. Forney, Jr., and I. Sunshine, *Biomed. Mass Spec., 2,* 206(1975).

83. R. M. Milberg, K. L. Rinehart, Jr., R. L. Sprague and E. K. Sleator, *Biomed. Mass Spec., 2,* 1(1975).

84. M. F. Schulman and F. P. Abramson, *Biomed. Mass Spec., 2,* 9(1975).

85. H. R. Sullivan, J. G. Page, and S. L. Due, *Biomed. Mass Spec., 2,* 53(1975).

86. G. Belvedere, L. Burti, A. Frigerio, and C. Pantarotto, *J. Chromatogr., 111,* 313(1975).

87. T. Nakagawa, M. Masada, and T. Uno, *J. Chromatogr., 111,* 355(1975).

88. T. Walle, J. Morrison, K. Walle, and E. Conradi, *J. Chromatogr., 114,* 351(1975).

89. G. Belvedere, A. Frigerio, and C. Pantarotto, *J. Chromatogr., 112,* 631(1975).

90. F. Marcucci, R. Bianchi, L. Airoldi, M. Salmona, R. Fanelli, C. Chiabrando, A. Frigerio, E. Mussini, and S. Garattini, *J. Chromatogr.,* 107, 285(1975).

91. E. E. Essien, D. A. Cowan, and A. H. Beckett, *J. Pharm. Pharmacol.*, *27*, 334(1974).

92. J. F. Nash, I. F. Bennett, R. J. Bopp, M. K. Brunson, and H. R. Sullivan, *J. Pharm. Sci.*, *64*, 429(1975).

93. W. F. Bayne, F. T. Tao, and N. Crisologo, *J. Pharm. Sci.*, *64*, 288(1975).

94. K. L. Hintze, J. S. Wold, and L. J. Fischer, *Drug Metab. Disp.*, *3*, 1(1975).

95. R. J. Murphy, T. L. Williams, R. E. McMahon, and F. J. Marshall, *Drug Metab. Disp.*, *3*, 155(1975).

96. C. C. Porter, B. H. Arison, V. F. Gruber, D. C. Titus, and W. J. A. Vandenheuvel, *Drug Metab. Disp.*, *3*, 189(1975).

97. W. G. Stillwell, P. A. Gregory, and M. G. Horning, *Drug Metab. Disp.*, *3*, 287(1975).

98. J. G. Allen, P. B. East, R. J. Francis, and J. C. Haigh, *Drug Metab. Disp.*, *3*, 332(1975).

99. K. B. Alton, R. M. Grimes, C. Shaw, J. E. Patrick, and J. L. McGuire, *Drug Metab. Disp.*, *3*, 352(1975).

100. D. J. Tocco, A. E. W. Duncan, F. A. Deluna, H. B. Hucker, V. F. Gruber, and W. J. A. Vandenheuvel, *Drug Metab. Disp.*, *3*, 361(1975).

101. E. J. Cone and C. W. Gorodetzky, *Drug Metab. Disp.*, *3*, 417(1975).

102. R. M. DeSagher, A. P. deLeenheer, A. A. Cruyl, and A. E. Clayes, *Drug Metab. Disp.*, *3*, 423(1975).

103. K. B. Alton, J. E. Patrick, C. Shaw, and J. L. McGuire, *Drug Metab. Disp.*, *3*, 445(1975).

104. D. J. Tocco, G. O. Breault, A. G. Zacchei, S. L. Steelman, and C. V. Perrier, *Drug Metab. Disp.*, *3*, .453(1975).

105. A. Karim, J. Hribar, W. Aksamit, M. Doherty, and L. J. Chinn, *Drug Metab. Disp.*, *3*, 467(1975).

106. B. K. Tang, T. Inaba, and K. Kalow, *Drug Metab. Disp.*, *3*, 479(1975).

107. M. Sano, Y. Yotsui, H. Abe, and S. Sasaki, *Biomed. Mass Spec.*, *3*, 1(1976).

108. D. Karashima, S. Takahashi, A. Shigematsu, T. Furukawa, T. Kuhara, and I. Matsumoto, *Biomed. Mass Spec.*, *3*, 41(1976).

109. G. C. Ford, N. J. Haskins, R. F.Palmerm, M. J. Tidd, and P. H. Buckley, *Biomed. Mass Spec.*, *3*, 45(1976).

110. H. Miyazaki, M. Ishibashi, M. Itoh, N. Morishita, M. Sudo, and T. Nambara, *Biomed. Mass Spec.*, *3*, 55(1976).

111. J. D. Baty, P. R. Robinson, and J. Wharton, *Biomed. Mass Spec.*, *3*, 60(1976).

112. R. T. Coutts, R. Dawe, G. W. Dawson, and S. H. Kovach, *Drug Metab. Disp.*, 4, 35(1976).

113. J. J. Rash and M. J. Lynch, *Drug Metab. Disp.*, *4*, 59(1976).

114. D. R. Knapp, N. H. Holcombe, S. A. Krueger, and P. J. Privitera, *Drug Metab. Disp.*, *4*, 64(1976).

115. T. R. Tyler, J. T. Lee, H. Flynn, and W. J. A. Vandenheuvel, *Drug Metab. Disp.*, *4*, 177(1976).

116. R. T. Coutts, G. W. Dawson, C. W. Kazakoff, and J. Y. Wong, *Drug Metab. Disp.*, *4*, 256(1976).

Chapter 7

THE MASS SPECTROMETER AS A DETECTOR FOR HIGH-PERFORMANCE LIQUID CHROMATOGRAPHY

Bobby G. Dawkins[*]

F. W. McLafferty

Department of Chemistry,
Cornell University,
Ithaca, New York

I. INTRODUCTION

A major limitation in the further application of high-pressure liquid chromatography
(LC) to important chemical and biomedical problems, such as the analysis of therapeutic

---

[*] *Presently at* E.I. DuPont de Nemours & Company, Aiken, South Carolina

TABLE 1. Performance of Continuous LC Detectors

Characteristic	RI	FID, EC, SID	UV[a]	Moving wire	Mem- brane	API	Direct CI
					MS interfaces		
Nondestructive	Yes	No	Yes	No	No	No	Yes
Sensitivity	10^{-6}g	10^{-9}g	10^{-9}g	10^{-7}g	10^{-9}g	10^{-9}g	10^{-9}g
Specificity, ident. ability: no. of resol. increments	2^{b}	1	~10		200-1000[c]		
no. of peaks/compound	1	1	~5	~50	~50	$\geq 1^{d}$	$\geq 1^{d}$
Sample limitations: volatility required vs	Minor[b]	Minor	UV		Volatility		
direct probe MS	None	None	None	Same	More	More	Less
LC solvent limitations	Gra- dient		UV		Non- polar		
Cost, complexity	OK	OK	OK		Expensive, complex		

[a]Fluorometric detectors are generally similar in performance, with higher ultimate sensitivity but more limited applicability.

[b]On the basis that the RI of the solute can be above or below that of the solvent, but not the same.

[c]Dependent on mol. wt.

[d]In a substantial proportion of cases the $(M + 1)^{+}$ or $(M - 1)^{+}$ ion is the only major peak, although the CI spectrum under certain conditions can provide detailed structural information.

agents, has been the availability of suitable detector systems [1-4]. The sensitivity, selectivity, and/or versatility of the common detectors are inadequate for many applications, and none give any detailed molecular structure information. Because of the well-known capabilities of gas chromatography (GC), the greatest interest in LC has been for those problems for which GC is not applicable, in particular for low-volatility or heat-labile compounds. The coupling of GC with the mass spectrometer (MS) has provided systems of unique analytical capabilities, so that LC/MS systems are an obvious possible solution to these LC detector problems.

The most widely used LC detectors (Table 1) [1-4] have been those utilizing ultraviolet (uv) absorption and refractive index (RI) measurements, which have maximum sensitivities of $\sim 10^{-9}$g and $\sim 10^{-6}$g, respectively. For the uv detector, the sample must (and the solvent must not) have a uv-active chromophore. Selectivity is limited because at most only a few different wavelengths in the uv can be used to differentiate sample components, and the relative absorptivities at these different wavelengths can provide only very general structural information. Although the RI detector will respond to almost any solute whose refractive index is different than that of the solvent, it thus provides almost no selectivity or structural information, and its sensitivity to temperature and solvent composition make it inapplicable for gradient separations. A variety of other detector systems have been proposed that show higher sensitivity than the RI detector, but most of these also give a single-parameter response. These include the spray impact detector (SID) [5], the electron capture (EC) detector [6], the flame ionization detector (FID) [7-9], and the piezoelectric quartz mass sensor [10]. Another promising system is the fluorescence detector; although this can give greater sensitivity than the uv detector, the types of substances showing suitable fluorescence are quite limited.

Interest in the mass spectrometer (MS) as an LC detector has primarily been motivated by its success as a continuous detector for GC. Although more selective and sensitive detectors are available for GC than for LC, the impressive success of MS as a GC detector has been due to its unusually high sensitivity and detailed spectral information which is characteristic of the molecular structure of the separated components. The main purpose of the most widely used GC/MS interfaces is the removal of the chromatographic mobile phase (usually helium) to increase the MS sensitivity for the separated sample components. A comparable LC/MS interface would require a separator for the liquid solvent phase, but such phases can exhibit a wide variety of properties (e.g., gradient elution) and are delivered to the interface at ~ 1 g/min versus GC carrier-gas flow rates of less than 1% of this value. One of the most important design considerations for LC/MS interfaces is the sample volatility requirement of mass spectrometry; this is not a factor in GC/MS, as the sample vapor pressure necessary for GC separation is even higher than for MS measurement. Despite the recent development of promising methods such as field desorption [11], direct chemical

ionization [12], and ionization with nuclear fission fragments [13], many important
high molecular-weight molecules, such as polynucleic acids, polypeptides and poly-
saccharides are not amenable to mass spectrometry. Fortunately, a large variety of
therapeutic agents give useful and characteristic mass spectra on evaporation from
a conventional direct probe introduced into the MS ion source.

This chapter will be devoted mainly to *continuous* LC detectors; a batchwise
detector is not only inconvenient and inefficient to operate, but cannot give a useful representation of the chromatogram. Non-MS detector systems will be reviewed
briefly first, followed by more extensive descriptions and discussions of the several
LC/MS systems that have been proposed.

II. NON-MASS SPECTROMETER DETECTOR SYSTEMS

A. Single-Parameter "Universal" Detectors

A number of detector systems have been proposed which give only a one-dimensional
response, but are sensitive to a wide range of compounds (Table 1). The most widely
used is the refractive index detector, whose response will indicate whether the solute
is higher or lower in RI than the solvent. It is thus sensitive to most solutes at
the microgram level but cannot be used in gradient elution. The spray impact detector
proposed by Mowery and Juvet [5] is basically a mass sensor in which the impact of the
solute on a target electrode causes a potential change; nanogram sensitivities have
been achieved. Similar sensitivities have been claimed for molecules possessing high
electron affinities using the electron capture detector of Willmott and Dolphin [6];
the design is similar to that of the corresponding GC detector. Flame ionization
detectors have been used in conjunction with moving wire transport systems which
effect solvent vaporization [7-9]. A piezoelectric quartz mass sensor [10] has been
reported to be capable of detecting 10^{-10}g; after solvent evaporation on the crystal
surface, the remaining solute mass is correlated with the crystal frequency. Electrochemical detectors [14] such as polarographic, voltammetric, and ion-selective
electrodes have limited applicability because of large volume requirements. Infrared
absorption utilizing lasers [15] appears to have promise for gel permeation and gradient elution chromatography, with sensitivities comparable to RI.

B. Several-Parameter Specific Detectors

The ultraviolet (uv) absorption spectrometer is the most widely used LC detector other
than RI, as it has much greater sensitivity for specific types of compounds when uv-
nonabsorbing solvents are used. For compounds with high molar absorptivities, such as
aromatics, detection of nanogram amounts is often possible. UV detection can also be
used if the compound sought can be converted into a uv-absorbing derivative. Some

selectivity and basic structural information can also be achieved with multiple wavelength detection.

Compounds (and derivatives) which can be made to fluoresce in the uv-visible region also can be detected conveniently in almost all common LC solvents. Subnanogram sensitivities are often possible, but the range of compound types which will undergo fluorescence is even more limited than those which absorb ultraviolet radiation, and such types often can be destroyed by the intense exciting radiation.

III. BASIC CHARACTERISTICS OF MASS SPECTROMETER DETECTORS

The unique advantages of the mass spectrometer in this application lie in its high detection sensitivity and the high information content of the resulting mass spectra. Modern commercial GC/MS systems can detect 10^{-12}g and obtain useable mass spectra on 10^{-9}g. The average mass spectrum contains ~75 peaks, each at one of hundreds of unique mass positions with abundances over a range of several orders of magnitude. Probably the most severe limitation of conventional mass spectrometry is that the sample molecules must be vaporized before ionization; samples must have vapor pressures of ~10^{-5} torr, or approximately 1000-2000 molecular weight for nonpolar organic compounds. Obviously this is not a limitation for compounds which can be separated by gas chromatography, but it is an important restriction for LC, which has no volatility requirements. A further serious drawback of mass spectrometry is the cost and complexity of the instrumentation involved. Although commercial instruments have become much more reliable, and the dedicated computer has substantially alleviated the data reduction problems, the major part of the investment and upkeep in any LC/MS system must be in the MS components.

A. Mass Spectral Data from Low-Volatility Compounds

Rapid heating, field desorption (FD), and direct chemical ionization (DCI) are methods by which mass spectral data have been obtained from particular compounds of volatility lower than that required for normal direct probe sampling. Although at present only DCI has been applied to the direct monitoring of LC, it is conceivable that continuous methods can be devised to utilize the first two also.

1. Rapid Heating

Friedman and his co-workers have shown [16] that it is possible to minimize thermal decomposition accompanying direct probe sample introduction by optimization of the heating rate and proper selection of the surface material from which the sample vaporizes. Sample heating rates of 12°/sec (complete sample evaporation in less than 30 sec) substantially reduce pyrolysis, as does the use of a Teflon sample probe in a Teflon-lined ion source. Optimum desorption is critically dependent on the absence of trace nonvolatile impurities.

Spectacular heating rates appear to be a key reason for the success of a unique method utilizing fission-fragment ionization proposed by MacFarlane and co-workers [13]. The sample is deposited on a thin foil behind which is a ^{252}Cf source. Thermalization of large nuclei fission fragments passing through the foil produce localized temperatures of ~25,000° C in 10^{-11} sec, vaporizing and ionizing the sample molecules. Spectra of low volatility molecules such as neurotoxins and large oligopeptides have been obtained, apparently because the rapid volatilization gives less time for thermal degradation. Sensitivity is limited by the radioactivity level of the ^{252}Cf source which can be tolerated; single-ion coincidence counting is used on the present time-of-flight instrument which is employed.

2. Field Desorption

Field ionization (FI) is a process in which an electron is removed from a gaseous molecule electrostatically in a high field gradient, usually about 1V/Å utilizing an overall field potential of ~10 kV at a sharp point or edge. Quantum tunneling allows the electron to be removed at minimum energy, so that the molecular ions produced exhibit a very low degree of fragmentation. Commonly used FI points are semiconducting organic microneedles grown by vacuum pyrolysis. In field desorption [17], the low volatility sample is deposited from solution directly on these emitter needles. When these are placed in the high potential field, the sample apparently is ionized by direct tunneling of the electron to the needle surface, the ions being repelled from the positive emitter toward the exit field. The degree of mass spectral fragmentation can be varied by changing the emitter temperature. A variety of examples have now been reported in which the sample volatility requirements have been greatly lowered by FD, such as mass spectral data on the underivatized nonapeptide bradykinin [18]. FD has been used for off-line monitoring of collected LC fractions [19].

3. Chemical Ionization

Ionization of sample molecules can also be effected by direct contact with an ion plasma. Such *chemical-ionization* (CI) [20] plasmas are commonly created in the ion source by electron bombardment of a reagent gas at a relatively high pressure (~1 torr). For example, CH_4 as a reagent gas produces CH_5^+ ions, which can protonate the sample molecules to give $(M + H)^+$ ions. Some reagent gases such as *n*-pentane cause hydride abstraction, giving $(M - H)^+$ ions. Although the sensitivity achieved by CI in comparison to normal electron ionization (EI) is highly dependent on the experimental configuration used, careful analysis of the two methods by Futrell and Wojcik [21] indicates that CI has the higher ionization efficiency and thus should have the higher ultimate sensitivity. Hunt and co-workers [22] have achieved higher ion densities in CI ionization with a Townsend discharge.

In *direct chemical ionization* (DCI) [12] the solid sample, preferably of high surface area, is exposed to the ion plasma directly in the ion source, which apparently lowers the thermal energy requirements of the vaporization process. The sample can be deposited from solution on the outside of the direct probe and then inserted in the ion source to accomplish this. Alternatively, the sample is dissolved in a solvent used as the ionizing reagent, and the solution is sprayed into the ion source, where solvent evaporation should produce microparticles of the solute. The volatility requirements for DCI are lower than conventional direct probe sample introduction, but do not appear to be as low as achieved by FD or fission fragment ionization. However, DCI with direct solution introduction has obvious possibilities for continuous LC monitoring.

IV. DESIGN PARAMETERS FOR LC/MS SYSTEMS

Modern trends in liquid chromatographic practice include fast separations with small-diameter columns ($\leq$2 mm i.d.) packed with small-diameter particles ($\leq$10 μm). As in GC/MS, any column bleeding would cause serious problems for a LC/MS system; chemically bonded packing materials, which have practically eliminated the problem of column bleeding, are an obvious choice for LC/MS operation. Sensitivity is a design parameter of major importance; any substantial increase in detector sensitivity will expand the possible applications in the analysis of therapeutic agents as well as pollutants, pheromones, metabolites, and so forth. Sensitivity will be dependent on the concentration of solute entering the ion source, on solute losses or split ratio in the interface system, and on the ionization method used. Because of the critical importance of gradient elution techniques, detector response should be relatively independent of solvent composition.

A. The Ion-Source Pressure Problem

The LC effluent is a liquid at atmospheric pressure or above; for the mass spectrometer ion source the sample must be a gas at <10^{-4} torr for normal EI operation, or ~1 torr for CI. Thus the LC/MS interface must vaporize the liquid effluent and remove sufficient effluent or solvent to maintain proper MS ion-source operating pressures continuously without serious fluctuations. Pumping requirements in both the interface and ion source region are obviously critical. The rate of solvent removal is dependent on the pumping system employed; cryogenic pumping (for example, liquid nitrogen cooled surfaces) is especially effective for the LC solvents usually employed. Vaporization requires high heat-transfer rates and care must be taken to avoid thermal degradation in local overheating.

B. LC Effluent Mixing

Minimum dead volume and low residence time for the interface are critical. Capillary tubing and fittings of the type normally used for LC sample transfer should minimize sample diffusion and loss of LC resolution. However, capillary transfer lines of the interface can give special problems. Preferential adsorption of polar compounds is possible [23]. Capillary plugging is common, but can be alleviated by in-line micro-filters. Evaporation from a capillary tip causes enrichment of the low volatility component, which can plug the capillary; a restriction at the exit can minimize this problem. Skurat [24] has done calculations to describe the liquid flow through capillaries from high pressure into a vacuum which are applicable for LC/MS systems.

The proposed types of LC/MS interfaces will be discussed under three major headings: those which employ solvent separation, those which vaporize the total effluent at atmospheric pressure, and those which introduce part of the effluent directly into the low-pressure MS ionization source. Noncontinuous methods have been proposed for evaporation of separate fractions on the direct probe for normal EI sample introduction [25], and by evaporation of such fractions on emitter tips for FD analysis [19]. All of these noncontinuous methods appear to require at least several minutes per spectrum, and thus would be useful mainly for structure determination of fractions which had been found in the LC chromatogram by some other method of detection.

V. SOLVENT SEPARATORS

Two techniques have been utilized in which a major portion of the solvent is removed from the effluent prior to MS introduction; these utilize solvent evaporation or preferential migration of the solute through a membrane. In cases in which LC is utilized because the samples are not amenable to GC separation, the differences in volatility and molecular size between solvent and solute should be favorable for such separations.

A. Wire Transport Introduction

Scott and co-workers [26] have developed an LC/MS interface which incorporates the wire transport of the LC/FID interface. The LC effluent passes over a moving wire which carries a portion of it into an evaporation chamber where the solvent is removed, leaving the solute deposited on the wire. The wire continuously moves through a suitable interface directly into the ionization chamber of the mass spectrometer where it is heated to vaporize the solute; either electron impact or chemical ionization spectra can then be obtained in the conventional manner. A similar interface utilizing a moving ribbon has been developed [27]. For the design in which an earthed wire is passed continuously through the ion source, a MS utilizing a low ion accelerating potential, such as the quadrupole, is advantageous; however, sample vaporization from the wire could also be done in a chamber just outside the source [27].

In the Scott interface [26], the wire passes through the dripping effluent at the LC column exit, carrying about 10 µl/min of solution into the vaporization region at the maximum wire speed of 15 cm/sec. In this way the interface also acts as a splitter, introducing only about 1% of the LC effluent into the MS. However, in the present design the remaining effluent apparently cannot be transferred to another detector or fraction collector without serious mixing, so that it has not been classified (Table 1) as a nondestructive detector. The wire passes through a differential pumping system with a 0.01-in. aperture in a ruby jewel which allows ion-source pressures of 3×10^{-6} torr to be maintained during operation. Electrical heating of the wire increases its temperature to as high as 250° inside the ion source, where the high-vacuum conditions minimize the cooling rate. Under these sampling conditions there appears to be little loss of LC resolution due to interface mixing or contamination. Sensitivity appears to be the most serious drawback of the present interface design. The minimum detectable amount of diazepam was 4×10^{-6} g/ml, measured as a 0.01% solution in a 10% tetrahydrofuran/chloroform solvent mixture. Although this obviously would be improved if more complete transfer of the effluent to the wire could be achieved, other substantial interface losses are indicated; even with a 100-fold improvement the sensitivity would still be substantially below the picogram sensitivities commonly achieved in GC/MS. Although mechanical difficulties have limited the acceptance for LC of the moving wire interface utilizing FID, dramatic improvements could come in several areas, such as localized wire heating in the ion source [26]. The theoretical capability of total transport of the solvent-free solute to the ion source, where either EI or CI spectra could be obtained, is certainly a powerful incentive for further research on this interface.

B. The Membrane Separator

One of the main types of interfaces used in GC/MS is the membrane separator, usually made of silicone rubber, designed by Llewellyn and Littlejohn [28]. This discriminates against the transmission of inert gases and many polar materials, and is selective for the transmission of relatively nonpolar materials which are sufficiently volatile at 250° C, the maximum working temperature of the silicone-rubber membrane. Westover and co-workers [29] have shown that enrichments of $^{\sim}10^{4}$ are possible for nonpolar organic materials in aqueous solution, and Jones and Yang [30] and Mieure and Dietrich [31] have shown examples of its use as an LC/MS interface. The device can be relatively simple and easy to construct, as, for example, when incorporated in the ion source end of the conventional MS direct probe [31]. A critical limitation is the rate of solute diffusion through the membrane, which can cause severe tailing of compounds of higher molecular weight (mol. wt.). For example [30], separation with a water/methanol gradient showed minimal tailing for anthracene (mol. wt. 178), substantial tailing for benzo[a]anthracene (mol. wt. 228) and serious tailing

for benzo[a]pyrene (mol.wt. 252). Subnanogram quantities of 3-methylcholanthrene were
detected by single-ion monitoring, although as a general rule the minimum useable
sample was about 25 ng. Despite the convenience of the membrane separator, with its
present capabilities as an LC/MS interface appear to be quite limited with applicability to few samples which cannot be examined by GC/MS.

VI. ATMOSPHERIC-PRESSURE VAPORIZATION

Two similar methods which have been proposed for LC monitoring involve vaporization
and ionization of the LC effluent at atmospheric pressure. In plasma chromatography
[32] the vaporized sample molecules are ionized by $\leq$66kV (mean = 17 kV) electrons
emitted from a ^{63}Ni foil; the resulting ions are accelerated and separated according
to their relative ion mobilities at atmospheric pressure. Subpicogram sensitivities
have been achieved, but this unique advantage is offset by the substantial variations
in sensitivity with the presence of other components and the observation that ion mobilities are not a strict function of mass/charge values [32]. Horning and co-workers
[33, 34] developed a system in which atmospheric-pressure ionization (API) is effected
with a ^{63}Ni source or a corona discharge, and the resulting ions are sampled through
a 25-μm pinhole into a mass spectrometer. Although subpicogram sensitivity has also
been demonstrated for trace components in air, a recent study applying this technique
as a continuous LC monitor gave a detection limit of 0.5 ng in the effluent solution.
Although this is a destructive detector, its sensitivity was shown to be comparable
to that of the uv detector for compounds such as polynuclear aromatic hydrocarbons
with high uv absorptivity, and far superior to the uv detector because of its specificity in monitoring for components of different molecular weight. The most severe
restriction of the API technique is the requirement of sample vaporization at atmospheric pressure; it appears that sample vapor pressure requirements are substantially
higher than those of normal direct probe introduction into the mass spectrometer,
limiting the number of compounds of insufficient volatility for GC separation which
can be run by LC/(API) MS.

VII. DIRECT ION SOURCE INTRODUCTION

As stated earlier, one of the basic problems of the LC/MS interface is the tremendous
disparity between the operating pressure at the LC exit and in the MS ion source.
Tal'roze and co-workers succeeded in continuously introducing volatile organic liquids
into a conventional EI ion source through a capillary leak system, but flow rates of
only ~10^{-6} ml/min were possible because of the <10^{-4}-torr sample pressure requirements
of the ion source [35]. Absorption of polar components cause difficulties which could
be alleviated by solvent rinsing [24], but cross contamination of samples, reproducibility of flow rates, and plugging were problems of the capillary interface system [23].

A. Solution Introduction With Direct Chemical Ionization [36-38]

MS sources utilizing chemical ionization operate at ~1 torr, making possible direct solution introduction at flow rates of ~10 µl/min. By using the LC solvent as the CI "reagent gas," approximately 1% of the LC effluent can be introduced continuously into the CIMS ion source. Although this obviously results in a reduction in sensitivity, the MS thus acts as an essentially "nondestructive" detector; the fractions which cannot be completely identified from the resulting CI mass spectra are available for examination by other techniques. For example, the solution could be dried down on the MS direct probe tip and a conventional EI spectrum obtained.

A low dead-volume interface utilizing DCI is shown in Fig. 1. The LC effluent flows past the end of a 0.07-mm i.d. glass capillary tube just inside the solids probe inlet of the MS. The other end of the tube terminates in the CI source, and the glass tip is pulled down to an orifice of about 5-µm i.d. restriction. The necessary heat of vaporization for the entering solution is supplied by electrical heaters in the ion source which heat the ion plasma. The turbulent ion plasma is thought to be especially efficient in effecting direct chemical ionization of the solute microcrystallites formed on evaporation of the solution; as noted above this makes possible the ionization of molecules whose vapor pressure is too low to obtain EI mass spectra by conventional direct probe introduction. There should be a very large pressure differential across the ion source; the volatile liquid solution flash vaporizes on entering, producing a gaseous volume of several ml/min even at atmospheric pressure, while the pumping is sufficiently rapid to produce pressures of ~10^{-4} torr in the ion source housing. The average sample residence time in the ion source should be <1 sec, so that the mixing here does not cause appreciable LC band spreading. If care is taken

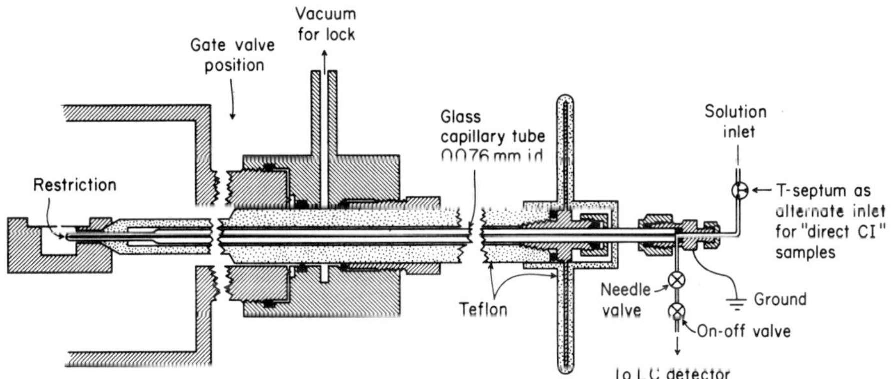

FIG. 1. Inlet probe for continuous introduction of solutions; replaces the direct introduction sample probe of the RMH-2, utilizing the same vacuum lock system [36].

to minimize dead volume in the splitter region at the entrance of the glass capillary, the LC peak shapes recorded with the CIMS detector are comparable to those found in duplicate runs using the uv detector.

For this system the maximum proportion of the LC effluent that can be introduced into the CIMS, and thus the detector sensitivity, is dependent on the instrument pumping speed. As mentioned, cryogenic pumping is efficient and relatively simple for the common LC solvents employed, but efficient operation is also possible if diffusion pumps alone are employed. Mechanical pumps should be vented, as unpleasant solvent breakdown products are often discharged, and the mechanical pump oil must be changed occasionally. In general, the ion source must be cleaned more frequently; with certain types of samples, this may be necessary every few weeks. The glass capillary tips are fragile and can be plugged if the solutions are not well filtered or if there are sudden concentration changes. However, the capillaries are relatively simple and easy to produce after a few trials.

Studies to date with this interface have been done with a magnetic mass spectrometer having a high potential on the ion source [36-39]. Many problems would be greatly reduced by using an instrument with a low source potential; J. D. Henion of Cornell has utilized such an LC/MS interface with a quadrupole mass spectrometer for analysis of a variety of low-volatility mixtures [41]. On the LC/MS pressure incompatibility problem, J. H. Futrell is constructing an interface of 1000x-higher pumping speed designed to take the full solvent load of a conventional LC.

B. LC/MS Operation

For the ideal chromatographic detector, all useful mobile phases (but no samples) should be "transparent". For GC/MS the carrier gases are generally of such low molecular weight (mol. wt.) that they cause no interference. Most useful LC solvents are of mol. wt. < 125; even those which tend to telomerize extensively in the CI source, such as H_2O, CH_3OH, and CH_3CN, give almost no peaks of m/e > 150 in their mass spectra at the operating pressures employed in this study. Fortunately, most samples for which LC is more appropriate than GC give the most abundant peaks in their CI spectra above m/e 150, so that almost all eluted sample components with sufficient vapor pressure for direct chemical ionization [12] are detectable.

The mass spectrum obtained depends on the nature of the solvent-solute couple, but with all the examples so far studied, a mass spectrum indicative of the solute could be obtained. In general, polar solvents such as tetrahydrofuran, acetonitrile, methanol, and water produce simple mass spectra where $(M + H)^+$ is often the most abundant ion. Saturated hydrocarbon solvents such as *n*-pentane give abundant hydride abstraction products, $(M - H)^+$. Chloroform gives mainly $CHCl_2^+$ and CCl_3^+ as solvent peaks, with peaks such as $(M - H)^+$, $(M + H)^+$, $(M + Cl)^+$, and $(M + CHCl_2)^+$ indicating the solute molecule.

The degree of molecular ion fragmentation (and thus the amount of molecular structure information) varies widely; as a general rule, the amount of excess energy transferred in ionization is dependent on the difference in proton affinities of the sample and ionizing reagent [20]. Behavior varies from cases in which the $(M + H)^+$ or $(M - H)^+$ is the major peak found to cases like oligopeptides, eluted with a H_2O/CH_3CN gradient [39], in which the DCI mass spectrum can give complete sequence information.

C. LC/MS/COM Systems

The described LC/CIMS system has also been coupled to a laboratory minicomputer (COM). The resulting LC/MS/COM system [38] shows many advantages now well established for GC/MS/COM systems, such as the real time preparation of reconstructed liquid chromatograms, mass chromatograms, and multiple ion detection [40].

For study of an unknown mixture using the LC/MS/COM system, mass spectra are collected repetitively during the LC run, using an MS scan time which is substantially less than the expected widths of the LC peaks. The simplest method of LC peak detection is to have the computer display a reconstructed chromatogram of the total ion

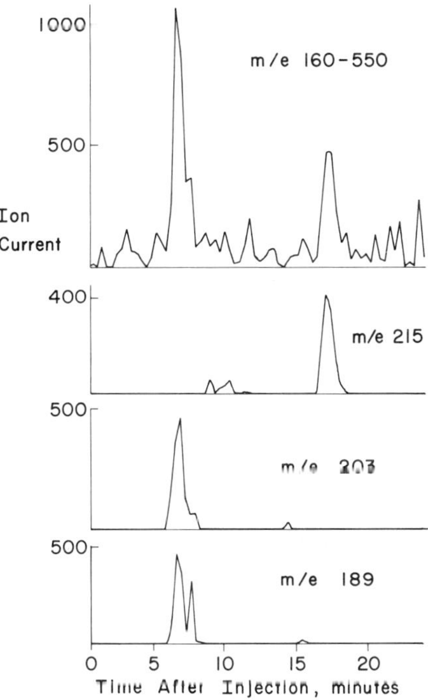

FIG. 2. Mass chromatograms for m/e 189, 203, 215, and the summed ions of m/e 160-550, in arbitrary units of ion current; LC separation of 50 x 10^{-9}g of trilaurin, ~1% of effluent to MS, signal threshold 100 units [38].

signal for all peaks of masses above those from the solvent. This is illustrated in
Fig. 2, a chromatogram resulting from the injection of a 50-ng sample of trilaurin,
mol. wt. 638, at the head of the LC column, eluted with methanol, with mass spectra
scanned repetitively (20 sec, m/e 70-700) on the ~1% of the effluent directed to the
MS. The CI spectrum of trilaurin with CH_3OH as the ionizing reagent shows a base
peak at m/e 215 and a smaller $(M + H)^+$ peak at m/e 639; the m/e 215 peak presumably
is protonated methyl laurate formed by reaction with the solvent ions. Figure 2
(m/e 160-550) indicates a number of eluted components from the supposedly pure sample. A "mass chromatogram" [40] using m/e 215 clearly shows the trilaurin eluting
with a retention time of 17 min. The m/e 215 peak eluting at ~9 min was found to be
reproducible, and is probably from lauric acid present in the sample as an impurity.
The largest peak in the reconstructed chromatogram was found to arise chiefly from
m/e 189 and 203 ions. These were found even with the injection of pure solvent, and
probably arise from impurities introduced from the septum. Many of the smaller peaks
also appear to be impurities from the inlet system, and scrupulous cleaning is necessary to achieve low noise chromatograms because of the generally high sensitivity of
CIMS to all compounds of sufficient volatility. Note that this sample gave no detectable signal on the uv detector.

D. Multiple Ion Detection

Improvement in sensitivity by monitoring only one or a few particular ions, a well-
known capability of GC/MS/COM systems [40] is also possible with the LC/MS/COM system. Injection of 10 ng of trilaurin gives a substantial m/e 215 chromatogram peak
using conventional analog recording (Fig. 3), with a substantial improvement in sig-
nal/noise with computer summation of the ion signal over 4-sec intervals (note also
the presence of the impurity of ~9-min retention time). A sample size of 0.5 ng of
trilaurin approaches the detection limit, although the integrated signal clearly
shows the presence of the peak. Note that this sensitivity is comparable or superior
to that shown by the uv detector for compounds of high molar absorptivities, despite
the fact that only 1% of the sample actually entered the mass spectrometer. If a
LC microcolumn could be developed requiring only ~0.01 ml/min solvent flow rates, the
total effluent could then go continuously to the CIMS. This should give picogram
sensitivities for LC/MS/COM, comparable to sensitivities now possible for GC/MS/COM.

E. Recent Developments

At the Seventh Triennial Mass Spectrometry Conference in Florence, Italy, September
1976, there were a number of papers on LC/MS coupling, and a round table discussion
on the subject chaired by V. L. Tal'roze with an attendance of several hundred. A
prototype model of the Finnigan moving ribbon system [27] was described by M. L. Story;

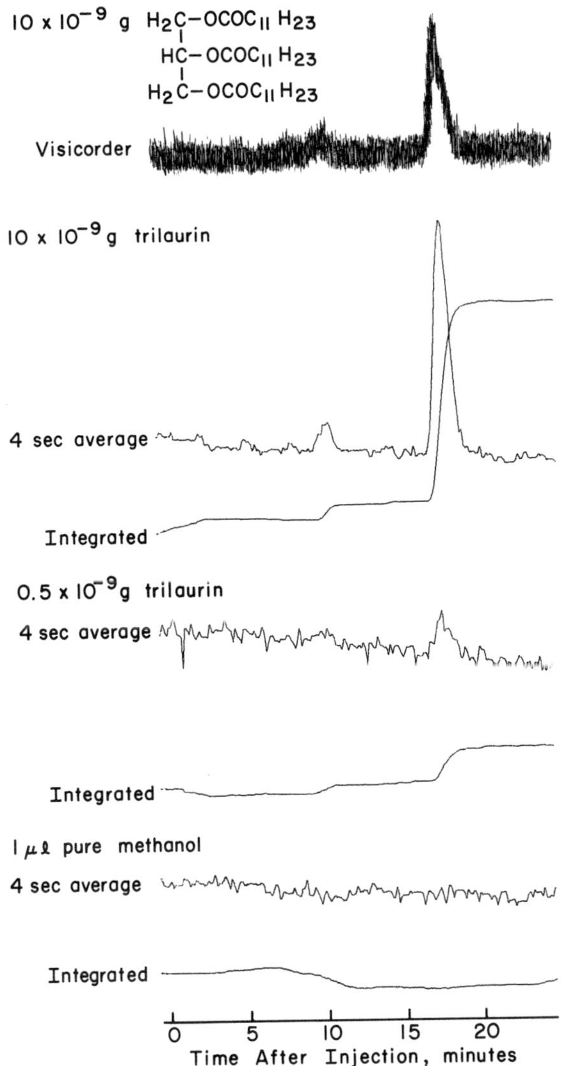

FIG. 3. Single ion detection of m/e 215 for LC separation of trilaurin, 1% of effluent to MS. "Visicorder" is an analog recording; "4-sec averaging" is computer summing of 50-kHz ion signals for 4-sec intervals; "integrated" is the running sum of these signals [38].

nanogram sensitivities have been achieved, but to date no compounds have been tried which have insufficient volatility for GC separation. Quite a number of LC users stressed the importance of the continuous performance of the LC/MS interface with compounds of marginal volatility. J. D. Henion of the School of Veterinary Medicine, Cornell University [41], described the application of the direct solution introduction/CI interface using a commercial quadrupole MS/COM system to natural product and metabolite mixtures. No other interfaces, including the moving wire and API systems, appeared to have been used in a routine fashion. The electrohydrodynamic ionization technique of C. A. Evans, Jr., and associates [42] might possibly be used as an LC/MS interface. Probably the most important conclusion to be gained from the round table session was the great utility foreseen for an LC/MS interface which would show performance characteristics comparable to those routinely achieved now with GC/MS systems.

ACKNOWLEDGMENTS

We are grateful to the following scientists who provided advice and/or special information on their own research results in the field: Drs. R. P. W. Scott, M. A. Baldwin, P. J. Arpino, R. Venkataraghavan, P. Irving, W. H. McFadden, and E. M. Chait. R. M. King and J. L. Waters of Waters Associates furnished helpful advice and the liquid chromatograph used in this study.

This work was supported by National Institutes of Health grant GM 16609 and the computer was purchased in part by funds supplied by National Science Foundation grant MPS 74-19871.

REFERENCES

1. L. R. Snyder and J. J. Kirkland, *Introduction to Modern Liquid Chromatography*, Wiley, New York, 1974.

2. B. L. Karger, M. Martin, and G. Guiochon, *Anal. Chem.*, *46*, 1640(1974).

3. P. J. Arpino, *La Recherche*, *6* (59), 769(1975).

4. R. C. Williams and J. T. Larmann, *Indust. Res.*, *17*, 60(August 1975).

5. R. A. Mowery, Jr. and R. S. Juvet, Jr., *J. Chromatogr. Sci.*, *12*, 687(1974).

6. F. W. Willmott and R. J. Dolphin, *J. Chromatogr. Sci.*, *12*, 695(1974).

7. R. P. W. Scott and J. G. Lawrence, *J. Chromatogr. Sci.*, *8*, 65(1970).

8. B. M. Lapidus and A. Karmen, *J. Chromatogr. Sci.*, *10*, 103(1972).

9. J. J. Szakasits and R. E. Robinson, *Anal. Chem.*, *46*, 1648(1974).

10. W. W. Shultz and W. H. King, Jr., *J. Chromatogr. Sci.*, *11*, 343(1973).

11. H. D. Beckey, *J. Mass Spec. Ion Phys.*, *2*, 500(1969).

12. M. A. Baldwin and F. W. McLafferty, *Org. Mass Spectrom.*, *7*, 1111, 1353(1973).

13. R. D. Macfarlane and D. F. Torgerson, *Science*, *191*, 920(1976).

14. B. Fleet and C. J. Little, *J. Chromatogr. Sci.*, *12*, 747(1974).

15. N. K. Freeman, F. T. Uphan, and A. A. Windsor, *Anal. Lett.*, *6*, 943(1973).

16. R. J. Buehler, E. Flanigan, L. J. Greene, and L. Friedman, *J. Amer. Chem. Soc.*, *96*, 3990(1974).

17. H.-R. Shulten and H. D. Beckey, *Org. Mass Spectrom.*, *6*, 885(1972).

18. S. Asante-Poku, G. W. Wood, and D. E. Schmidt, Jr., *Biomed. Mass Spectrom.*, *2*, 121(1975).

19. H.-R. Shulten and H. D. Beckey, *J. Chromatogr. Sci.*, *83*, 315(1973).

20. F. H. Field, in *Mass Spectrometry* (A. Maccoll, ed.), Butterworths, London, 1972, p. 133.

21. J. H. Futrell and L. H. Wojcik, *Rev. Sci. Instrum.*, *42*, 244(1971).

22. D. F. Hunt, C. N. McEwen, and T. M. Harvey, *Anal. Chem.*, *47*, 1730(1975).

23. V. L. Tal'roze, G. V. Karpov, I. G. Gorodetskii, and V. E. Skurat, *J. Phys. Chem.* (USSR) *43*, 198(1969).

24. V. E. Skurat, *J. Phys. Chem.* (USSR) *46*, 570(1972).

25. R. E. Lovins, S. R. Ellis, G. D. Tolbert, and C. R. McKinney, *Anal. Chem.*, *45*, 1553(1973).

26. R. P. W. Scott, C. G. Scott, M. Monroe, and J. Hess, Jr., *J. Chromatogr.*, *99*, 395(1974).

27. W. H. McFadden, Finnigan Instruments, private communication, October 1975.

28. P. M. Llewellyn and D. P. Littlejohn, U. S. Patent 3,429,105, February 1969.

29. L. B. Westover, J. C. Tou, and J. H. Mark, *Anal. Chem.*, *46*, 568(1974).

30. P. R. Jones and S. K. Yang, *Anal. Chem.*, *47*, 1000(1975).

31. J. P. Mieure and M. W. Dietrich, First North American Chemical Congress, Mexico City, December 1975, ENVT 59.

32. F. W. Karasek, *Anal. Chem.*, *46*, 710A(1974).

33. D. I. Carroll, I. Dzidic, R. N. Stillwell, K. D. Haegele, and E. C. Horning, *Anal. Chem.*, *47*, 2369(1975).

34. M. McKeown and M. W. Siegel, *Amer. Lab.*, *7*, 89(November 1975).

35. V. L. Tal'roze, G. V. Karpov, I. G. Gorodetskii, and V. E. Skurat, *J. Phys. Chem.* (USSR), *42*, 1658(1968).

36. P. J. Arpino, M. A. Baldwin, and F. W. McLafferty, *Biomed. Mass Spectrom.*, *1*, 80(1974).

37. P. J. Arpino, F. W. McLafferty, and B. G. Dawkins, *J. Chromatogr. Sci.*, *12*, 574(1974).

38. F. W. McLafferty, R. Knutti, R. Venkataraghavan, P. J. Arpino, and B. G. Dawkins, *Anal. Chem.*, *47*, 1503(1975).

39. F. W. McLafferty and B. G. Dawkins, *Biochem. Soc. Trans.*, *3*, 856(1975).

40. R. A. Hites and K. Biemann, *Anal. Chem.*, *42*, 855(1970).

41. J. D. Henion, *Adv. Mass Spectrom.*, 7(1977).

42. D. F. Simons, B. N. Colby, and C. A. Evans, Jr., *Int. J. Mass Spectrom. Ion Phys.*, *15*, 291(1974).

Chapter 8

ISOLATION OF SAMPLES PRIOR TO CHROMATOGRAPHY

Howard Ko

Edgar N. Petzold

The Upjohn Company
Kalamazoo, Michigan

I. INTRODUCTION

The chromatographic analysis of a crystalline drug often is simple straightforward
procedure. However, a sample of this same compound in a biological milieu where
hundreds of other substances are present, frequently overwhelm the capabilities of
present day GLC and HPLC systems. Only in rare instances is no cleanup required and
the drug concentration simply adjusted before the drug is injected directly into the
chromatograph [1]. These substances might interfere by interaction with the drug
itself, or possibly with the column packing or detector of the GLC or HPLC system.
Thus, additional procedures are nearly always required prior to chromatography to
reduce the number of interfering substances so that GLC and HPLC can resolve the remaining components efficiently with a lower "noise" level and detection limit.

An interaction between the drug and the sample constituents can take place by
encapsulation, by adsorption, or by reaction. Except in cell-free fluids (e.g.,
plasma, urine, or fermentation beer), drugs must be isolated from biological matrices
such as tissue, cells, and mycelia, and thus be freed of their encapsulating environ-
ment. This is usually done by homogenizing and extracting. However, certain compounds are impervious to cellular membranes and consequently must be freed by disrupting them ultrasonically, osmotically, or enzymatically. Even when freed of the
biological matrix, the drug can become bound or solvated by other sample constituents,
e.g., proteins or oils, or adsorbed to the container walls. . Thus, serious losses
can occur in those isolation steps where these constituents are removed or when the
sample is being transferred.

Other undesirable effects are where sample impurities interact with the column
packing or detector of an HPLC or GLC system. Some detectors easily become saturated
and even poisoned. To the extent that the detector is not exclusively specific for
the drug, interfering materials will tend to raise the noise level and, hence, the
threshold detection limit; or a peak may even appear in the chromatogram as a "pseudo
drug." Excessive interfering substances can reduce and even nullify the resolving
power of column packings by overloading. Thus, to lower the detection limit and to
efficiently resolve the drug from its metabolites or concomitant impurities in the
sample, purification is usually required prior to chromatography.

The purification of the drug, once it is freed from its encapsulating envirom-
ment, depends upon the chemical nature of the drug and its contaminants. As a general rule, those extraction conditions should be chosen which will selectively extract
the drug, leaving unextracted as much of the endogenous material as possible while
providing quantitative recovery of the drug. The problems encountered in the subsequent purification steps will be related to the amount of extracted nondrug materials.
Some of the properties that can be utilized for separating compounds into classes
are polarity, affinity for other molecules in solvents or on surfaces, charge, and

molecular size. These properties determine their distribution between immiscible
solvents, solubility, volatility, and chromatographic properties.

Some procedures for isolating drugs from an extract include extraction into an
immiscible solvent, volatilization, and chromatography. Alternatively, the contaminating compounds can be removed by extraction and/or precipitation, thereby enriching
the drug concentration in the extract. Absorbents used in open tubular columns or on
TLC plates, e.g., charcoal, florisil, alumina, silica gel are often useful for perform-
ing class separations. Although these isolation systems do not have the resolving
power of those in HPLC/GLC, they can be heavily loaded with sample and thus serve well
in cleaning up the sample prior to HPLC/GLC. After elution of the drug from the adsorbent, the sample containing the enriched drug is often taken to dryness and derivatized to improve its volatility for GLC or uv absorbance for HPLC. The following
sections describe some of these techniques and their applications.

II. INTERNAL STANDARDS

To help correct for recovery of a compound, an internal standard similar to the compound of interest can be added to the media or test tube before isolation is begun.
A good internal standard will negate bias in the assay from losses of drug in the
isolation steps and improve the precision of the result by compensating random errors
associated with aliquot taking, derivatization, and instrumental performance. An internal standard ideally should accompany the drug in a constant ratio throughout all
isolation steps, and then be separable on the GLC or HPLC system.

Homologs or analogs of the compound differing by the addition, e.g., of a methyl
group or halogen will be excellent internal standards when the differences in polarity
between drug and internal standard are negligible. The reasons for this are as follows. Isolation steps involving extraction and TLC are most responsive to differences
in polarity, but are insensitive to changes in molecular weight. GLC, however, is
sensitive to differences in molecular weight and, hence, will often separate homologs
having like polarity.

III. ISOLATION OF A DRUG FROM AN ENCAPSULATING ENVIRONMENT

A. Osmotic Shock

Fragile cell membranes often can be disrupted by osmotic shock; the shocking reagent
can then also function in precipitating the proteins before extraction of the drug.
As an example, nucleosides, nucleotides and purine and pyrimidine bases [2] were isolated from erythrocytes by adding a suspension of the erythrocytes dropwise to cold,
vigorously stirred trichloroacetic acid (12%, TCA). The TCA osmotically shocks and
disrupts the cells, and precipitates the protein.

B. Disruption of Cells by Boiling

Another way of disrupting blood cells and precipitating proteins is to heat blood to
the point of denaturing the protein [3]. As part of the procedure for isolating
clonazepam from blood, a 1-ml sample in 2 ml borate buffer (pH 9.0) was heated intermittently for no more than 30 sec or until a rust-brown precipitate formed. The drug
was extracted after cooling the sample to room temperature.

C. Tissue Grinders

Soft tissues such as those in brain can be homogenized in a tissue grinder (e.g.,
Potter-Elvehjem size 19 homogenizing tube with an AAA pestle, medium ground glass,
1-ml capacity) in a small amount of extracting medium [5]. To avoid metabolism or
degradation during homogenization, the tissue can be frozen in aluminum pans with
solid CO_2 or liquid nitrogen prior to grinding. A solids-free extract can usually
be obtained optimally from the homogenate by centrifugation (e.g., 12,000g at $4^{o}C$
in a Sorvall RC2-B centrifuge equipped with an SM-24 rotor).

More resistant tissues or organic particles (up to 5mm) can be treated by homo-
genizers such as the Polytron series which can handle volumes ranging from 1 to 5000
ml depending upon the model (Brinkmann Instruments, Inc.). Thermally labile samples
should be chilled during processing to avoid degradation.

D. Ultrasonic Disruption

Plant and animal cells have been disrupted by means of ultrasonic waves. The collaps-
ing cavitation bubbles which are generated tend to break most easily covalent bonds
of polymers having molecular weights of greater than 0.5 million and thus can selectively rupture tissue cell walls while leaving the drug intact. The disruption of
tissue in order of increasing difficulty [4] is shown in Table 1. There are, however,
sono-chemical reactions which might potentially lead to drug degradation [4]. Ultrasonic waves can generate temperatures estimated at 10,000K at the instantaneous point
when the microscopic bubbles collapse. The localized heat energy induces free radical
formation and "sonic luminescence," which, in the presence of air and water, produce
hydrogen atoms, hydroxyl radicals, hydrogen peroxide, molecular oxygen and hydrogen,
and nitrous and nitric acid. In addition, the rates of the following phenomena increase: ester hydrolysis, hydration of acetylene, dissolution of powders, decalci-
fication of bone, oxidation of unsaturated oils, and aging of alcohols. One way to
reduce or avoid free-radical formation and sonic luminescence generated by ultrasonic
waves is to degas the water or saturate it with hydrogen or carbon dioxide. Another
way is to add a small quantity of a volatile liquid heat sink, e.g., acetone or methanol, to rapidly dissipate the heat.

TABLE 1. Disruption of Tissue in Order of Difficulty[a]

1.	Brain	11.	Kidney
2.	Mucous	12.	Lung
3.	Tissue Culture	13.	Fibrin
4.	Fat	14.	Skin
5.	Feces	15.	Muscle
6.	Necrotic Tissue	16.	Bone
7.	Tumor Cells	17.	Heart Muscle
8.	Liver	18.	Green Leaves
9.	Bladder	19.	Hair
10.	Aorta	20.	Renal Papilla

[a]Copyright held by International Scientific Communications, Inc.

E. Hydrolysis by an Acid, a Base, or Enzymes

The drug may combine in various ways with components or compounds in the biological milieu. Protein-drug complexes may be formed with drugs containing a free amino group. Drugs with a carboxyl group may form sugar esters such as glucuronides. One way to free the drug moiety is by acidic or alkaline hydrolysis, provided it is stable under these conditions. Some fat soluble substances, e.g., chlorinated hydrocarbons can be freed of fatty tissues by digestion with perchloric and acetic acid [9, 50].

The use of enzymes is preferred to acids or bases when milder hydrolysis conditions are needed to avoid drug degradation or an undesirable hydrolyzate which would complicate the isolation steps to follow, and when selective hydrolysis is desired.

One may choose from a variety of sources of enzymes to select the type of hydrolysis or conditions desired. The pH optima for different β-glucuronidases vary from pH 3.8 to pH 7 depending upon the source of the enzyme [6]. Good specific sources of β-glucuronidase are the female rat preputial gland [6] and beef liver [45], which are both devoid of sulfatase activity. Good sources of sulfatase in addition to β-glucuronidase are extracts of the snail, *H. pomatia*, and of the limpet, *Patella vulgata* [6]. Known inhibitors of β-glucuronidases are glucaro-(1 ﹐ 4)-lactone, heparin, chondroitin sulfate, acid mucopolysaccharides, hyaluronic acid and ions such as Hg^+, Hg^{2+}, and Cu^{2+}. Substances such as albumin, chitosan and ethylenediaminetetracetic acid activate or stablize the enzyme, presumably by protecting it from heavy-metal surface denaturation [6].

Typical examples of drug hydrolysis in urine with β-glucuronidases are the following. Glucuronides of tolmetin and its carboxylic acid derivative [7] were hydrolyzed in 16 hr (pH 6.8, phosphate buffer) with bacterial β-glucuronidase. Conjugates of minoxidil [8] and conjugates of a metabolite of mefexamide [52] were hydrolyzed in 24 to 64 hr at 37°C (~pH 5.5) employing 1000 Fishman units of beef liver glucuronidase.

IV. REMOVAL OF INSOLUBLE MATERIALS

A. Centrifugation

Insoluble materials can be quickly removed by centrifuging the sample at 1000-3000g for 10 sec to 10 min, depending upon the sample (e.g., see Refs. 12, 15, and 31). Centrifuges such as the International PR-6000 or the Sorvall RC-2 have rotors which can centrifuge seventy-two 12- x 75-mm tubes simultaneously.

B. Filtration

Although filtration may appear superficially to be a simple straightforward procedure, it can be a tedious step when filtration becomes difficult. When filtration problems occur, one should seriously consider alternative procedures for removing the drug from the homogenate, e.g., centrifugation, steam distillation, or liquid extraction (batchwise or continuous). When quantitative recovery of the entire sample is unnecessary, taking an aliquot of the filtrate will cut down the time required for filtration. Several techniques for improving filtration rates are

1. The use of organic solvents or precipitating agents, e.g., solutions of trichloroacetic acid or lead acetate, which will coagulate proteins.
2. Addition of a filter aid, e.g., diatomaceous earth or cellulose pulp, batchwise.
3. Imbedding the sample in a filterable material, e.g., cellulose powder, and extracting in a chromatographic tube [42].
4. A low pH (2-5), if compatible with the drug, usually provides a faster filtration rate than that under neutral or alkaline conditions.

Some molecular filters have recently become available which provide selective filtration of low molecular-weight substances from high molecular-weight materials. Ultrafiltrates from molecular filtration of plasma can be injected directly into a liquid chromatograph for the determination of theophylline [43].

V. EXTRACTION

Liquid-liquid extraction is a well-used tool for cleaning up samples prior to analysis for several reasons. First, the drug can usually be extracted away from high concentrations of contaminating materials. This separation step is often employed at an early step in the procedure to enrich the drug and reduce the solids in the sample.

Second, this procedure is simple, rapid and several samples can be extracted simultaneously. Third, phasic separations can be enhanced by centrifugation. Fourth, the sample can often be rapidly concentrated by extraction into a smaller volume of a solvent having a larger solubility of the drug.

A. Choice of Solvents

The choice of solvents is dependent upon two main considerations:

1. The solvents after equilibration must have formed two immiscible phases.
2. The substance(s) to be analyzed must strongly favor distribution into one phase, while the impurities must strongly favor distribution into the other phase.

The distribution ratio D of a drug between the two phases is defined to be [10]

$$D = \frac{\text{Stoichiometric concentration of all species of the drug in the organic phase}}{\text{Stoichiometric concentration of all species of the drug in the aqueous phase}} \quad (1)$$

Inefficient extraction (< 50%) usually leads to a larger variation in recovery of the drug since it is more susceptible to nominal fluctuations in conditions, e.g., sample composition, temperature, and pH. Reproducibility will generally improve with efficiency of extraction.

Compounds have been classified by their ability to enter into different types of intermolecular interactions, viz., dispersion, dipolar, and hydrogen-bonding interactions. The generality that solutes tend to dissolve into like solvents can be applied with the help of the interaction parameters shown in Table 2 [10].

Compounds tend to dissolve best in solvents that have the same polarity or solubility parameters (SP), where $(SP)^2$ is the intermolecular interaction energy per unit volume in the pure liquid [10], i.e., the solubility parameter (SP) increases with the solvent polarity. As the difference in the solubility parameters between two solvents increases, the mutual solubility of the solvent pairs decreases until they become partially immiscible, a condition mentioned above which is required for separating compounds. Likewise, solutes which have large dispersion parameters (DD) tend to interact more with solvents with large dispersion parameters, and solutes with large dipole parameters (DP) interact more strongly with solvents with large dipole parameters. Hydrogen-bonding parameters follow the principle that a proton donor solute (DA) will tend to interact with a proton-acceptor solvent (DB), and vice versa.

These specific intermolecular interactions can be used as a basis for finding solvents that will preferentially separate components which differ in their interactions with the solvents.

In order to obtain a solvent of a given polarity, two miscible solvents which approximate the desired polarity can be mixed. The best separation for a given

TABLE 2. Interaction Parameters[a] of Different Solvents[b]

Solvent	SP	DD	DP	DA	DB
Perfluoroalkanes	6.0	6.0	0	0	1.0
$CFCl_2$-CF_3	6.2	5.9	1.5	0	0
Isooctane	7.0	7.0	0	0	0
Diisopropyl ether	7.0	6.9	0.5	0	0.5
n-Pentane	7.1	7.1	0	0	0
CCl_3-CF_3	7.1	6.8	1.5	0	0.5
n-Hexane	7.3	7.3	0	0	0
n-Heptane	7.4	7.4	0	0	0
Diethyl ether	7.4	6.7	2.0	0	2.0
Triethylamine	7.5	7.5	0	0	3.5
Cyclopentane	8.1	8.1	0	0	0
Cyclohexane	8.2	8.2	0	0	0
Propyl chloride	8.3	7.3	3.0	0	0
CCl_4	8.6	8.6	0	0	0.5
Diethyl sulfide	8.6	8.2	2.0	0	0.5
Ethyl acetate	8.6	7.0	3.0	0	2.0
Propylamine	8.7	7.3	4.0	0.5	5.0
Ethyl bromide	8.8	7.8	3.0	0	0
m-Xylene	8.8	8.8	0	0	0.5
Toluene	8.9	8.9	0	0	0.5
$CHCl_3$	9.1	8.1	3.0	0	0.5
Tetrahydrofuran	9.1	7.6	4.0	0	3.0
Methyl acetate	9.2	6.8	4.5	0	2.0
Benzene	9.2	9.2	0	0	0.5
Perchloroethylene	9.3	9.3	0	0	0.5
Acetone	9.4	6.8	5.0	0	2.5
CH_2Cl_2	9.6	6.4	5.5	0	0.5
Chlorobenzene	9.6	9.2	2.0	0	0.5
Anisole	9.7	9.1	2.5	0	2.0
1,2-Dichloroethane	9.7	8.2	4.0	0	0
Methyl benzoate	9.8	9.2	2.5	0	1.0
Dioxane	9.8	7.8	4.0	0	3.0
Methyl iodide	9.9	9.3	2.0	0	0.5
Bromobenzene	9.9	9.6	1.5	0	0.5
CS_2	10.0	10.0	0	0	0.5
Propanol	10.2	7.2	2.5	4	4.0
Pyridine	10.4	9.0	4.0	0	5.0

TABLE 2.   (Continued)

Solvent	SP	DD	DP	DA	DB
Benzonitrile	10.7	9.2	3.5	0	1.5
Nitromethane	11.0	7.3	8.0	0	1.0
Nitrobenzene	11.1	9.5	4.0	0	0.5
Ethanol	11.2	6.8	4.0	5	5.0
Phenol	11.4	9.5			
Dimethylformamide	11.5	7.9			
Acetonitrile	11.8	6.5	8.0	0	2.5
Methylene iodide	11.9	11.3	1.0	0	0.5
Acetic acid	12.4	7.0			
Dimethylsulfoxide	12.8	8.4	7.5	0	5.0
Methanol	12.9	6.2	5.0	5	5.0
1,3-Dicyanopropane	13.0	8.0	8.0	0	3.0
Propylene carbonate	13.3				
Ethanolamine	13.5	8.3	Large	Large	Large
Ethylene glycol	14.7	8.0	Large	Large	Large
Formamide	17.9	8.3	Large	Large	Large
Water	21.0	6.3	Large	Large	Large

[a]Interaction parameters: (SP) solubility parameter (calculated from the boiling point), (DD) dispersion solubility parameter, (DP) orientation (polar) solubility parameter, (DB) proton-acceptor solubility parameter, and (DA) proton donor solubility parameter.

[b]Copyright held by John Wiley and Sons, Inc.

solubility parameter will usually be obtained when the difference between the two solubility parameters is maximal and yet the mixture is still monophasic [10].

The optimum separation between two substances A and B occur [10] when

$$C_a C_b = 1 \tag{2}$$

where

C_a = capacity factor of A = $\dfrac{\text{total amount of A in the organic phase}}{\text{total amount of A in the aqueous phase}}$

C_b = capacity factor of B

A similar definition applies to the capacity factor of B. To exemplify with a situation where the drug has a distribution ratio, D, which is 10x that of the contaminant, the most favorable solvents would extract 90% drug and 10% contaminant. The drug would be extracted with too great an efficiency, *over-capacity*, if 99.9% of the drug and 99% of the contaminant were extracted.

B. Extracting Drugs with Ionizable Groups

Drugs having ionizable groups will be most efficiently extracted into the organic phase from the aqueous phase when uncharged. At a given pH near the pK_a of an acidic drug, there will be a protonated species, HA (uncharged), and an unprotonated species, A^- (charged). For the basic drug, the protonated species will be BH^+ (charged) and the unprotonated species, B (uncharged). The predominantly uncharged species of the drug will exist in the pH region below the pK_a for acids, above the pK_a for bases and at the isoelectric pH for amphoteric compounds. Let us consider first extraction of the acidic drug [10].

The distribution of HA between the organic and aqueous phases may be represented by the equation,

$$(HA)_o = (HA)_w \, K_d \tag{3}$$

where $(HA)_o$ represents the concentration of HA in the organic phase, $(HA)_w$ is the concentration in the aqueous phase, and K_d is the distribution coefficient. When the acidic drug dissociates in the aqueous phase, the total concentration will be the sum of the concentrations of the ionized (A^-) and unionized species (HA). Using Eq. (1), the distribution ratio

$$D = \frac{(HA)_o}{(HA)_w + (A^-)_w} \tag{4a}$$

Considering that

$$K_a = \frac{(H_3O^+)(A^-)_w}{(HA)_w}$$

we find

$$D = \frac{(H_3O^+)K_d}{(H_3O) + K_a} \tag{4b}$$

This equation for the distribution ratio shows that at relatively low pH values [high hydronium ion concentrations, where $(H_3O^+) \gg K_a$], the distribution ratio approximates the distribution coefficient, i.e.,

$$D = K_d \tag{5}$$

When the hydronium ion concentration equals the dissociation constant, i.e., $pH = pK_a$, then

$$D = \frac{K_d}{2} \tag{6}$$

When the hydronium ion concentration is much smaller than the dissociation constant, i.e., pH > pK_a, so that the contribution of the $(H_3O)^+$ to the divisor of Eq. (4b) becomes negligible, then

$$D = \frac{(H_3O^+)K_d}{K_a} \tag{7}$$

We see that for the pH values much higher than the pK_a, the relative amount of HA in the organic phase becomes vanishingly small.

For a basic drug, analogous equations would be obtained except that higher ratios of uncharged to charged species would be obtained at pH values greater than the pK_a and, consequently, higher distribution ratios. Alternatively, a lower fraction of the drug would be uncharged at pH values lower than the pK_a of the basic group and consequently lower distribution ratios would be obtained.

1. Isolation of Organic Acids

Recently published methods for free fatty acids [11] and tolmetin [12] illustrate the utility of extracting from the aqueous phase into the organic phase, and vice versa, simply by manipulating the pH of the aqueous phase.

In the method for free fatty acids in plasma, the sample containing internal standard in a stoppered centrifuge tube was protonated into the uncharged species with a mineral acid, and extracted with heptane-isopropanol by vigorously mixing and then allowing the phases to stand. The fatty acids in the heptane phase were sufficiently pure for derivatization and GLC analysis.

Essentially the same rationale was used for extracting tolmetin [12] from plasma into ethyl ether, but the extract still contained other materials, presumably neutral and basic lipoid compounds, which could also be removed by liquid-liquid extraction. After the aqueous phase was frozen with a dry ice acetone bath, the drug in the ether extract was transferred into aqueous base, and again after acidulation, was extracted back into ethyl ether. At high pH, the neutral lipids and any basic compounds remained with the ether phase, and hence were removed in the aqueous base extraction step. The drug in the final ethyl ether phase could be concentrated under mild conditions and was then derivatized and analyzed by GLC.

2. Isolation of Basic Drugs

Equations similar to those given above for acidic drugs define the distribution of basic drugs between aqueous and organic phases, where the essential difference is that the uncharged species of the base is obtained in the pH region above its pK_a. Thus, the distribution coefficient increases with increase in pH, i.e., the distribution of drug is more favorable to the aqueous phase below the pK_a and more favorable to the

organic phase above the pK_a. Basic drugs are often extracted from alkaline aqueous
solution into an organic solvent such as diethyl ether. Thus, the basic drug is extracted away from aqueous soluble compounds, e.g., proteins, amino acids, and salts of
some organic acids, but not from neutral lipids which are also extracted. The extent
of the purification depends upon the extraction efficiency of the organic solvent for
the various components. In some cases, basic drugs [13], such as cocaine, codeine,
ciclazocine, levallorphan, levorphanol, methadone and pentazocine, nalorphine, and
naloxone, were recovered in yields of > 80% from plasma with a single extraction into
benzene-isopropanol (9:1). After removal of the solvent, the drugs were ready for
dissolution, silylation where required, and GLC analysis.

Other methods [14] required more extensive cleanup and the drug was back extracted
from the organic solvent into an aqueous acidic solvent, e.g., 1 N HCl, at a pH below
the pK_a of the drug, leaving behind most of the neutral lipoid compounds and more of
any remaining acidic compounds. The basic drug was reextracted into an organic solvent, e.g., dichloroethane, in a pH region above the pK_a of the basic group. Extraction in and out of the aqueous phase in this manner provided an enriched extract of
organic bases which was compatible with GLC analysis. This procedure, for example,
gave an 83% recovery of thioridazone from plasma [14]. In another method [15], amines
($5 \times 10^{-3} M$ to $6 \times 10^{-7} M$) were extracted into benzene from 0.1 M NaOH with a recovery
of more than 95% in a single extraction, when a phase-volume ratio of the aqueous phase
V_{aq} to organic phase V_{org}, V_{aq}/V_{org} = 1/3 was used. Solvents more polar than benzene
were not suitable for extraction because they increased the extraction of extraneous
material from biological samples. Extraction of the amines from benzene into an acidic
water phase, and then reextraction of the water phase at high pH, reduced the back-
grounds in GLC-EC chromatograms.

3. Amphoteric Drugs

For amphoteric compounds having one acidic and one basic group, the isoelectric point
pI is midway between the pK_a values of the two groups, viz.,

$$pI = \frac{pK_{a1} + pK_{a2}}{2} \tag{8}$$

If the pK_a of the base, e.g., an aromatic amine, is less than the pK_a of the acid,
e.g., a phenolic group, the molecules will be more favorably distributed into the organic phase in the pH region near the isoelectric point. In the reverse situation
where the pK_a of the base is greater than that of the acid, e.g., the amino acids,
the molecule will contain separated but equally opposite changes at the isoelectric
point. These charges will increase the molecule's aqueous solubility, despite its
electrical neutrality; thus it becomes more difficult to extract into organic solvents.

C. Ion-Pair Extraction

When compounds are hydrophilic and charged, extraction into the organic phase will be inefficient, i.e., the distribution ratio will be small. The extraction efficiency can often be improved if the compound is extracted as a complex that is more soluble in the organic phase [22]. For example, anions of sulfonic or carboxylic acids can be extracted as complexes with cationic dyes or phenothiazines and organic ammonium ions as complexes with anions such as picrate, dipicrylamine, chloride, nitrate, or iodide [22].

Suppose a cation HA^+ complexes with B^- to form an ion pair, HAB. Then

$$K_{HAB} = \frac{HAB}{HA^+ + B^-} \tag{9}$$

where K_{HAB} is the extraction constant, HAB is the concentration of the ion pair in the organic phase, and HA^+ and B^- are the concentrations of the ions in the aqueous phase. If there are no side reactions, the distribution ratio D_{HAB} is

$$D_{HAB} = \frac{HAB}{HA^+} = K_{HAB}(B^-) \tag{10}$$

Thus, the relative amount of the complex in the organic phase can be adjusted by varying the type or concentration of the counterion, B^-.

For some compounds, the partition ratio increases rapidly with decreasing concentration because of dissociation of the ion pair in the organic phase. For example, in the extraction of choline into CH_2Cl_2 from an aqueous phase at pH 11.5 as an ion pair with dipicrylamine, the partition ratio increases rapidly with decreasing concentration. Thus, when the concentration is below 10^{-5} mole/liter, choline is extracted quantitatively (i.e., $D_{choline} > 100$). Thus, in investigating ion pair extraction as a means for isolating trace amounts of drugs in biological systems, the possibility of the partition ratio increasing with decreasing concentration of drug should be investigated.

When components of an ion pair undergo protolysis, the uncharged species that results can distribute between the organic and aqueous phases. The distribution ratio of the drug between the two phases should therefore be measured as a function of pH in order to determine where the optimal distribution occurs, as e.g., in the extraction of piribenzil, dextropropoxyphene, nicotine, and tryptamine as an ion pair with bromothymol blue into methylene chloride [23].

Distribution of ion pairs into the organic phase increases with solvation of the ion pair. The ability of molecules of the organic phase to solvate increases with the polarity of both the solvent and solute and with the ability of the solvent to form hydrogen bonds with the solute. The extraction constants of picrate ion pairs [24] decreases with the following solvents (dielectric constant): 1-pentanol (13.9) > methylisobutylketone (13.1) > ethyl acetate (6.0) > methylene chloride (8.9) > chloroform (4.8) > benzene (2.28) > carbon tetrachloride (2.24).

A good example on the use of ion-pair extraction as an efficient "cleanup step" is in the analysis of spectinomycin, an antibiotic containing two secondary amine groups, from various biological samples [44]. The antibiotic is first extracted from tissue in dilute trichloroacetic acid and lead acetate. The normally water-soluble antibiotic is partitioned into methylene chloride containing sodium dinonyl sulfonates (NaDNNS), resulting in the formation of the spectinomycin$^+$-DNNS$^-$ ion pair. The antibiotic is back-extracted into water by addition of a fatty amine, viz., tricaprylyl monomethyl ammonium chloride (Aliquot 336), which forms a stronger ion pair, Aliquot 336$^+$DNNS$^-$.

D. Drugs with No Ionizable Groups

Lipophilic drugs which remain unionized and stable over a wide pH range (for example, 2-12) can be purified by extracting contaminating compounds into the aqueous phase as their salts. The drug might be extracted into the organic solvent first at a pH of 12, which would leave anionic salts of organic acids unextracted in the aqueous phase. A second extraction of the organic layer might be made at pH 2, which would remove cationic salts of organic bases.

Nonpolar contaminants can be removed by dissolving the drug in a relatively polar solvent and extracting the nonpolar contaminants into a nonpolar solvent. For example, a procedure [16] was described for extracting nanogram amounts of steroids from plasma and urine. Ammonium carbonate (2.5 g) was added to plasma (2 ml) and water (3 ml) to increase the dielectric constant so that the neutral steroids would be more efficiently extracted into ethyl acetate (3 times with 2 ml.). The combined extracts were washed with water (0.3 ml) and dried with a stream of nitrogen. Fatty acids and triglycerides were removed by dissolving the residue in methanol-10% aqueous acetic acid (8:2, v/v) and extracting with equilibrated hexane. The steroids in the methanol solution were subsequently derivatized prior to analysis by GC-MS. Recoveries of [^{14}C]steroids (cholesterol, cortisol, cortisone, dexamethasone, testosterone, and norethindrone) and [^{3}H]steroids (estradiol and estriol) were 85-100%. A similar procedure [16] was used to extract steroids from urine except that the washing steps with water and hexane were omitted.

E. Aqueous Soluble Drugs

Aqueous soluble drugs can be purified to some extent by extracting away lipid interferences and precipitating the proteins. For example, 10 volumes of 40% heptane in ethanol, when mixed with 2 volumes of plasma, provides about 70% ethanol in the aqueous phase. Under these conditions, triglycerides, cholesterol, and cholesteryl esters [17] are extracted into the heptane phase and most of the major proteins, such as albumin and γ-globulins, are precipitated [18].

TABLE 3. Adsorption of Phenmetrazine in Benzene Solution to 40 ml Glass Centrifuge Tubes[a]

Tube	Pretreatment	Notes	Recovery % (N = 3)
1	Ethanol washing	Tube stored was exposed to laboratory air one week prior to test	21
2	Ethanol washing		51
3	Ethanol washing	Test solution contained 5×10^{-2} moles/l trimethylamine	100
4	Ethanol washing	Test solution contained 5×10^{-2} mole/liter isobutyl alcohol	100
5	Silanization and ethanol washing		89
6	Silanization and ethanol washing	Centrifuge tube contained 0.5 ml of 0.1 M NaOH	100

[a]Copyright held by the Swedish Academy of Pharmaceutical Sciences.

F. Reduction of Glass Adsorption Losses

Polar molecules, e.g., certain amines in nonpolar solvents, are adsorbed to some extent to glass [15]. Three ways of eliminating adsorption have been used: (1) the polarity of the solvent is increased by adding a small amount of an alcohol, (2) the polarity of the glass is decreased by silanization, and (3) a tertiary amine is added as an adsorption suppressor where alcohols interfere in a reaction.

As an example [15], phenmetrazine, when shaken in benzene (113 ng/ml) under anhydrous conditions, was extensively lost by adsorption to the glass unless certain precautionary steps (Table 3) were taken. The adsorption of phenmetrazine to glass was completely prevented by addition of either 0.5% isobutyl alcohol or 0.5% trimethylamine or by the presence of alkaline aqueous phase. Presilanization of the tubes also improved recovery (though not to 100%) of phenmetrazine from the tube.

Several losses of amines to glass will occur, even during the brief time of contact, while transferring from a pipette. This problem can be bypassed by substituting polypropylene for glass wherever possible, by wetting the pipettes with isoamyl alcohol, or by silanizing them with 5% dimethyldichlorosilane in pyridine overnight, rinsing with methanol, and drying at 105° C [15].

G. Reduction of Losses Due to Coprecipitation or Encapsulation

In biological such as blood, urine, and spinal fluid, proteins serve as carriers of lipophilic compounds. Although many lipophils can be quantitatively extracted away from the protein and cellular debris into organic solvents, some, when coprecipitated, cannot be completely extracted, e.g., the prostaglandins [21]. A drug can often be more efficiently extracted by avoiding protein precipitation through using 40% ethanol (final concentration by volume) or diethyl ethyl ether. If precipitation of the proteins cannot be avoided because of the choice of solvent, the use of an internal standard in the medium prior to precipitation of the proteins will often adjust for these losses, where the extent of adjustment depends largely upon the similarity in distribution ratio and protein binding between the substance analyzed and the internal standard.

H. Drying Agents

Agents such as magnesium sulfate or sodium sulfate are added to organic extracts in order to remove dissolved water [19]. The removal of water by these agents increases the rate of evaporating an organic solvent by a stream of nitrogen or by boiling prior to dissolving the residue in a suitable solvent for injecting into HPLC.

VI. CONCENTRATION OF SAMPLES

A. Enrichment by Extraction

Some drugs, e.g., methindione, while poorly recovered when attempting evaporation of the solvent [19], can be concentrated simply by extraction into a relatively small volume of organic phase. The polarity of the aqueous phase can be increased by increasing the ionic strength, and doing so increases the distribution coefficient. For example [19], the extraction efficiency of methindione (from 11 ml aqueous phase into 0.5 ml toluene) was increased from 89% to more than 99% by increasing the ionic strength of the aqueous phase from 0.1 to 2.9.

B. Enrichment by Evaporation

Although the evaporation step is conceptually a simple one, its practical application in drug analysis is often problematical with respect to the time and attention required, and the loss of sample incurred. Most procedures leading to GLC/HPLC analysis require a 10- to > 1000-fold reduction in sample volume, or even complete solvent removal where derivatization is necessary. When the sample is enriched by means of evaporation, the sample may be lost principally in three ways, viz., degradation, volatilization, and incomplete transfer.

The drug is often degraded because of its exposure to an overwhelming excess of solvent or its contaminants. Degradation might be avoided, then, by either choosing another solvent or purifying the solvent. When the drug can be degraded by a free-radical mechanism, the addition of a free-radical scavenger, e.g., hydroquinone, and the purging of the solvent with an inert gas will often stabilize the drug. There are several techniques which reduce the temperature and time of heating the sample. Since overheating will most likely occur when most of the solvent has evaporated, a special apparatus was designed so that the heating of the samples stopped when the final volume was attained [46]. Another practical way to avoid degradative loss by overheating is to heat moderately while the sample is swept with a stream of inert gas or is evaporated at a reduced pressure using a vacuum source.

Among the varieties of vacuum pumps available, the water aspirator and mechanical oil pump are used the most. The water aspirator is inexpensive and mechanically simple, but has a low evacuation rate and pressure limit of 15 mm Hg. The vacuum pump, although having a higher evacuation rate and lower vapor pressure, is more expensive and easily damaged by contamination of the oil and corrosion of its mechanism. Specially designed dry ice traps have so reduced the oil contamination problem that several liters of solvent can be evaporated from six rotary-evaporation units operating simultaneously [47].

A drug might be lost by volatilization through coevaporation of the sample with the solvent. This will happen, e.g., with low molecular-weight nonpolar compounds, phenols, and amines. Volatile samples, however, can be quantitatively recovered by evaporation when the more volatile component, i.e., the solvent, is removed at a nominal rate and high reflux ratio. Some condensors designed specifically for this purpose are the Kaderna-Danish evaporator column [20], Micro Snyder column [48, 49], Vigreux column [46], and the Stoner column [51].

Sample may be lost during or after the evaporation step due to "bumping" or "spattering." The boiling chip is commonly used in large-scale evaporations to avoid this problem, but better recovery can be attained on a small scale by using an ebullator [46] or rotary-evaporator unit [47]. Losses occurring after evaporation are caused by incomplete recovery of sample which often has spread over a large surface area. To attain quantitation, the vessel walls must be thoroughly rinsed and the sample then reconcentrated. However, the sample will not be completely recovered, even when using good transferring techniques, when there is adsorption on the glass (see Sec. V.F.).

VII. THIN-LAYER CHROMATOGRAPHY

When extraction has not sufficiently reduced the amount of interfering materials in the sample, thin-layer chromatography (TLC) can frequently separate the substance analyzed from other classes of compounds. TLC separates the sample into zones containing groups of compounds similar in chemical type but varying in degree of alkyl substitution i.e., the molecular-weight selectivity of TLC is lower than its compound-type selectivity.

A. Selection of Adsorbent, Solvent System, and Internal Standard

The most important considerations for developing a good TLC system are the choice of adsorbent, the solvent system, and an internal standard.

The adsorbents which have been most frequently used for cleaning up extracts are silica gel and, to a lesser extent, alumina. The activity of these adsorbents is inversely related to the water content of the adsorbent [53]. The adsorbent is activated on the plate by heating in an oven at 110-150° C. The degree of activation can be adjusted by equilibrating the TLC plate in a constant humidity chamber. The relative humidity desired can be obtained by equilibrating a mixture of water and sulfuric acid with the atmosphere in an enclosed jar or a Vario-KS chamber [54].

The adsorbent can be impregnated with transition-state metal ions for specific separations of olefins. For example, TLC plates impregnated with silver ion were used to separate fatty acid esters according to their degree of unsaturation as well as separating them from other classes of compounds [28]. The silver ion may be removed in a later step by washing the sample in a water-immiscible solvent with water. In a later study [29] on complexing prostaglandins with transitional-metal salts such as cobalt (II), iron (III), chromium (III) and silver (I), PGA_2 and PGB_2 were best resolved on ferric chloride-impregnated silicic acid plates [29]. Another study [30] showed that ferric chloride impregnated plates separated the prostaglandins PGA_1, PGB_1, and PGC_1, which differ only in their position of the double bond in the cyclopentanone ring. The resolution of these prostaglandins is attributed [29, 30] to differences between two competing processes: (a) deactivation of the adsorbent that tends to increase the mobility of the prostaglandins, and (b) complexing of the prostaglandins with the ferric moiety that tends to decrease their mobility.

Solvent systems for TLC of various classes of compounds have been described by Randerath [25], Bobbitt [26], and Snyder [27, 55]. Those solvent systems used on the same class of compounds as the substance being analyzed are usually a good starting point in finding a suitable system for the separation desired. Table 4 [27] may be used as a guideline for modifying a given system. The solvents in this table are listed in increasing order of solvent strength, i.e., according to their ability to elute compounds from adsorbents. Trial runs with different solvents will reveal the solvent strength required to move the compounds of interest as zones which have migrated various fractions (RF) of the distance from the origin to the solvent front, i.e., the mobility of a substance under a given set of conditions may be expressed by the RF value:

$$RF = \frac{\text{Distance of the zone center from the starting point}}{\text{Distance of solvent front from starting point}} \tag{11}$$

In practice, when the RF values range from 0.3 to 0.7, the substance being analyzed is usually optimally resolved from the contaminating materials. When previously

TABLE 4. Solvent Strength with Alumina as Absorbent [27]

Solvent	Polarity index
Fluoroalkanes	-0.25
n-Pentane	0.00
Isooctane	0.01
Petroleum ether, skellysolve B, etc.	0.01
n-Decane	0.04
Cyclohexane	0.04
Cyclopentane	0.05
Diisobutylene	0.06
1-Pentene	0.08
Carbon disulfide	0.15
Carbon tetrachloride	0.18
Amyl chloride	0.26
Xylene	0.26
1-Propyl ether	0.28
1-Propyl chloride	0.29
Toluene	0.29
n-Propyl chloride	0.30
Chlorobenzene	0.30
Benzene	0.32
Ethyl bromide	0.37
Ethyl ether	0.38
Ethyl sulfide	0.38
Chloroform	0.40
Methylene chloride	0.42
Methyl-1-butylketone	0.43
Tetrahydrofuran	0.45
Ethylene dichloride	0.49
Methylethylketone	0.51
1-Nitropropane	0.53
Acetone	0.56
Dioxane	0.56
Ethyl acetate	0.58
Methyl acetate	0.60
Amyl alcohol	0.61
Dimethyl sulfoxide	0.62
Aniline	0.62
Diethyl amine	0.63

TABLE 4. (Continued)

Solvent	Polarity index
Nitromethane	0.64
Acetonitrile	0.65
Pyridine	0.71
Butyl cellosolve	0.74
1-Propanol, *n*-propanol	0.82
Ethanol	0.88
Methanol	0.95
Ethylene glycol	1.11
Acetic acid	Large

reported solvent systems are not applicable, one can begin by trying solvents in
Table 4 having strengths of about 0.25, 0.5, and 0.75. Solvent systems of intermediate strength can be obtained by mixing two solvents whose strengths bracket the value
desired. Sample streaking, which affects both isolation and recovery of the drug,
sometimes occurs even when the RF is in the range of 0.3 to 0.7. Three principle
causes of sample streaking are

    1. Overloading of the sample on the plate.

    2. The existence of diverse binding forces between solute (drug) and adsorbent.

    3. Partial dissociation or protonation of the drug in the solvent system employed.

The sample overloading problem is usually solved by applying a smaller amount of sam-
ple/unit of adsorbent. Some ways of doing this are to apply a smaller amount of sample, apply the sample as a stripe rather than a spot, and apply the sample on a thicker
layer, e.g., on specially channeled plates [56]. Streaking due to bonding forces between sample and adsorbent will often disappear by partial or complete deactivation of
the adsorbent sites with moisture, i.e., by exposing the adsorbent to water vapor or
by adding water to the solvent. Streaking due to dissociation or protonation of the
drug usually occurs with acids, phenols, and amines in solvent systems where they will
exist, and consequently migrate, in both the ionized and unionized forms. This problem
is usually resolved by adjusting the acidity or basicity of the solvent system to keep
the drug completely in its unionized form. Alternatively, polar groups of the compound may also be derivatized into nonpolar groups, e.g., by derivatizing the amino
group into an amide group and the carboxyl group into an ester. The nonpolar derivative will also be more volatile and usually more applicable for GLC.

For more difficult separations, secondary solvent effects, e.g., carbon branching,
aromaticity, and hydrogen bonding may be utilized. The solvent providing secondary effects should be combined in a proportion with the existing solvent mixture which will

bracket the desired solvent strength. Secondary solvent strengths can also be increased by making up a solvent mixture of two or more components having widely different solvent strengths. Thus, if a solvent strength of 0.5 is found to be optimum
but the separation is still not satisfactory, solvent mixtures of 0.6 and 0.4 (or
0.7 and 0.3) might be tried to increase secondary solvent effects and thus improve
resolution.

An internal standard should be chosen which will be separable from the drug on
the GLC because of differences in molecular weight, but inseparable from the drug by
TLC. In TLC, the migration of the substances in the zones is more sensitive to changes
in the classes of the compounds and less sensitive to changes in molecular weight of
the compounds. Thus, for example, the fluorine group in a compound can be exchanged
for a methyl group, or a chlorine group exchanged for a methyl in the internal standard [31, 32]. These functional group changes provide internal standards that have
the same RF on the TLC plate but different retention times on the GLC column. Thus,
such internal standards provide not only corrections for extraction efficiency but
also for biases resulting from application and isolation from the TLC plate.

B. Isolation of Zones from TLC Plates [31, 32]

The TLC chromatograms are thoroughly air dried at room temperature and the zones are
scraped into separate glass-stoppered centrifuge tubes. A polar solvent (for example,
2 ml methanol) is added and the tubes are gently agitated with a mixer. After the
tubes are centrifuged (1000g for 5 min) to sediment the silica gel, a portion of the
solvent is transferred for direct analysis or for derivatization and then analysis.

VIII. ISOLATION OF DRUGS FROM URINE BY XAD-2 RESIN

Basic (e.g., amphetamine), and amphoteric (e.g., morphine) drugs have been rapidly
and efficiently extracted from urine using XAD-2 resin [33]. Only one adsorption and
desorption step was required. These compounds were bound most efficiently when the
compounds were electrically neutral. Amphoteric compounds, whose acidic group has a
higher pK_a than its basic group, will be maximally adsorbed at an intermediate pH
and less well at higher and lower pH. Thus, morphine exhibited an amphoteric character and was adsorbed best at about pH 8.5 and was less well adsorbed at pH values
different from this value. The polarity of the eluant affects the efficiency of elution to a considerable degree. Thus, the efficiency of eluting amphetamine and phenobarbital from XAD-2 resin increased as the percentage of isopropanol in ethyl acetate/
dichloroethane (3:2) was increased from 0% to 40%. Most of the urinary pigments were
left on column when there was 20-25% isopropanol in ethyl acetate/dichloroethane.
Amphetamine was extracted with an efficiency greater than 92% between pH 6.5 and 12.5.
Phenobarbital was extracted with an efficiency greater than 96% between pH 2.5 and 5.

These drugs were efficiently extracted at concentrations ranging from 0.5 to 500 mcg/ml of urine. The drugs bound to XAD-2 resin tended to be more efficiently desorbed when the flow rate of the eluant was slower and volume of the eluant larger. High concentrations of one drug did not reduce the adsorption of low concentrations of the other drugs. However, other components in urine can saturate the adsorption sites on the resin and thus reduce drug adsorption. Hydrolyzed urine samples reduced the recovery of drug 5-15%. Thus, the total amount of sample must be adjusted for the total number of adsorptions sites on the amount of XAD-2 being used.

IX. VOLATILIZATION TECHNIQUES FOR CLEANING UP SAMPLES

Distillation is a well-known technique for purification of liquid samples from other substances which usually have a lower vapor pressure and are often called *sludge* or *residue*. Direct distillation techniques fail, however, when the substance in question is a trace component and consequently of insufficient quantity to pass through the apparatus and be recovered in good yield in the condensate. This problem may be overcome in some instances by sweeping the sample from the other substances with another gas into which it will mix and thereafter be condensed. A classical example of this approach is the steam distillation cleanup step used in some assays for phenols and amines.

More recent elaborate procedures called *forced volatilization* [34] and *sweep codistillation* [34] have been described which applies the foregoing principle by using a mixture of an inert gas, e.g., H_2 and He, and a gaseous organic substance to remove the desired substance from the extractive. An available apparatus [35] to perform this function bears a functional resemblance to the gas chromatograph, but differs by its use of a larger sample, macroscale support (sand, glass wool, glass beads), and a sweeping solvent. Important considerations for optimization of this technique are

1. Volatility and stability of the sample when vaporized
2. Sample size
3. Type of support used in separation tube
4. Separation tube temperature
5. Sweeping rate of gas
6. Conditions for adding sweeping solvent

The best applications of sweep codistillation have been in the analysis of organo-phosphate and chlorinated hydrocarbon pesticides where several time-consuming cleanup steps were eliminated, recoveries improved, and assay time shortened [34, 36-40]. It was particularly useful in analyzing for these pesticides in edible oils [37]. Gram quantities of oil were injected into the separator containing glass beads, then swept with petroleum ether, and collected in a Teflon tubing condenser. Although this technique has been used mostly on pesticides, its use for analysis of drugs, i.e., benzodiazepines, phenothiazines and bromureides, has been documented [41]. However, in these applications the recoveries ranged from 70 to 100%.

Theoretically, this technique should be applicable to any sample which is stable enough to pass through a gas chromatograph in good yield and indeed, the method has been applied extensively. Other considerations which may have limited its use are

1. The complex conditions that must be optimized for obtaining good results

2. Pyrolysis of the extractives and contamination of the condensate

3. Losses of the desired substance by reaction with other extractives

4. Instability or irreversible adsorption of the sample

5. Low volatility of the sample

6. Low recoveries of 50% or less when the prior conditions occur.

REFERENCES

1. M. H. Penner, *J. Pharm. Sci.*, *64*, 1017(1975).

2. P. R. Brown, S. Bobick, and F. L. Hanley, *J. Chromatogr.*, *99*, 587(1974).

3. J. A. F. DeSilva and I. Bekersky, *J. Chromatogr.*, *99*, 447(1974).

4. H. Alliger, *Amer. Lab.*, *7*, 75(1975).

5. H. Ko, R. Lahti, D. J. Duchamp, and M. E. Royer, *Anal. Lett.*, *7*, 243(1974).

6. G. A. Levvy and J. Conchie, in *Glycuronic Acid Free and Combined* (G. J. Dutton, ed.), Academic Press, New York, 1966.

7. M. L. Selley, J. Thomas, and E. J. Triggs, *J. Chromatogr.*, *94*, 143(1974).

8. R. C. Thomas and H. Harpootlian, *J. Pharm. Sci.*, *64*, 1366(1975).

9. D. Lisk, *Science*, *154*, 93(1966).

10. B. L. Karger, L. R. Snyder, and C. Horvath, *An Introduction to Separation Science*, John Wiley and Sons, New York, 1973.

11. H. Ko and M. E. Royer, *J. Chromatogr.*, *94*, 143(1974).

12. M. L. Selley, J. Thomas, and E. J. Triggs, *J. Chromatogr.*, *94*, 143(1974).

13. F. Medzihradsky and P. J. Dahlstrom, *J. Pharmacol. Res. Commun.*, *7*, 55(1975).

14. R. G. Muusze and J. F. K. Huber, *J. Chromatogr. Sci.*, *12*, 779(1974).

15. T. Walle and H. Ehrsson, *Acta Pharm. Suecica*, *8*, 27(1971).

16. W. G. Stillwell, A. Hung, M. Stafford, and M. G. Horning, *Anal. Lett.*, *6*, 407(1973).

17. V. P. Dole and H. Meinertz, *J. Biol. Chem.*, *235*, 2595(1960).

18. R. B. Pennell, in *The Plasma Proteins* (F. W. Putnam, ed.), Vol. 1, Academic Press, New York, 1960, p. 9.

19. J. Vessman, S. Stromberg, and G. Freij, *J. Chromatogr.*, *94*, 239(1974).

20. F. A. Gunther, R. C. Blinn, M. J. Dolbezen, J. H. Barkley, W. D. Harris, and H. S. Simon, *Anal. Chem.*, *23*, 1835(1951).

21. W. G. Unger, I. F. Stamford, and A. Bennett, *Nature*, *233*, 336(1971).

22. G. Schill, in *Ion Exchange and Solvent Extraction* (J. A. Marinsky and Y. Marcus, eds.), Vol. 6, Marcel Dekker, New York, 1974, p. 1.

23. K. O. Borg, H. Holgersson, and P. O. Lagerstrom, *J. Pharm. Pharmacol.*, *22*, 507(1970).

24. R. Modin and S. Back, *Acta Pharm. Suecica*, *8*, 585(1971).

25. K. Randerath, *Thin-Layer Chromatography*, Academic Press, New York, 1966.

26. J. M. Bobbitt, *Thin-Layer Chromatography*, Rheinhold, New York, 1963.

27. L. R. Snyder, *Principles of Adsorption Chromatography*, Marcel Dekker, New York, 1968.

28. L. J. Morris, in *New Biochemical Separations* (A. T. James and L. J. Morris, eds.), Van Nostrand, New York, 1964.

29. R. L. Spraggins, *J. Org. Chem.*, *38*, 3661(1973).

30. J. A. F. Wickramasinghe and S. R. Shaw, *Prostaglandins*, *4*, 903(1973).

31. D. G. Kaiser, S. R. Shaw, and G. J. VanGiessen, *J. Pharm. Sci.*, *63*, 567(1974).

32. J. A. Wickramasinghe and S. R. Shaw, *J. Pharm. Sci.*, *60*, 2451(1971).

33. M. P. Kullberg, W. L. Miller, F. J. McGowan, and B. P. Doctor, *Biochem. Med.*, *7*, 323(1973).

34. R. W. Storherr and R. R. Watts, *J. Ass. Offic. Agr. Chem.*, *48*, 1154(1965).

35. A. F. Plant, U. S. Patent 3,496,068, February 17, 1970.

36. R. R. Watts and R. W. Storherr, *J. Ass. Offic. Agr. Chem.*, *48*, 1158(1965).

37. R. W. Storherr, E. J. Murray, I. Klein, and L. A. Rosenberg, *J. Ass. Offic. Agr. Chem.*, *51*, 662(1968).

38. R. R. Watts and R. W. Storherr, *J. Ass. Offic. Agr. Chem.*, *50*, 581(1967).

39. R. W. Storherr and R. R. Watts, *J. Ass. Offic. Agr. Chem.*, *51*, 662(1968).

40. P. Maini and A. Collina, *J. Ass. Offic. Agr. Chem.*, *55*, 1265(1972).

41. R. H. Drost and R. A. A. Maes, *Forensic Sci.*, *3*, 175(1974).

42. A. Arbin, *Acta Pharm. Suecica*, *12*, 119(1975).

43. L. C. Franconi, G. L. Hawk, B. J. Sandmann, and W. G. Haney, *Anal. Chem.*, *48*, 372(1976).

44. A. R. Barbiers, D. D. Chapman, E. V. Hubert, E. N. Petzold, L. J. Smith, and T. J. Vangorder, unpublished results.

45. S. L. Cohen, *J. Biol. Chem.*, *192*, 147(1951).

46. M. Beroza and M. C. Bowman, *Anal. Chem.*, *39*, 1200(1967).

47. R. E. Gosline, L. F. Krzeminski, and A. W. Neff, *J. Agr. Food Chem.*, *17*, 2114(1969).

48. J. A. Burke, P. A. Mills, and D. C. Bostwick, *J. Ass. Offic. Agr. Chem.*, *49*, 999(1966).

49. F. A. Gunther and R. C. Blinn, *Analysis of Insecticides and Arcaricides*, Interscience, New York, 1955.

50. R. L. Stanley and H. T. Lefavoure, *J. Ass. Offic. Agr. Chem.*, *48*, 666(1965).

51. E. Stoner, D. Cowburn, and L. C. Craig, *Anal. Chem.*, *47*, 344(1975).

52. A. A. Forist, A. L. Pulliam, and D. G. Kaiser, *Res. Commun. Chem. Path. Pharmacol.*, *8*, 385(1974).

53. R. A. DeZeeuw, in *Progress in Thin-Layer Chromatography and Related Methods*

(A. Niederwieser and G. Pataki, eds.), Vol. III, Ann Arbor Science Publishers, Ann Arbor, Mich., 1972.

54. G. Geiss and H. Schlitt, *Chromatographia*, *1*, 392(1968).

55. L. R. Snyder, *J. Chromatogr.*, *63*, 19(1971).

56. M. J. Matherne, Jr. and W. H. Bathalter, *J. Ass. Offic. Agr. Chem.*, *49*, 1012(1966).

Chapter 9

PREPARATIVE HIGH PERFORMANCE LIQUID CHROMATOGRAPHY:
SMALL SCALE AND TRACE COLLECTION

Richard A. Henry[*]

Virginia M. Smith

Spectra-Physics, Inc.
Autolab Division
Santa Clara, California

I. INTRODUCTION

Classical liquid chromatography (LC) has been largely preparative and typically has involved separation of 1-10 g of material on very large columns having low efficiency. It has been a tedious and time-consuming process that often requires hours or even days when gravity flow is employed. Recent improvements in packings and hardware allow more difficult separations to be made in much shorter times. Many recent articles [1-7] have shown that columns containing high-efficiency packings can be used with larger sample sizes without loss of performance. Figure 1 shows efficient high-speed

* *Presently at* Altex Scientific Inc., Berkeley, California

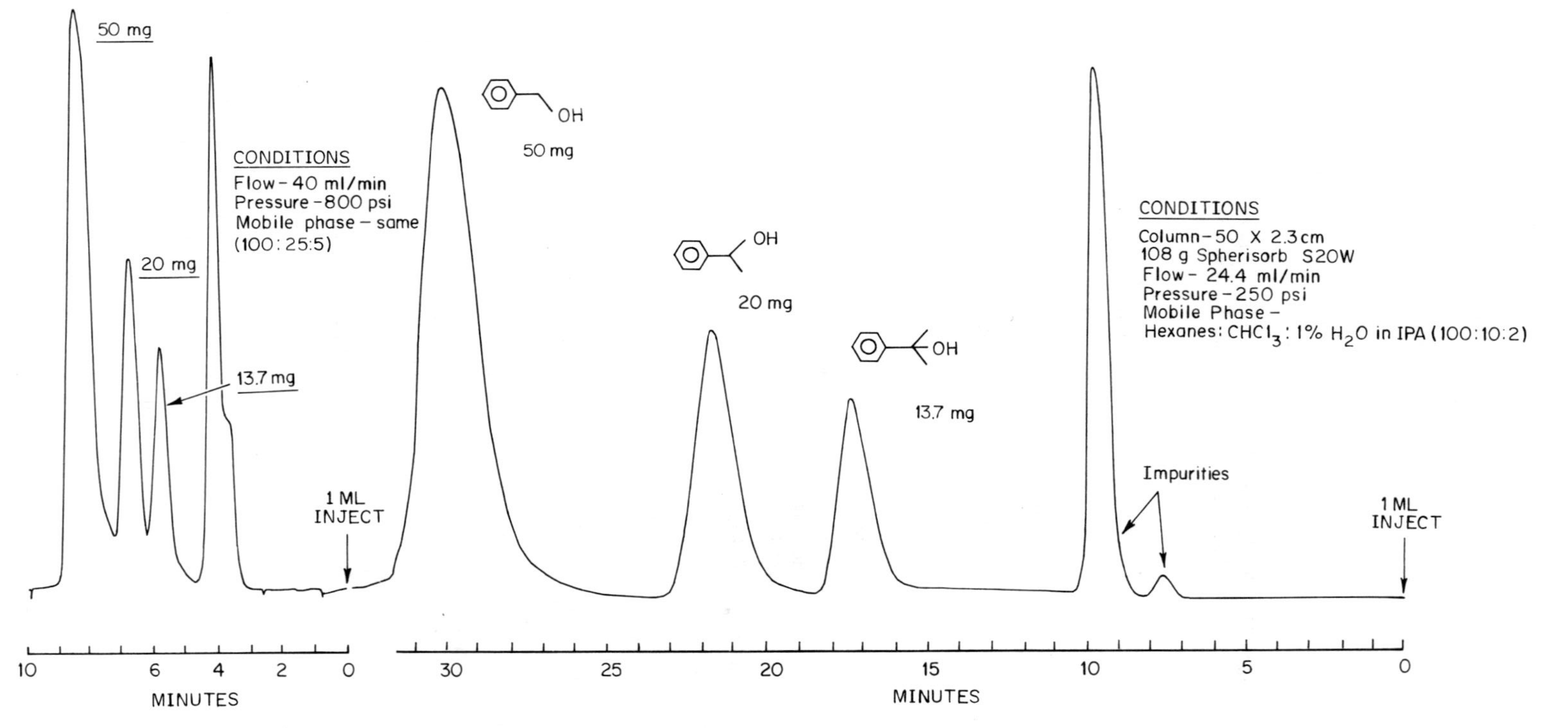

FIG. 1. High Speed Separation on a 1-inch column [8].

separations on a 1-in. i.d. column packed with spherical, totally porous, 20-µm dia-meter silica particles [8]. The column used in Fig. 1 was prepared by a simple dry-packing method. Modern preparative LC offers high speed, high resolution, and low risk of compound degradation.

The ability to quantitatively recover purified solutes in undegraded or unchanged form is particularly useful in biological and pharmaceutical research, where positive identification is critical to the success of new drug development or disease diagnosis, and pure compounds enhance the prospect of a conclusive experiment. Figure 2 depicts several uses for preparative LC and typical amounts of material required.

It is the intention of this chapter to provide practical information for the chromatographer who finds it necessary to collect eluent fractions of very high purity. More specifically, this chapter will deal with the area sometimes referred to as small-scale preparative LC in which milligram rather than gram quantities of purified material are collected. It is also our intention to point out how experimental techniques should be modified when the compound to be isolated is a minor or trace component of a sample rather than a major component.

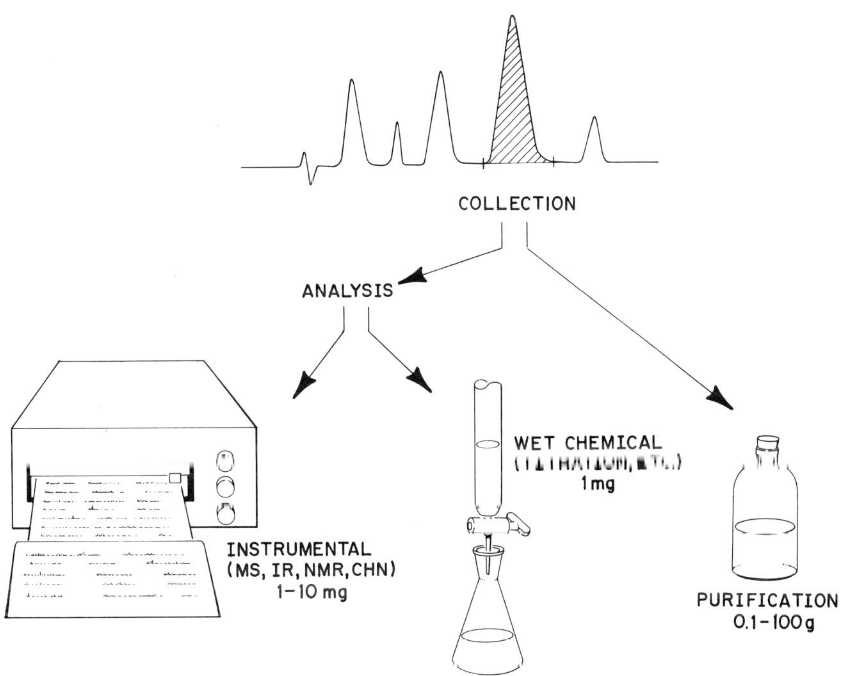

FIG. 2. The preparative experiment.

II. OPTIMIZING THE SEPARATION

If small-scale preparative LC, including trace collection and identification, is to
be a frequent requirement in the laboratory, it will be desirable to establish an
area for this type experiment and make certain initial financial investments that will
quickly be returned in the form of pure material and positive identification. Investments should include special pumps that have higher-than-usual flow-rate capability,
large-diameter columns and fittings, large amounts of modern LC packings, and the
necessary equipment to make columns. Special detectors and cells to handle high flow
rates, solvent stills for reclaiming the expensive high-purity solvents usually required in LC, rotary evaporators to recover solute after collection, and an excellent
ventilation system that minimizes exposure to solvent vapors should also be included.
More commonly, small-scale preparative LC will be an occasional requirement in the
laboratory and it will not be possible to justify a large investment in a special
area. In this case, it will often be sufficient to develop a separation on a column
having analytical dimensions and make repeated injections of the largest sample amount
that can be introduced without excessive overload.

The first question that must be answered is how much pure material is required
in order to accomplish the objective. As mentioned earlier, it is seldom possible to
obtain positive identification with less than a milligram of material and therefore,
a milligram will be the minimum objective of most preparative LC experiments. If the
compound of interest makes up 10% by weight of the sample and excellent resolution is
obtained so that essentially all the compound can be collected on one pass through the
column, then 10 mg of total sample must be injected. Even on typical analytical columns containing microparticle porous packings this can usually be accomplished in one
or two injections. Figure 3 illustrates the collection of a large fraction on a 2-mm
i.d. column packed with spherical, totally porous 5 μm particles. However, if more
pure compound is required or if the desired compound is present as a trace constituent
(less than 1% by weight), scale-up to larger diameter columns may be necessary.

Before beginning any preparative LC experiment, it is important to list the facts
that are known about the sample. How many other compounds are present and what are
their chemical characteristics? It is often desirable to preface a high resolution
LC experiment with a simple solvent extraction or class separation on a less expensive
packing. A high resolution LC column is best suited for separating compounds of a
similar chemical nature. If very complex mixtures must be introduced to the high resolution column, conditions should be adjusted so that most of the unneeded compounds
pass through the column unretained. Unneeded material that is more strongly retained
than the compound of interest must be regularly cleaned from the column by step elution with a very strong solvent to avoid contamination during subsequent collection
experiments. It is valuable to perform a material balance experiment in order to

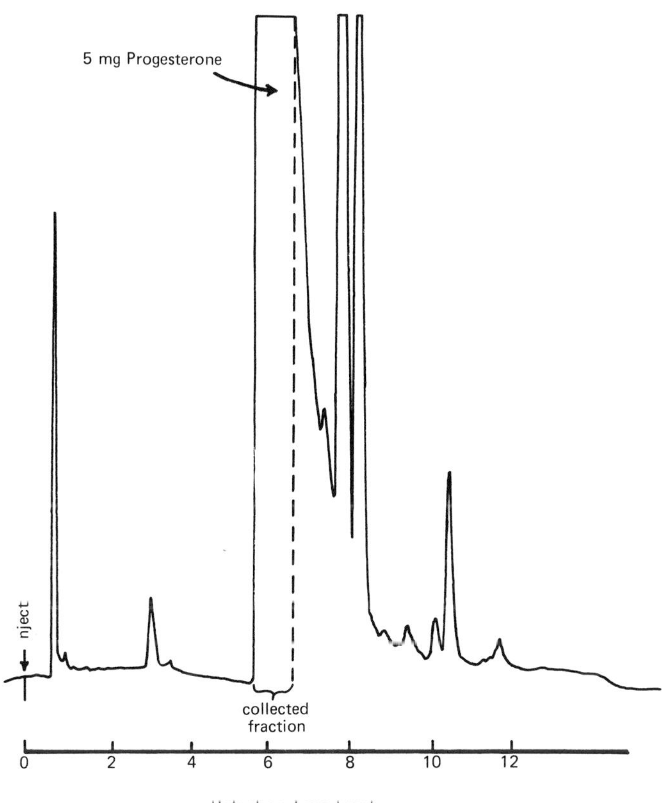

FIG. 3. Chromatogram of Impure Progesterone. Sample size, 5 mg; column, 0.25 m x 2.1 mm i.d. packed with Zorbax-SIL; temperature, ambient; mobile phase, linear gradient from 100% dichloromethane to 90% dichloromethane; 10% ethanol at 2%/min; column pressure 1500 p.s.i.g.; detector sensitivity 0.02 absorbance units full scale. (D. R. Baker, R. A. Henry, R. C. Williams, D. R. Hudson, and N. A. Parris, *J. Chromatogr.*, 233(1973). Reprinted with permission of copyright owner.)

establish conditions where all compounds are being eluted from the column. Two common methods for doing this employ gravimetric and spectrophotometric techniques.

There have been many attempts in the literature to describe the optimum experimental conditions for various objectives in chromatography. Two recent examples [9-10] have dealt specifically with preparative high performance LC. Fortunately, there is a practical approach to chromatography that does not require a detailed knowledge of theory. In fact, when all careful theoretical arguments have been employed, it will be discovered that preferred conditions will closely resemble the following:

1. A microparticle (5 to 30 μm) porous packing, probably silica

2. A stainless steel column, 25 cm in length and 3- to 8-mm internal diameter

3. A k' equal to 5 or less for the compound to be collected

4. A linear velocity of 1 cm/sec

5. A sample injection volume of 100-1000 ml and a sample concentration of
   1% wt./vol

In addition, it will be found that the number of theoretical plates N is not important
as long as N is greater than one thousand. Although it will occasionally be necessary
to scale up to columns greater than 8-mm i.d. and flow rates greater than 10 ml/min,
the arguments against it for small-scale preparative work are very compelling:

1. Pumps that are optimized for high flow rates are usually not designed for precise analytical work and cannot be justified on the basis of total capability.

2. Large columns and the large amounts of high performance packings to fill them
   are expensive and not readily available. When a packed column fails, replacement cost is high.

3. Large columns are more difficult to pack than small ones by slurry methods.

4. Dilution factors on larger columns are greater for equal sample weights,
   therefore, recovery of isolated material is more expensive and time consuming. Dilution also makes detection and collection of trace components
   more difficult.

5. High flow rates tend to waste solvents and are more likely to create hazardous spill situations when left unattended.

6. A single injection of a large sample is more risky than repeated injections
   of smaller samples.

7. High flow rates usually require special detector cells or a stream splitter
   before the detector.

A. Speed/Capacity/Resolution Compromises

Although excellent results for most objectives can be obtained by following the general guidelines set forth above, a knowledge of chromatographic theory can be helpful
in solving difficult problems and in explaining unexpected results. The following
discussion is excerpted from Ref. 11:

> *Resolution is the objective of all chromatography*. It results from a column's
> selectivity and efficiency. Selectivity moves zones centers apart and depends
> primarily upon chemical properties such as the polarity of the packing and mo-
> bile phase. Efficiency keeps zones narrow and depends primarily upon physical
> properties such as packing particle size.
>
> *An important factor in efficiency is mobile phase flow inequalities*. Stream
> paths with different velocities cause zone spreading and result from hetero-
> geneity of the packed bed structure, as illustrated in Fig. 4. Efficiency is
> always improved by increasing packing homogeneity. This factor affects the ef-
> ficiency of both retained and unretained peaks.
>
> *Another factor in efficiency is non equilibrium of solute between the two phases*.
> This is also termed resistance to mass transfer. The faster the transfer between

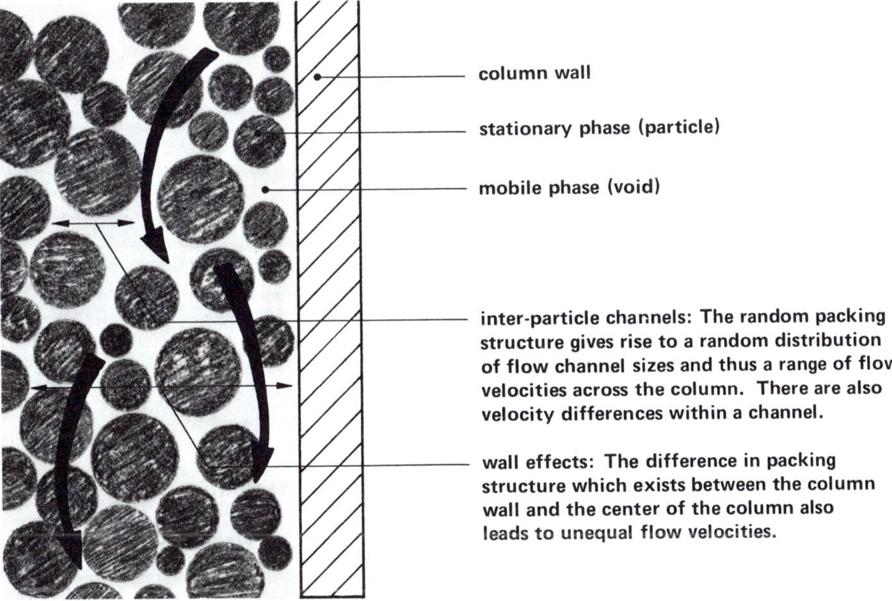

FIG. 4. Illustration of column-bed structure.

phases, the better the efficiency. Reducing the flow rate improves efficiency because more time is allowed for the transfer. There are two components to this phenomena: mass transfer in the stationary phase (i.e., packing particle), and mass transfer in the mobile phase (i.e., flowing solvent surrounding the particle). Efficiency is improved by increasing transfer rates in either of these phases, preferably in both. This factor primarily affects the efficiency of only retained peaks.

Efficient columns permit high flow rates while maintaining adequate resolution. High flow rates reduce analysis time but require pressure at the column inlet:

$$\Delta P = \frac{v l \eta}{d_p^{\,2}}$$

where

ΔP = pressure drop

v = flow rate

l = column length

η = viscosity

d_p = packing particle diameter.

It is desirable for a column to offer minimum resistance to flow, a characteristic referred to as permeability.

Column efficiency is quantitatively described by N, the number of theoretical plates. N is proportional to column length, so longer columns have more plates if all other variables are equal. A measure of efficiency independent of length

is H, the plate height. It describes the degree of spreading per unit length
of column and is related to N by the expression, H = L/N.

A 250-mm long column which has 5,000 plates thus has a plate height of 0.05 mm
or 50 μm. This is very good performance since the plate height is comparable
to the diameter of one particle, usually 5-20 μm.

Efficiency, analysis time, and permeability are all factors of column performance.
Since plates can be increased by reducing the flow rate, at the expense of analysis time, it is meaningful to compare plates per second, i.e., the number of
plates that can be generated per unit time. Similarly, since plates can be increased by increasing column length, at the expense of a greater pressure requirement, another useful comparison is the performance factor. It is an expression which takes into account both efficiency and permeability of a column.

Resolution occurs as a sample travels along the column and some components of the
sample spend more time associated with the packed bed than others. When compounds are
present in excess in the vicinity of an active site, an overload condition exists and
peaks become non-Gaussian in shape and skewed toward shorter retention times. Obviously, overload occurs early in every separation, but dilution and resolution quickly
permit equilibrium to take place for normal sample sizes (1-100 mg per gram of packing). In preparative experiments, larger samples are usually required and overload
conditions exist during most of the time that the sample is on the column. The amount
of overload that can be tolerated will be determined by resolution between the peak of
interest and neighboring peaks, as well as by the purity requirement of the collected
fraction.

Much has been written about the relationship between speed, capacity, and resolution in chromatography. Resolution that has been achieved on a given column under
a set of experimental conditions can be lost by increasing the linear velocity (speed)
or sample size (capacity). Conversely, resolution can be improved by decreasing the
linear velocity and sample size. Resolution is lost at higher velocities, primarily
due to nonequilibrium factors which cause sample zone spreading. Small-diameter
packings are much better for maintaining column efficiency (narrow zones) at high
carrier velocities than those with large diameter. Keep in mind that it is not flow
rate but linear velocity that affects resolution. When scaling up to large-diameter
columns, linear velocity can be maintained by increasing the flow rate in proportion
to the cross-sectional areas.

Generally, if modern LC packings are employed for small scale preparative and
trace compounds collection, the compromises are not serious and milligrams of very
pure material can be obtained in a few minutes under a wide variety of column conditions.

B. Selecting the Packing

A separation should always be developed on a small diameter column to conserve sample,
mobile phase, and packing. Only when the separation has been accomplished and the
identity of peaks verified, should scale-up to larger diameter columns be attempted.

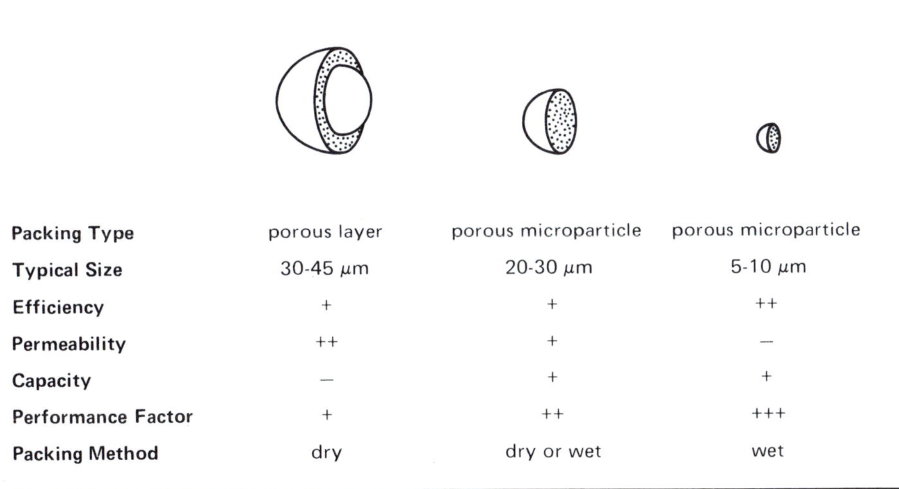

Packing Type	porous layer	porous microparticle	porous microparticle
Typical Size	30-45 μm	20-30 μm	5-10 μm
Efficiency	+	+	++
Permeability	++	+	—
Capacity	—	+	+
Performance Factor	+	++	+++
Packing Method	dry	dry or wet	wet

FIG. 5.   Comparison of LC packings.

With many excellent packings now available, it is not difficult to get a good separation if certain guidelines are followed. Choose a packing that has an affinity for the compound of interest so that it can be selectively retarded and removed from undesired components. A packing should also be in a physical form that provides adequate performance, yet is convenient to handle. Figure 5 compares the characteristics of the three popular physical shapes available. Figure 6 shows a chromatographic comparison of particle size on resolution with pellicular and totally porous packings. Note that the low capacity of Vydac makes it less retentive than Spherisorb under the same mobile-phase condition. While other chapters of this book deal specifically with methods for selecting the right packing for a particular separation problem, it is nevertheless useful to discuss briefly some of the advantages and limitations of common packings for preparative LC.

1. Adsorption

Silica is the most widely used packing for both analytical and preparative LC because it is readily available and much is known about its chromatographic behavior. Silica can be used with a wide variety of solvents and compounds. It has best selectivity for compounds having polar functional groups that can hydrogen bond with silica hydroxyl groups. Alumina is a much neglected adsorbent because it has not been as readily available as silica, but it exhibits better selectivity for many classes of compounds and is often easier to handle than silica because it is more dense and less hygroscropic.

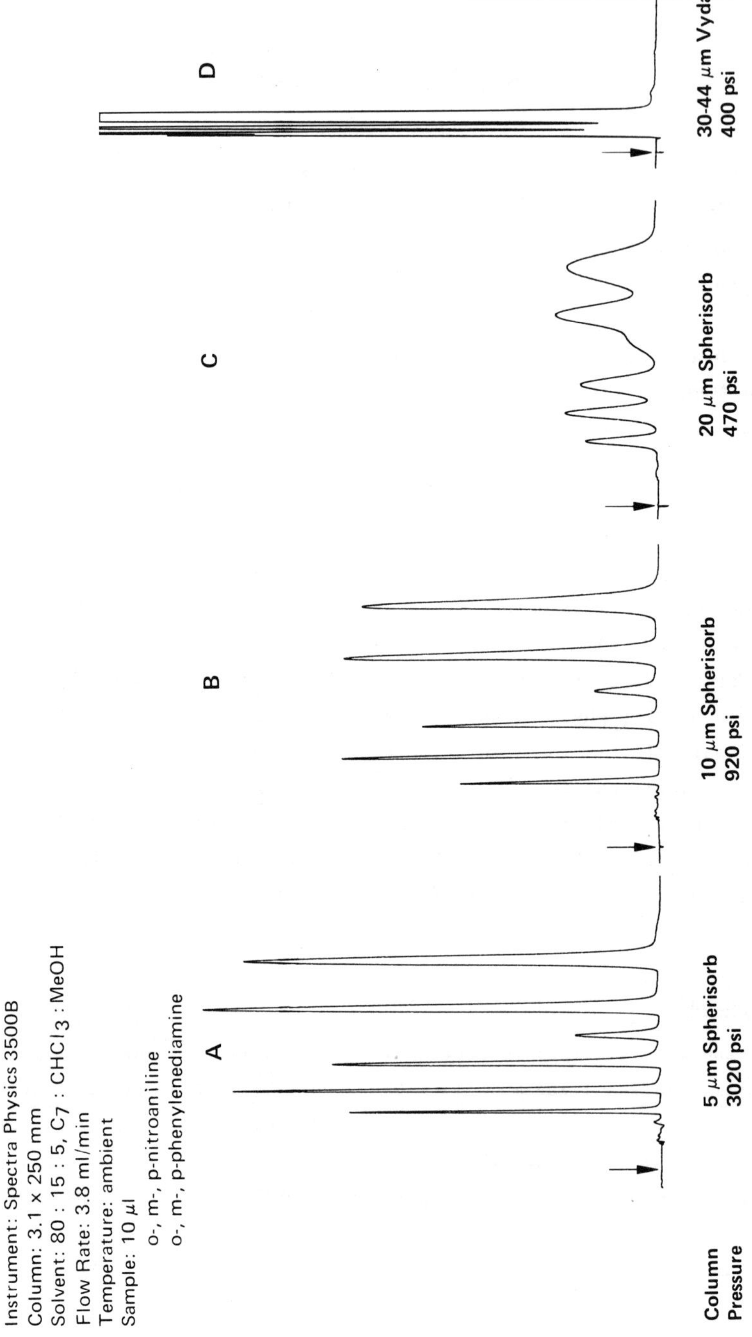

FIG. 6. Chromatographic performance of porous and pellicular LC packings.

In adsorption chromatography, it is desirable to avoid packings which have a wide range of very small pore diameters because of their tendency toward irreversible adsorption. Newer packings with uniform pore size in the 100-$\overset{\circ}{A}$ range and greater are becoming readily available and show little of the irreversible adsorption behavior of earlier packings. It is especially important to avoid irreversible adsorption if trace collection is necessary.

Silica and alumina should be the first choices for doing analytical or preparative LC, but they are not optimum for either very polar, or very nonpolar compounds. An exception might be alumina, which has good selectivity for aromatic hydrocarbons.

Recently, there has been a rapid increase in the popularity of silica adsorbents which have been modified with a bonded organic hydrocarbon to alter selectivity. Problems that have prevented the bonded-phase packings from becoming widely used in preparative LC include high cost; chemical instability, especially in polar solvents; and lower capacity than is usually observed for uncoated adsorbent. The recent availability of porous, microparticle, totally organic adsorbents which overcome these stability and capacity problems with silica bonded phases promises to be important for separations in polar solvents.

2. Ion Exchange

Ionic compounds are usually best separated on organic packings which have bonded ionic functional groups. Due to difficulties in handling classical ion-exchange resins, there has been an increase in the popularity of silica with ionic bonded phases. Bonded-phase silica for ion-exchange chromatography, however, can only be used over a narrow pH range and has serious stability problems in aqueous solution. The technique of reacting an ionic compound with an organic ion of opposite charge to form a nonpolar ion pair that can be separated on adsorbents or bonded-phase silica should be kept in mind as an alternate to ion-exchange chromatography. Continuing improvements in ion-exchange resins and methods for handling them should make these packings preferred for the separation of polar, ionic compounds, on a preparative scale.

The biggest problem associated with preparative ion-exchange LC is eliminating buffer compounds and other additives present in the mobile phase. Buffer compounds present in the collected fraction may be removed by (1) a "desalting" technique using aqueous steric exclusion chromatography or (2) solvent extraction. The use of inorganic salts and salts of low molecular-weight organic acids, and bases can facilitate later removal from higher molecular-weight organic compounds. Avoid selecting buffer components that are similar in size and structure to the compounds that are to be collected. Evaporation of aqueous solvents is difficult; however, and the risk of alteration or loss of collected compound is great. Low molecular-weight organic acids can often be removed by evaporation along with the aqueous solvent.

3. Steric Exclusion

Separation by size difference is an excellent high capacity technique; however, it is seldom possible to obtain adequate purity by steric exclusion alone. More often, it has merit as a prepurification or cleanup technique that can drastically reduce the complexity of difficult samples and prolong the life of a column. Cleanup by steric exclusion improves the chances of getting resolution and purity on a high performance column by eliminating the possibility that high and low molecular-weight compounds are co-eluting. Column cleanup techniques should be especially valuable in collection of trace amounts where interferences from other components is a difficult problem.

Currently, the best packings for steric exclusion work are soft or semirigid organic gels which are useful primarily in organic solvents. New developments in rigid inorganic packings and gels which are optimized for polar and aqueous solvents promise to make this technique even more popular in biochemical and drug applications.

Although it is common to refer to adsorption, partition, ion exchange, and steric exclusion, most separations take place under chemical conditions, where more than one mechanism is operating. Future understanding of these mechanisms will lead to the development of improved packings and better guidelines for their use.

C. Special Considerations

Once a separation has been accomplished, it is relatively simple to scale up, if necessary, to a larger column that contains enough packing to permit collection of the quantities of pure material that are needed. As stated earlier, a 2- to 3-mm i.d. analytical column will usually suffice when only a few milligrams are required. The use of a sample valve with flexible loop size permits scale-up to large injection volumes. Injection volumes of 100 µl or greater can be employed in LC without serious loss of resolution, even on short, 2- to 3-mm i.d. analytical columns [12].

A special technique that has been employed when collection of a trace or very dilute component is desired uses the column to concentrate the compound of interest. A column is selected that is very retentive for the compound even when the sample solvent is the eluting mobile phase. It is desirable to adjust conditions, if possible, so that compounds which are not of interest pass through the column unretained. When enough solvent has passed through the column to concentrate the compound near the inlet, a mobile-phase change is made to elute the compound for detection and collection. This technique has obvious value for waste-water analysis and for the study of dilute extracts of body fluids, where harsh concentration steps such as evaporation can cause alteration or loss of compound.

The subject of trace LC analysis has been treated extensively by Kirkland [13]. The major problems in collection of trace components are achieving resolution and detector sensitivity. To avoid excessive sample dilution, most trace collection will be carried out on short, 3- to 8-mm i.d. columns at relatively low flow rates. In any case, isolation of large quantities of a trace component will be a tedious experiment that will involve multiple injections. The most expedient route to isolation of a trace component may be to collect under heavily overloaded conditions where visual resolution is lost, and then to reinject the now concentrated component under high resolution conditions.

III. HARDWARE IMPLEMENTATION

Most manufacturers of LC packings and instruments have suitable valves, columns, and fittings available; however, many specialty companies also have excellent technology. Some commonly used specialty companies include

1. Handy and Harmon Tube Company, Whitehall and Township Line Road, Norristown, Pennsylvania
2. Haskel Engineering and Supply Company, 100 E. Graham Place, Burbank, California
3. Mott Metallurgical Corporation, Farmington Industrial Park, P. O. Drawer L, Farmington, Conneticut (porous metal frits, filters, and fittings).
4. Rheodyne, 2809 Tenth Street, Berkeley, California (high- and low-pressure valves).
5. Scientific Systems, Inc. 1120 West College Avenue, State College, Pennsylvania (high-pressure valves and fittings).
6. Valco Instrument Company, P. O. Box 12032, Houston, Texas (high-pressure valves).

A. Columns

The two main choices of column materials are glass and stainless steel. Glass is the usual selection for classical low-pressure chromatography because it is relatively inexpensive and easily adjusted to varying column bed lengths with special fittings; however, glass is not mechanically stable and cannot withstand the higher pressures normally required in modern LC. Stainless steel is readily available and is the material of choice for most high performance LC, including preparative.

Column fittings should accomodate a small porosity frit to hold the packing material in the column. These fittings should be replaceable without disturbing the column bed. This is of particular importance in preparative chromatography since the larger columns are more costly and difficult to pack. Also, frits may plug more in preparative chromatography due to the higher throughput of solvents required. Figure 7 shows a typical column and fitting configuration.

Preparative columns may either be dry-packed or slurry-packed depending on the type of material selected [11]. Particles larger than 20 μm that are not the totally

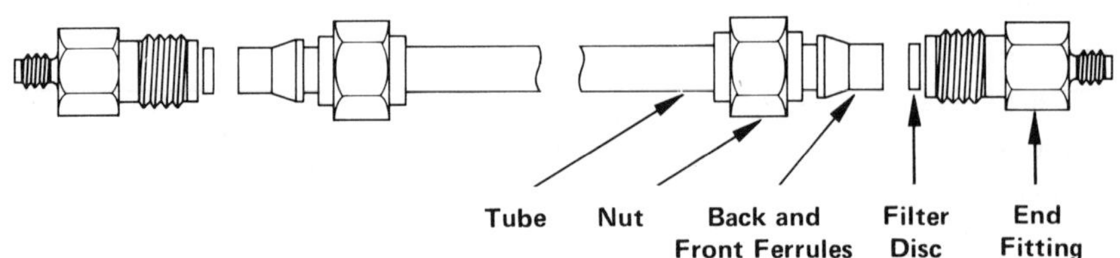

FIG. 7. Column and fitting configuration.

organic, swellable type are easily dry-packed and should be selected for preparative
LC when very high resolution is not required.

B. Pumps

A desirable feature of a pumping system for preparative LC is that it be capable of
delivering a high flow rate from a large reservoir. A compromise will have to be
made at some point because the price for high flow-rate capability usually is lower
pressure limit. Higher flow rates are necessary to maintain linear velocity of mobile phase through larger diameter columns. Because the permeability of a column is
determined by the characteristic of the packed bed, the same pressure is required to
maintain equal linear velocity on any diameter column.

In actual practice, preparative liquid chromatography is usually carried out at
somewhat lower pressures than analytical chromatography. This is due not only to the
limitations of pumping systems, but also to pressure limitations of larger diameter
columns and fittings.

Resolution of the component of interest in preparative LC can generally be accomplished under isocratic conditions; however, a step or continous gradient may be
needed for the collection of two or more components which have widely different capacity factors (k'). Gradient conditions may also be required for the cleanup of a
column prior to subsequent injections. One special technique which requires gradient
elution has already been discussed earlier in this chapter; that is, the concentration
of very dilute samples on the head of the column prior to elution with a stronger mobile phase. Because of the extra cost associated with most gradient generating devices, it is desirable to employ isocratic or step gradient methods in preparative
LC. A step gradient can be most conveniently employed by placing a low-pressure valve
between the pump and reservoirs and switching at the appropriate time. Figure 8 shows
an example of a preparative experiment carried out on a 1-in. diameter column under
step-gradient conditions.

C. Sample Introduction

Although syringe injection is still popular in many types of liquid chromatography,
sample valves have become preferred for convenient, reproducible introduction of

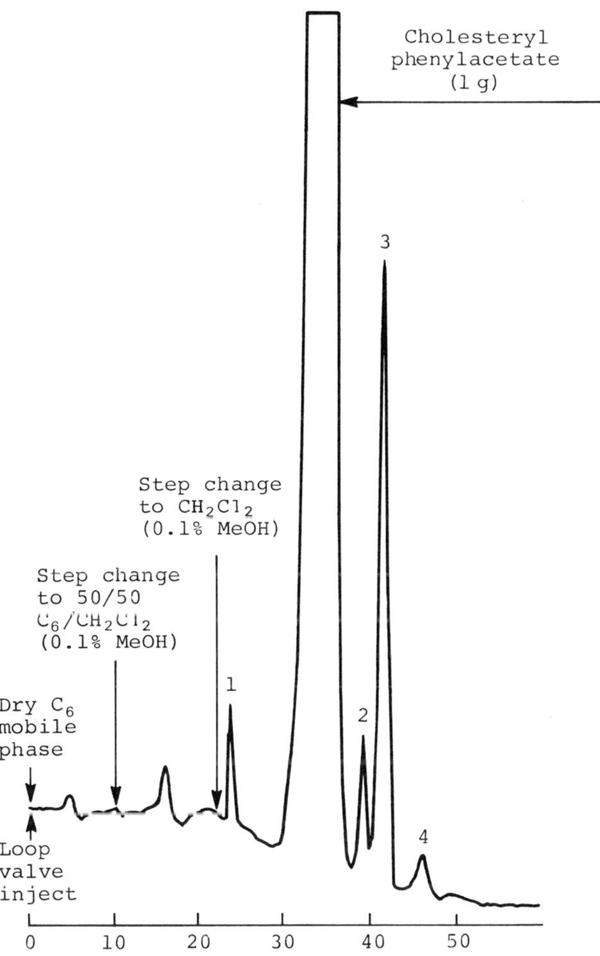

FIG. 8. Chromatogram of a 1-g sample of cholesteryl phenylacetate. Column, two 0.5 m X 23 mm I.D. columns packed with Spherosil XOA-400 connected in series; mobile phase, step elution from dry hexane to dichloromethane containing 0.1% methanol; temperature, ambient; column pressure, 1000 p.s.i.g; flow rate, 30 ml/min; detector sensitivity, 0.32 absorbance units full scale. (D. R. Baker, R. A. Henry, R. C. Williams, D. R. Hudson, and N. A. Parris, *J. Chromatogr. 83*, 233(1973). Reprinted with permission of copyright owner.)

large samples. Figure 9 shows a typical sample valve arrangement. The sample loop can be easily changed for different sample-size requirements or the loop can be syringe-loaded for efficient, no-waste injection of valuable materials. As modern preparative liquid chromatography becomes a routine laboratory procedure, automatic injection of large samples will become a necessity. Figure 10 shows a diagram of an automatic sampling system, in which the electronic control unit operates both the sample mechanism and the injection valve.

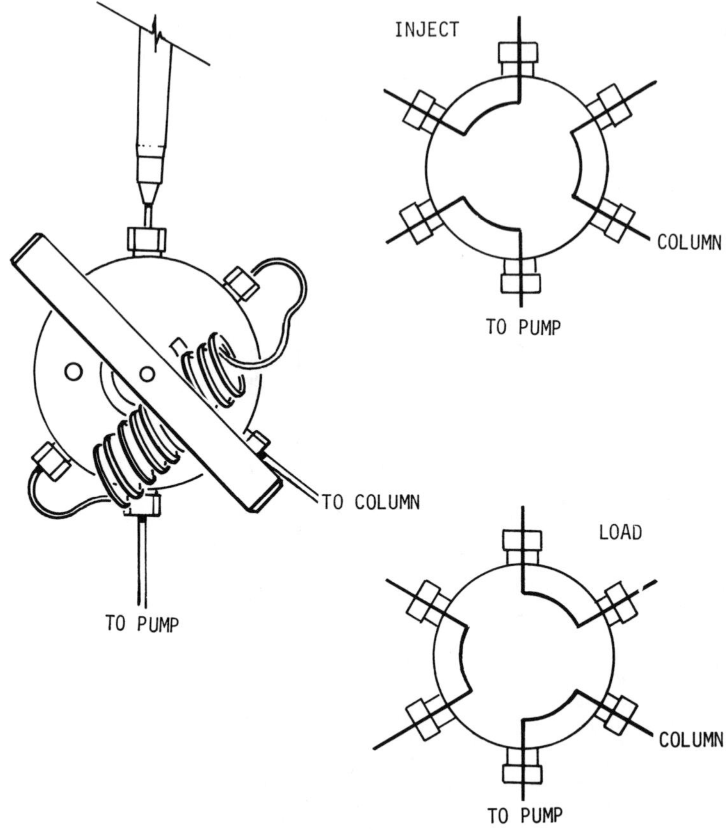

FIG. 9.   Sample injection valve.

D. Detectors

Because of the large sample sizes and high flow rates often employed, special detector
cells or stream splitters may be necessary to avoid overloading the detector or exceeding the flow-rate capacity of the detector cell. Special cells may have larger volume
and shorter path length, and they should also have large-bore inlet tubing to minimize
pressure drop. The low dead-volume splitter is an excellent technique for employing
an analytical detector in preparative work. The flow stream can be recombined after
the cell if desired so that the entire peak can be collected. Splitting the flow
stream will not change the response of detectors such as the photometric and refractive index which are sensitive to concentration of sample in the mobile phase; therefore, a stream splitter would be preferred for collection of a trace component at high
flow rates.

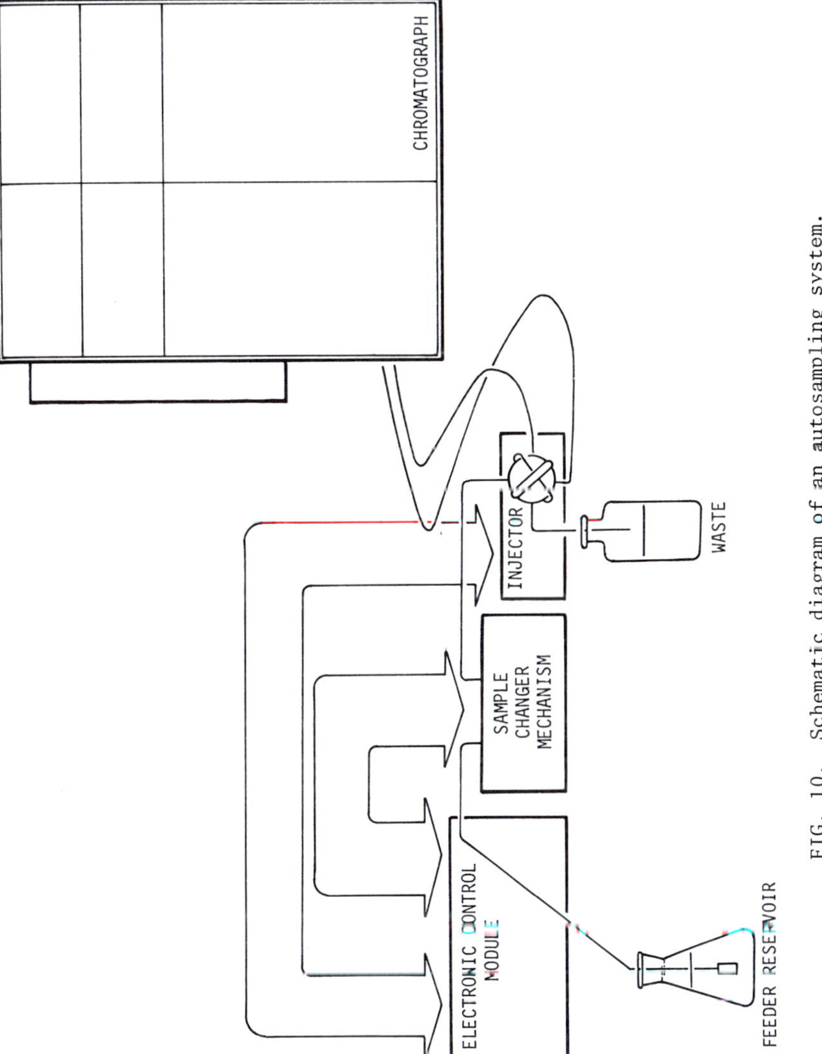

FIG. 10. Schematic diagram of an autosampling system.

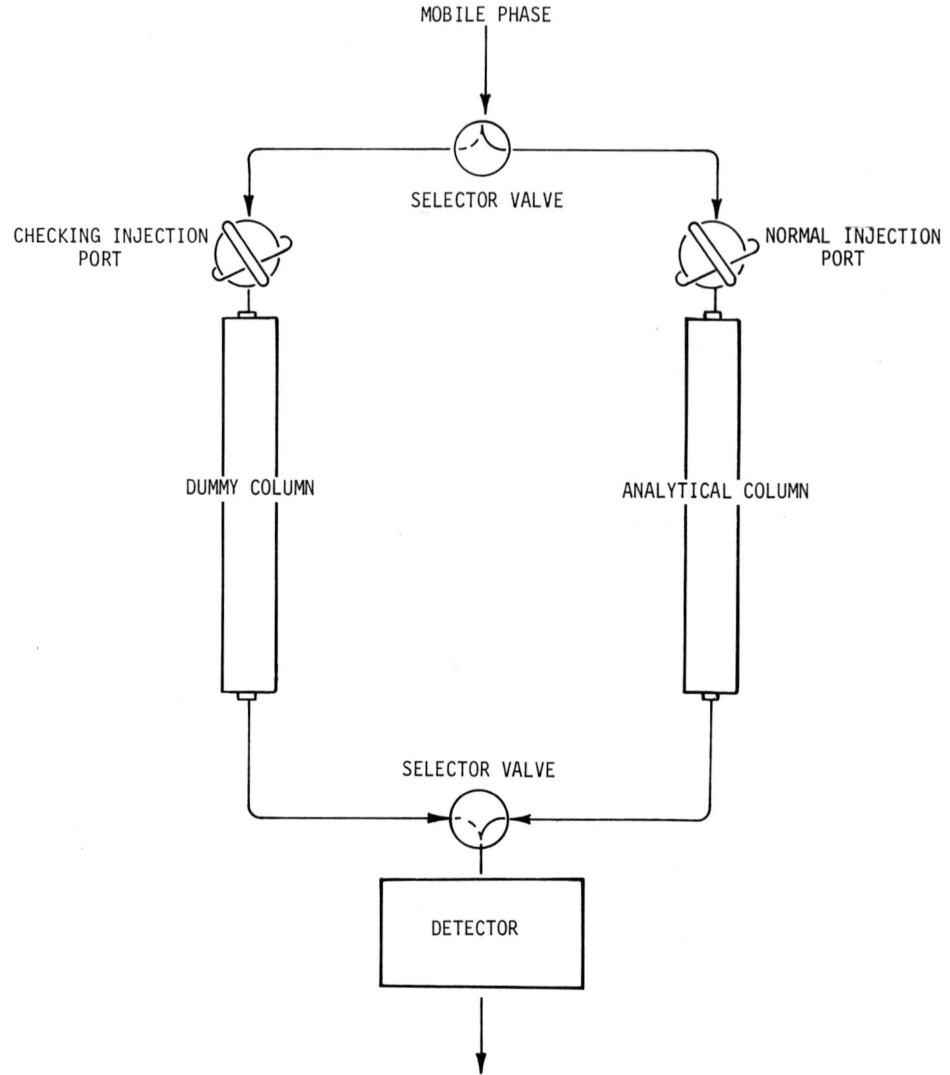

FIG. 11. Detector calibration using a bypass line.

Another special technique that can be very important in preparative LC detection is the use of a bypass column to calibrate the detector. Figure 11 shows an arrangement for detector calibration using a bypass line. By using a dummy column that will not retain the sample, it is easy to estimate whether the detector will be overloaded by the sample and whether all of the sample is being eluted from the separating column.

Ultraviolet/visible photometric detectors are probably the most widely used today for all types of liquid chromatography. In addition to special cells and splitting

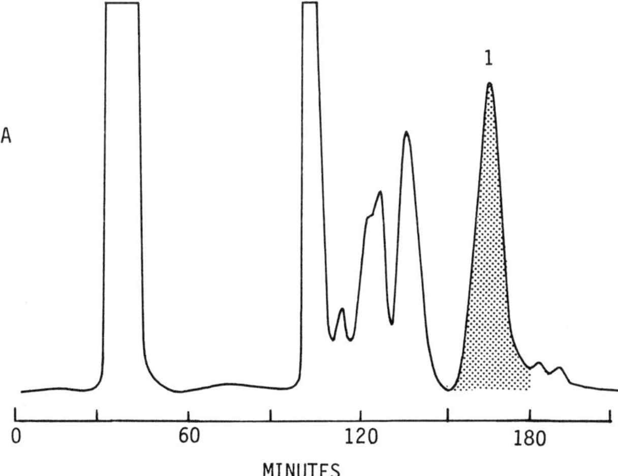

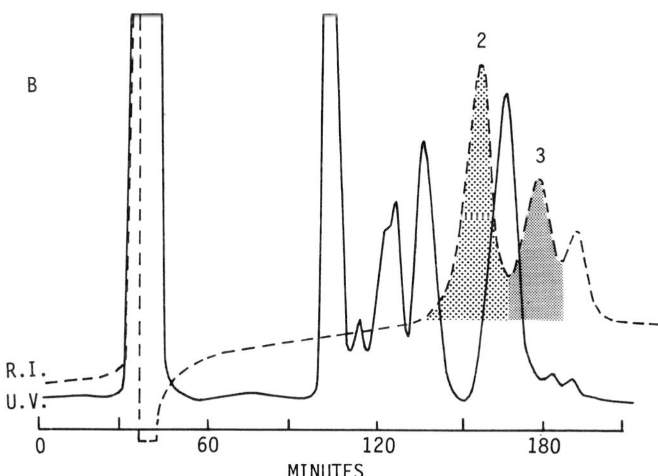

FIG. 12. Detector anomaly in preparative steroid separation. Chromatogram A, UV dotector only. Chromatogram B, dual UV/RI dotoctors; Column, 88 X 2.3 cm I.D.; Spherosil XOA-400; mobile phase, pentane/dioxane (1:1) with 1.5% water; pressure, <300 p.s.i.g; flow-rate, 14 ml/min; sample, 5 ml of 200 mg/ml in mobile phase. (J. J. DeStefano and J. J. Kirkland, *Anal. Chem. 47*, 1193A(1973); *47*, 1103A(1973). Reprinted with permission of copyright owner.)

techniques, several other methods can be used to improve the performance of the photo-motric dotoctor for proparativo applications. Onc mcthod is to usc thc dctcctor in a percent-transmittance mode rather than an optical density or absorbance mode. In the %-T mode, the detector is sensitive at low concentrations, but uses a compressed scale at high concentrations. A variable-wavelength photometric detector can be positioned to a wavelength where components of interest do not have strong absorption, so that peaks can be kept on scale in a linear response region.

The refractive index detector, which is less sensitive than the photometric, gains utility in preparative LC where sensitivity is not usually a requirement. The RI detector's universal nature allows all components of the sample to be observed and isolated if desired. Also, because it is less sensitive, the detector is less likely to saturate. The response of the refractive index detector is generally more uniform for compounds of similar structure so that the relative size of peaks is a good indication of the concentration of those components in the original sample. Figure 12 shows a separation monitored by both uv and RI detectors, which indicates the complimentary information that can be obtained by using these detectors in series. The chromatograms show that steroid isomers (peaks 2 and 3) elute on either side of a minor impurity (peak 1) which was strongly uv-absorbing. Collection of peak 1 in chromatogram A only yielded 39% recovery of the total sample injected, while collection of peaks 2 and 3 in chromatogram B yielded 37% and 29% recovery, respectively. Figure 12 is an excellent example of the earlier statement that relative peak sizes using the RI detector are a good indication of compound concentration in the total sample, while peak sizes can be very misleading with photometric detectors. Many other detectors such as fluorescence, conductivity, flame ionization, etc., can be employed in preparative LC; however, they are less useful than the photometric and refractive index detectors and will not be covered. Techniques for using these other detectors are very similar to those described.

E. Collection

At present, most collection in preparative LC is done manually by diverting flow into a collection vessel as the peak of interest passes through the detector. Many small valves are available to facilitate or even automate this operation. There will be increased use of automated valves for collection, although automating a complex chromatogram will be difficult due to the need for column cleanup between samples.

If trace quantities are being collected, it may be important to reduce to final volume in a narrow-bottom vial to concentrate the trace compound and avoid unnecessary loss.

F. Special Uses of Valves

When multiple columns are needed in order to obtain adequate resolution of two closely related compounds, a method known as recycling can be a practical alternative to adding columns in a series. Recycle can be accomplished in either of two ways. One method recycles sample through the pump, injection system, and column and is diagramed

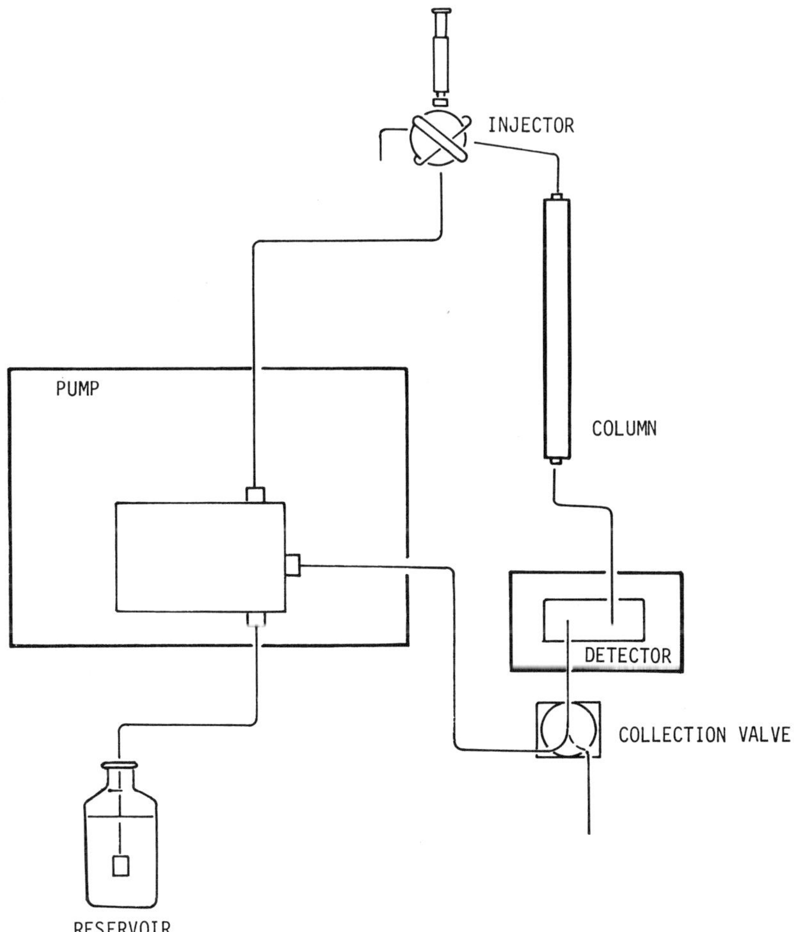

FIG. 13.   Flow configuration for closed-loop recycle.

in Fig. 13. This approach can be employed with normal, low-pressure detector cells, but requires special attention to extra-column dead volume and special design of the pumping system and injection port. It can be employed best in preparative work where collection of large amounts of major components is the objective and is less desirable for high-resolution preparative work or collection of trace components in a sample. Another method, illustrated in Fig. 14, uses alternate columns with a low dead-volume switching valve to keep the sample moving through the two-column system. This method can be used with virtually any pump and inlet system, but requires high-pressure detector cells as it is diagramed in Fig. 14. The detector in alternate-pumping recycling can also be placed in the low-pressure drain line and the valve switched just as the peaks begin to emerge from the column system. Steps which are common to both recycle methods are as follows:

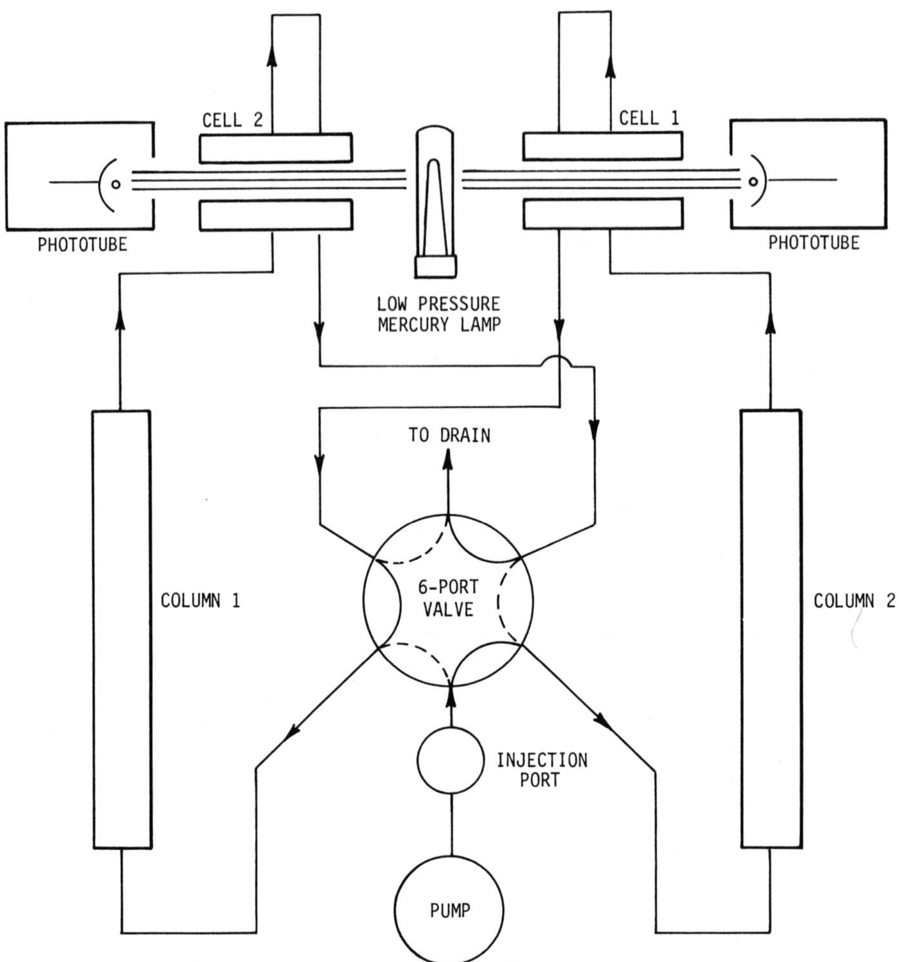

FIG. 14. Flow configuration for alternate-pumping recycle.

1. Collect the portion of the chromatogram that needs further resolution.
2. Clean up the column by switching to a strong solvent.
3. Reduce the volume of the collected fraction and reinject.
4. Recycle the sample until adequate resolution has been obtained or until the peaks broaden to fill the column system.

Another important use of valves in preparative LC will be for column switching. Column-switching techniques may be used in lieu of gradient elution when a complex sample containing both strongly and weakly held components must be investigated. Column-switching techniques also permit non-uv-absorbing compounds to be monitored using the refractive index detector, which cannot be used in conjunction with

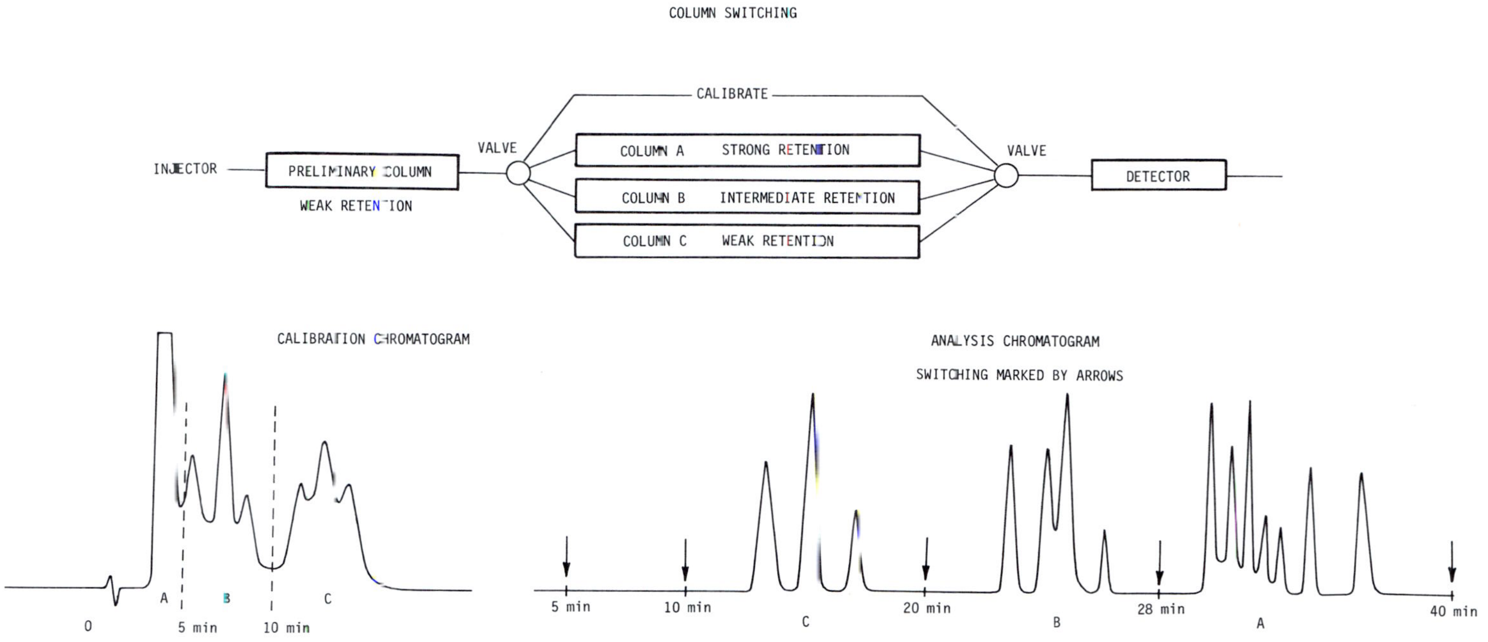

FIG. 15. Diagram of a column switching arrangement.

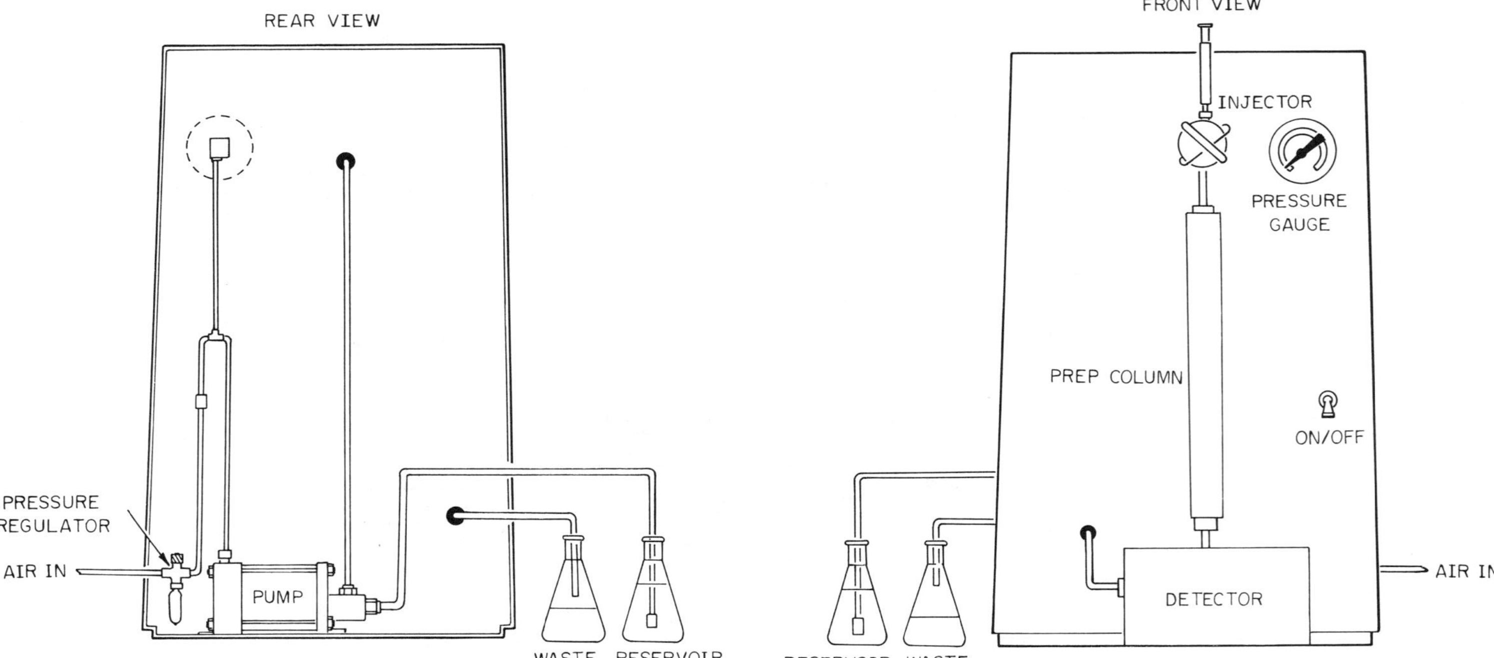

FIG. 16.   Design of a low-cost preparative LC.

gradient elution. Figure 15 illustrates one configuration that can be employed for
column switching, sometimes referred to as column or stationary-phase programming.
In Fig. 15, the sample is first passed through the preliminary column, which is selected to retain the most strongly held components, then directly to the detector
where resolution is measured. After the calibration run, the partially resolved
sample is divided into fractions on the basis of time. The second and all subsequent
injections are then switched at the appropriate time to adequately resolve the entire
sample without making a mobile phase change. Figure 15 also shows an example of a
calibration and analysis chromatogram in which the sample is divided into fractions
A, B, and C by the precolumn and then further resolved in the parallel-column arrangement. Fraction A is unretained and is allowed to pass into column A which is strongly
retentive; fraction B is weakly retained and is diverted into column B for further resolution; fraction C is most strongly retained and is switched into column C for additional resolution, if necessary. When fraction C has eluted, the flow is diverted to
column B to elute fraction B and finally to column A to elute fraction A.

G. Suggested Design for a Low-Cost Preparative Liquid Chromatograph

While any commercially available liquid chromatograph can be adapted for small scale
preparative work, it is useful to describe the design of a low-cost instrument that
can be dedicated to preparative analysis with minimum investment. Figure 16 illustrates the configuration of a preparative LC built-around a pneumatic-amplifier
pump.[†] This pump is recommended because of its low cost, high-pressure limit, and
large flow-rate capability. The price of the pumping system with 50 ml/min and
4500-psi capability, valve injector, and detector, should be in the $2,000-$3,000
range. Flow rate can be selected with this design by adjusting the inlet pressure
to the pump. Flow-rate precision and detector baseline stability will be adequate
for all but the most demanding analysis.

IV. SUMMARY AND PROJECTION

High performance LC is an excellent method for rapidly isolating reasonable quantities
of very pure material for a variety of end uses. Because separation takes place in
solution at or near ambient temperature, fractions may be isolated easily by diverting
the eluent stream to a collection vessel as the peak of interest emerges from the column. For this reason, preparative LC is also rather simple to automate using low-
pressure, low dead-volume solenoid-actuated valves.

Liquid chromatography is also an excellent tool for isolating trace quantities
for identification or further investigation. As improvements are made in detectors

---

[†]Haskel minipump, Model MDST-36C (designation C includes inlet filter/regulator/
valve manifold).

and interfaces to other analytical instruments, this application of high performance LC should increase dramatically.

Preparative LC will benefit in the near future from

1. Advances in automation techniques such as autosamplers, auto-collectors, and a variety of automated valves.

2. Advances in column technology which will result in lower cost to the user. The most important will be the discovery that packing an LC column, even with microparticles, is not difficult.

3. Improvements in detectors that will allow more sensitive detection of a wider variety of compounds.

REFERENCES

1. S. T. Sie and N. J. van den Hoed, *J. Chromatogr. Sci.*, *7*, 257(1969).

2. J. J. DeStefano and H. C. Beachell, *J. Chromatogr. Sci.*, *8*, 434(1970).

3. H. C. Beachell and J. J. DeStefano, *J. Chromatogr. Sci.*, *10*, 481(1972).

4. J. J. DeStefano and H. C. Beachell, *J. Chromatogr. Sci.*, *10*, 654(1972).

5. R. A. Henry, in *Gas Chromatography 1972* (S. G. Perry, ed.), Applied Science Publishers, Essex, England, 1973, p. 199.

6. J. P. Wolf, *Anal. Chem.*, *45*, 1248(1973).

7. D. R. Baker, R. A. Henry, R. C. Williams, D. R. Hudson, and N. A. Parris, *J. Chromatogr.*, *83*, 233(1973).

8. Unpublished data supplied by Bill Ja, Stauffer Chemical Co., Richmond, California.

9. K. J. Bombaugh and P. W. Almquist, *Chromatographia*, *8*, 109(1973).

10. J. J. DeStefano and J. J. Kirkland, *Anal. Chem.*, *47*, 1193A(1975); *47*, 1103A(1975).

11. S. R. Bakalyar, R. A. Henry, and J. Yuen, How to Pack Liquid Chromatography Columns, Spectra-Physics Technical Bulletin 114-176(1976).

12. B. L. Karger, M. Martin, and G. Buiochon, *Anal. Chem.*, *46*, 1640(1974).

13. J. J. Kirkland, *The Analyst*, *99*, 859(1974).

Chapter 10

PREPARATIVE HPLC: MULTIGRAM
QUANTITIES

William H. Pirkle

School of Chemical Sciences
University of Illinois
Urbana, Illinois

I. INTRODUCTION

High pressure liquid chromatography (HPLC) has found rapid and widespread acceptance
as a means for qualitative and quantitative analysis of chemical mixtures. Moreover,
the technique is rapidly being demonstrated as a powerful method for effecting preparative scale separations. To date, most such separations have been on a small
scale and have involved subgram quantities of materials. This does not mean that
large scale HPLC separations cannot be carried out, but simply that analytical techniques have outpaced the preparative aspects of HPLC. Even so, it is clear that
multigram HPLC will be a popular, practical, and generally applicable purification
method.

Historically, liquid chromatography developed as a preparative rather than an
analytical technique. Nevertheless, liquid chromatography is seldom used for multigram separations unless more conventional purification methods such as crystallization

or distillation are ineffective or inapplicable. This bias against large-scale open-column liquid chromatography may be attributed to the low separation efficiency generally afforded. Since recent advances in analytical HPLC have resulted in separation efficiencies undreamed of 10 years ago, it is understandable that this success has tantalized synthetic chemists and has stimulated research on the preparative aspects of HPLC. This research is late, perhaps because of initial pessimism concerning the performance levels that might be expected from "large" HPLC columns, stemming from the fact that an increase in the diameter of a gas chromatography column markedly lowers the efficiency of the column. Nevertheless, it is now evident that large HPLC columns can offer convenient, rapid, and efficient means of obtaining multigram amounts of pure materials.

It is anticipated that considerations of cost, practicality and convenience will frequently dictate a unique approach to multigram HPLC that will be rather different from those presently employed for analytical HPLC. This chapter considers the ways in which multigram HPLC is apt to differ from present analytical and subgram preparative HPLC techniques. Additionally, the subject of automated multigram HPLC is covered and its advantages and limitations are discussed.

II. ANALYTICAL AND SUB-GRAM HPLC: A SYNOPSIS

Analytical HPLC units consist of the column, a means of introducing the sample onto the column, a pump to force the eluting solvent through the column and means of detecting and recording the elution of the various components of the sample. Vast and continuing effort has gone into the external design of the column, methods for packing columns, and the development of high-efficiency packings. An excellent and well-documented discussion of high performance liquid chromatography packings has recently appeared [1]. The use of small (5 to 10 μm) uniformly-sized particles of packing is becoming standard in analytical HPLC systems owing to the rapidity and efficiency of the separations thus afforded. It should be noted at this point that the efficiency of a column of given dimensions (as judged by the number of theoretical plates it affords) is quite dependent upon flow rate, sample size, packing techniques, and many other experimental parameters as well as the size and nature of the column packing material, the solvent employed and the solute itself.

Although most analytical separations are effected on a sub-milligram scale, larger samples can be accomodated to analytical HPLC instruments simply by increasing somewhat the diameter of the columns. Several manufacturers (e.g., Varian Associates, Palo Alto, Calif.; Waters Associates Inc., Milford, Mass.; and Whatman Inc., Clifton, N. J.) now offer columns of about 10-mm bore that are prepacked with high-efficiency 5- to 10-μm packings. Albeit rather expensive, these columns equal or exceed the number of theoretical plates offered by their small bore analytical counterparts.

Columns of this type thus allow preparative separation of tens or hundreds of milligrams of material on analytical instrumentation. Thus, apart from the trivial complication introduced by having to collect fractions, relatively small-scale preparative HPLC offers few difficulties beyond those encountered in achieving an analytical separation. While not the topic of this chapter, sub-gram separations of the preceding type are important and will become increasingly common. Often, however, an investigator will require more material than can be purified on an 8- to 10-mm diameter column which is fundamentally "analytical" in origin.

III. MULTIGRAM HPLC

There are two basic approaches to the problem of obtaining multigram amounts of a pure material via HPLC techniques. The first is to use a modestly scaled-up "analytical" system to repetitively effect however many subgram separations are required to afford the necessary amount of material. The second approach is to use a much larger multigram HPLC system. The first approach is simple in that the presently widespread analytical technology is adequate and no special "multigram" apparatus is necessary. However, this approach can be rather inefficient in terms of the expense, time, and labor involved. A number of practical and strategic considerations relevant to subgram preparative HPLC have been discussed by Fallick [2] and by DeStefano and Kirkland [3]. These discussions are also relevant to multigram HPLC although some differences in strategy may occur, especially if automated systems are employed.

A. Design and Packing of Large-Bore Columns

The key issue concerning the practicality of multigram HPLC is the question of whether "giant" HPLC columns can achieve the performance levels necessary to routinely separate multigram samples into their individual components. One might simply suppose that further scale-up of high-efficiency analytical columns would solve this problem. Unfortunately, the cost of filling a rather large column with the very expensive packings used for analytical applications effectively precludes this approach. Purely from the standpoint of practicality, most large-volume multigram HPLC columns will require relatively inexpensive packings to be economically feasible. Owing to their relatively low cost and ability to accommodate higher sample loadings, porous silica gel and alumina will be more widely employed for multigram HPLC than will be pellicular, "bonded-phase," or other specialized packings [1]. Actually, the exclusion of expensive ultrahigh-efficiency packings from large-bore columns is not as great a loss as might be supposed. While these packings can afford analytical columns exhibiting as many as 40,000 theoretical plates per meter, inspection of HPLC chromatograms published in the literature or in manufacturers advertisements shows that most columns fall far short of this performance level in actual usage.

Further advances in gel-permeation-type packings could render these materials especially valuable for multigram HPLC. Among other advantages that include indefinite lifetimes and the ability to accommodate unusually high sample loads, gel permeation packings are chemically inert and allow chromatography of reactive compounds which would not survive exposure to silica gel or alumina. Despite their potential, gel permeation type packings will not soon surplant silica gel and alumina as the packing predominantly used for multigram HPLC due to their cost.

Contrary to early expectations, increasing the diameter of an HPLC column does not necessarily degrade its efficiency. Indeed, Wolf [4] and others [5-8] have shown that large bore columns can equal or exceed the efficiency of identically packed small bore analytical columns. It should be noted that in general, when very high column efficiencies are reported, they are attained by using low sample loadings of essentially unretained solute at low flow rates through small particle packings. Small particle packings afford higher efficiency than do large particle packings not only in theory but also in practice. However, the sizing techniques that must be employed for the smaller sized particles cause them to be rather expensive. Moreover, they are difficult to pack and their use requires pumps and system components capable of generating and sustaining rather high pressures. Consequently, the ultrasmall (5-10 µm) particles used for analytical HPLC will seldom be encountered in multigram applications. It is to be expected that most multigram HPLC columns will compromise performance with practicality, not only in terms of choice of packing, but also in terms of optimum flow rates and sample sizes. Therefore, most multigram HPLC packings will generally fall between 40µ to 200µ in average diameter with the 60- to 100-µm average diameters being most popular.

Considerable effort has gone into methods of packing large bore HPLC columns so as to optimize the level of performance when intermediate sized particles are used. Workers from Merck have concluded [9] that a bumping technique works best for silica gel particles larger than 30 µm in diameter. Particles less than 30 µm in diameter are apparently best packed by the balanced-density method [10, 11], a technique that would be rather troublesome to employ in the case of very large columns. Other packing techniques for large HPLC columns include dry packing under vibration and wet packing as a slurry. Packing techniques become much less critical as particle diameters exceed 60 µm but still influence column efficiency. A narrow range of particle sizes improves column efficiency (and increases cost of the packing), regardless of the size of the column or the average size of the particles. Prepacked multigram HPLC columns are commercially available. EM Laboratories, Inc. offers three sizes of glass columns prepacked with silica gel particles having a narrow size distribution. The largest of these has a length of 44 cm and a bore of 3.7 cm. While the supplier makes no quantitative concerning the efficiency of these columns, on the basis of a chromatogram appearing in the manufacturer's brochure, the larger column is estimated to afford about 800 theoretical plates. This level of efficiency has been attained in the author's laboratory for an EM column of this size employing 50- to 100-mg sized samples. Lower efficiencies are obtained for larger samples.

Surprisingly high efficiencies have been obtained on large (6 cm bore, 120 cm length) homemade glass columns [12] dry packed (by bumping) with inexpensive 58-μm silica gel having a surface area of 600 m^2/g. Using methylene chloride as an eluent, one of these columns has exhibited over 7000 theoretical plates for a 100-mg sample of pyrene at a flow rate of 2.5 liter/hr. However, this efficiency degrades rapidly as multigram samples are applied. For example, under the previous conditions, samples of 500 mg, 5 g, and 10 g afford 3800, 700, and 300 theoretical plates, respectively. While plots of sample size vs column efficiency will vary with the nature of the sample (especially for retained solutes) and the adsorbent, the rapid loss of efficiency with increasing sample size strongly suggests that efficient multigram HPLC columns will have to be quite large. For retained solutes (which of course covers most compounds of interest), heavy sample loadings tend to saturate the active sites on the surface of the adsorbent and give rise to rapid spreading of the sample band. In general, the greater the surface area of the adsorbent, the less susceptible it is to this effect for a given sample size. This overloading effect cannot be totally overcome by more efficient packing techniques, since it is intrinsic to the chromatographic process. The factors affecting solute band spreading have recently been reviewed by Grushka [13].

Attempts to increase multigram HPLC column efficiency through mechanical compression of the packing have been fruitful. French workers [14] have developed an 8-cm diameter metal column in which the lower end contains a pneumatically driven piston that compresses the packing material along the length of the column. Such compression is claimed to double the number of theoretical plates in the column when 10-40 μm silica is compressed to 8-10 atm. This innovation has been incorporated into a commercial multigram HPLC system, the Chromatospac-Prep 100 (J-Y Optical System, Metuchen, N. J.). Although the system is advertised to afford 7000 theoretical plates per meter and to accommodate samples of up to 100 g, the preceding paragraphs should make it clear that both claims cannot be met simultaneously. Published [14] chromatograms obtained on this system clearly demonstrate the efficiency of this column for small samples. Apart from column design, the Chromatospac-Prep 100 is fairly conventional and utilizes a 10-liter solvent reservoir pressurized with inert gas in lieu of a pump.

Even more recently, Waters Associates (Milford, Mass.) has developed a multigram HPLC unit, the PrepLC/System 500, employing disposable 5.7-cm x 30-cm columns. These cartridge like columns with thin walled polyethylene sleeves containing 80-110 μm silica gel are loaded into a tube-like steel chamber where end plugs are sealed in place with hydraulic pressure. The chamber is then pressurized with fluid, and the

thin-wall cartridge undergoes radial compression forcing the silica gel into a tightly
packed homogeneous column. The radial compression improves column efficiency markedly
and efficiencies of 1000-1400 theoretical plates per foot are claimed for unretained
300- to 450-mg samples at flow rates of 80-100 ml/min. These efficiencies are very
high in view of the sample size, high flow rates and large size of the silica gel
particles (80-110 μm) used. Apart from reduced cost, the use of moderately large
silica gel particles allows high flow rates to be achieved with relatively low pressures. High flow rates are a prerequisite for rapid separations when large columns
are employed. This new Waters preparative unit utilizes a twin-pistoned pump capable
of affording flow rates of up to 0.5 liter/min at a pressure of 500 psi.

B. Solutions to Some Problems That Accompany Multigram HPLC

Large multigram columns will frequently require many liters of solvent per chromatography. The purchase, storage, handling, and disposal of large quantities of solvents constitute economic, safety, and ecological barriers to large scale HPLC. To
date, no commercial preparative HPLC unit has been designed to overcome these particular problems associated with large scale operation. Large scale HPLC units occasion
other problems as well. For example, a common practice in preparative HPLC is the
use of a fraction collector to collect numerous periodic fractions. However, most
commercial fraction collectors lack the capacity to accommodate the large volumes of
eluent that multigram HPLC will entail. Relevant to this discussion is the fact that,
as previously noted, column efficiency is degraded rapidly as sample loading is increased. Hence, large samples of materials moderately difficult to separate may have
to be chromatographed in portions even when large HPLC columns are used. Accordingly,
a very desirable feature of a multigram HPLC system is that it be capable of automatically effecting repetitive separations once an initial set of operational parameters
has been defined. Repetitive automatic operation of a multigram HPLC unit compounds
the problem of fraction collection (the receivers soon overflow) and also generates
problems in the areas of solvent supply and sample introduction.

1. Automated Fraction Collection

Generally, liquid chromatographic fractions are collected manually or by a fraction
collector which actuates on a regular time or volume basis. Instrumentation Special-
ties Company (Lincoln, Nebraska) offers a "peak separator" accessory which advances
the fraction collector to the next tube if slope detection circuits note slope inflec-
tions on the chromatographic recorder trace. However, even this variation on the
usual approach of collecting many small periodic fractions is not suited for a repetitive system inasmuch as several runs would exhaust the storage capacity of the fraction collector. In any event, the collection of a large number of fractions to be

subsequently pooled or discarded (depending upon content) is pointless when, as will frequently be the case, the compound(s) of interest can be cleanly separated from the remainder of the mixture. In such cases, collection of the entire desired chromato- graphic band in one receiver is permissible and preferable. This principle has been employed on commercial automated preparative GLC instruments and has been applied to preparative liquid chromatography by Aitzetmüller [15] and by Pirkle and Anderson [16]. The latter workers have avoided the problem of exhausting the storage capacity of their receivers during protracted repetitive runs by continuously distilling the volatile eluting solvent from the receiver as it is collected. Nonvolatile solutes remain in the receiver. This approach not only substantially reduces the number of fractions which must be collected and avoids the pooling and concentration of frac- tions as separate steps, but also solves two major remaining problems in large-scale HPLC, the consumption of overwhelming quantities of organic solvents and the tedium associated with routine repetitive operation. As the solvent is distilled from the receiving kettles, it is returned to the main pump. Hence, a closed-loop solvent system is obtained. This conceptually simple innovation makes automated repetitive multigram HPLC feasible from both the practical and economic viewpoints. The closed- loop solvent system ensures that a relatively small amount of solvent suffices for a protracted series of large scale runs; the receivers never overflow and the pump never runs out of solvent.

2. Description of a Prototype Automated Multigram HPLC System

Operationally, the automated multigram HPLC system developed by Pirkle and Anderson at the University of Illinois works as follows. A sample pump introduces the sample onto the column through a one-way check valve. Sample size is controlled by the length of time the pump runs (programmable), the rate of pumping (programmable), and the sample concentration. At the conclusion of sample introduction, the sample pump is shut down by a counter-sequencer-timer unit (CST). This CST unit then starts the main pump that draws solvent from a reservoir (a gallon solvent bottle) and introduces it into the column through a second one-way valve. Eluent from the column passes through a short path length flow cell in an ultraviolet detector, through one of four solenoid valves and into one of four steam-heated boiling flasks. The volatile elut- ing solvent is flash distilled as the eluent is introduced into the boiling flask. Solvent vapor is conducted into an efficient horizontal manifold-like stainless steel condenser that serves all four boiling flasks. After condensation, the distilled and degassed solvent is returned to the original reservoir. Nonvolatile solutes remain in the boiling flask. The CST unit counts peaks (up to nine) and preprogramming of a 4 x 10 array of toggle switches sequentially directs desired components into dif- ferent boiling kettles and undesired components into a refuse kettle. After a pro- grammable length of time great enough for completion of the run, the CST unit stops

the main pump, resets to the start position, and activates the sample pump to introduce yet another sample. Thus, the system runs repetitively and unattended once
started.

It should be noted that preparative columns of any type or design can be used
in this automated system. Thus, improvements in column technology such as Waters'
"radial compression" concept, could be easily incorporated into an automated closed
solvent-loop HPLC system.

a. Limitations of the Closed Solvent-Loop Approach

As presently constituted, the automated system has two operational limitations brought
on by use of the closed-loop solvent system. First, solutes must not be volatile or
thermally unstable or they will be lost from or decomposed in the heated receiving
flasks. However, a larger quantity of solvent would then have to be initially supplied to the pump and the receivers would have to be manually emptied to prevent
overflow. Apart from the faculties for automatic collection of the peaks and repetitive operation, the system would thereby be reduced to what is now standard practice.
The second limitation is that gradient elution schemes are not readily compatable
with a closed-loop solvent system. Fortunately, the major fraction of multigram HPLC
applications can very likely be done isocratically. Stepwise gradient elution, a
common practice for preparative liquid chromatography, is likewise not accommodated
to an automated HPLC system without considerable intervention of the operator.

An alternate approach that has somewhat the same desired effect as gradient elution but one that is better suited to the closed solvent loop concept is that of temperature programming of the column. This technique, common to gas chromatography, has
been applied [17] to HPLC but has not been widely adopted. Technically, the application of temperature programming to multigram columns seems feasible and easily con-
trollable on an automated basis. Typically, increasing the temperature of the column
hastens the elution of the solute. Since this technique would require no change in
the composition of the eluting solvent during the run, no "reactivation" of the column would be required apart from allowing it to cool. Similar effects can be achieved
by programming an increase in flow rate during the run to assist in the elution of the
slower moving components of the mixture [18].

b. Applications of Automated Multigram HPLC

Although used for diverse preparative separations, the most extensive application of
the Illinois system has been in the resolution of optical isomers via the chromatographic separation of diastereomeric derivatives [19]. In particular, the resolution
of 2,2,2-trifluoro-1(1-naphthy)ethanol has been detailed [12] and chromatograms illus-
trative of the performance of the automated repetitive system have appeared [12, 16].
As originally described [16], 6-10 g of diastereomeric carbamates could be separated

in 24 hr. This value has since been raised to 40-100 g per 24 hr and further scale-up is possible. Interestingly, this performance has been obtained on Brinkmann 63-200 μm alumina, a relatively inefficient (in terms of theoretical plate height) packing owing to the large average-particle size and the large range of particle sizes. However, the material is inexpensive, easily packed, and allows high flow rates (75-150 ml/min) to be obtained with relatively low pressures (20-40 psi) that, in turn, allow the use of large glass columns and inexpensive pumps.

C. Pumps for Multigram HPLC

To be suitable for multigram HPLC, a pump should be capable of affording high flow rates. Pressure requirements are dictated by the column being used but typically will be substantially less than those encountered in analytical HPLC. The flow need not be pulseless, since detectors will generally be operated at low sensitivities. However, large pressure surges are undesirable. Waters Associates has a newly designed twin-piston pump capable of affording flow rates of up to 30 liter/hr at 500 psi. At present, this pump is available only with the PrepLC/System 500. Haskel Engineering and Supply Co. offers a pneumatically driven pump designed for HPLC that can afford flow rates of up to six liters per hour at pressures up to 4500 psi. This pump is being used by several commercial HPLC instruments.

Pumps comprised of sturdy solvent reservoirs pressurized with an inert gas are relatively inexpensive and simple to operate. This type of pump is used in several commercial HPLC systems and can provide high flow rates until the reservoir is emptied. Refilling the reservoir requires intervention by the operator and possible interruption of the run. Since the reservoir cannot be refilled by gravity flow while pressurized, this type of pump is not suited for use with a closed-loop solvent system. Doubtless, an auxiliary pump could be used to continuously refill the pressurized reservoir with reclaimed eluent. Such an arrangement hardly seems worthwhile since a mechanical pump obviates the problem entirely.

As previously noted, the Illinois automated multigram HPLC system [12] usually employs large columns that require but low pressures to attain high flow rates. For these applications, a stainless steel rotary pump produced by the Eco Pump Corporation (Chicago, Ill.) has proven to be especially suitable. In this pump, a driven spur gear engages an idler gear and solvent is trapped between the meshed gears and transported a short distance from the low-pressure inlet to the high-pressure outlet. Conceptually simple, the pump has no valves, has but one pressure seal (around a rotating shaft), is very compact, and yet easily affords pulseless flow rates of from 0 to 1 liter/min at pressures up to 100 psi.

Other metering pumps capable of delivering high flow rates at modest pressures are available from Fluid Metering, Inc. (Oyster Bay, N. Y.), and the Milton Roy

Company (St. Petersburg, Fla.). In some cases, the operating temperature of large
metering pumps may be high enough to vaporize volatile solvents such as methylene
chloride and pentane during the suction portion of the pumping cycle. A water-cooled
jacket or heat-transfer plate solves this problem.

D. Detectors for Multigram HPLC

Virtually any commercial detector system can be used for multigram HPLC provided it
responds to the solutes of interest. Lack of sensitivity will not often be a serious
problem for preparative work owing to the large samples employed. Refractive index
detectors are perhaps more universally applicable than ultraviolet-absorption detectors but are less convenient to use if the character of the elution solvent changes
during the run as it would when gradient elution or temperature programming were employed. The ability of RI detectors to be kept "on scale" by attenuation during elution of highly concentrated solute bands is an added advantage. Ultraviolet detectors
can be blinded by high solute concentrations and this sometimes occurs even with relatively low concentrations of strongly adsorbing solutes. This problem can be partially offset by reducing the length of the optical path of the flow cell although
this may require a stream splitter to reduce the volume of eluent flowing through
the cell. Alternatively, a different monitoring wavelength may be chosen so as to
reduce the absorbance of the solute. Gradient elution or temperature programming
poses no problems for ultraviolet detectors.

IV. CONCLUSION

Although high-efficiency multigram HPLC is still in its infancy and is not yet widely
used, it is clear that the method is technically and economically feasible. Indeed,
it can be confidently predicted that, within a few years, multigram HPLC techniques
will largely surplant crystallization and distillation as methods for purifying research derived compounds not only in laboratories, but in certain industrial applications as well. For example, multigram HPLC will soon be the separation method of
choice in many instances involving isolation and/or purification of intermediates
in protracted synthetic schemes, the purification and isolation of natural products
and the final purification of pharmaceuticals or fine chemicals. These predictions
take cognizance of the fact that multigram HPLC separations can be quantitative as
well as rapid and convenient. Moreover, there are very few limitations on the types
of compounds amenable to HPLC separation.

ACKNOWLEDGMENT

The literature survey essential to the writing of this chapter was conducted via a
computer search of the Chemical Abstract tapes. The extent and thoroughness of this

search, conducted under the direction of Professor Thomas Isenhour of the University of North Carolina, is perhaps not readily apparent, inasmuch as this chapter is not intended to be an exhaustive survey of the HPLC literature. The author is much indebted to Professor Isenhour for his assistance.

REFERENCES

1. R. E. Majors, *Amer. Lab.*, *7*, 10:13(1975).

2. G. Fallick, *Amer. Lab.*, *5*, 7:19(1973).

3. J. J. DeStefano and J. J. Kirkland, *Anal. Chem.*, *47*, 1103A, 1193A(1975).

4. J. P. Wolf III, *Anal. Chem.*, *45*, 1248(1973).

5. J. J. DeStefano and H. C. Beachell, *J. Chromatogr. Sci.*, *8*, 434(1970).

6. H. C. Beachell and J. J. DeStefano, *J. Chromatogr. Sci.*, *10*, 481(1970).

7. S. T. Sie and N. van den Hoed, *J. Chromatogr. Sci.*, *7*, 257(1969).

8. J. H. Knox and J. F. Parcher, *Anal. Chem.*, *41*, 1599(1969).

9. D. Randau and W. Schnell, *J. Chromatogr.*, *57*, 373(1971).

10. R. E. Majors, *Anal. Chem.*, *44*, 1722(1972).

11. R. M. Cassidy, D. S. Legay, and R. W. Frei, *Anal. Chem.*, *46*, 340(1974).

12. W. H. Pirkle and M. S. Hoekstra, *J. Org. Chem.*, *39*, 3904(1974).

13. E. Grushka, *Anal. Chem.*, *46*, 510A(1974).

14. E. Godbille and P. Devauz, *J. Chromatogr. Sci.*, *12*, 564(1974).

15. K. Aitzetmüller, *J. Chromatogr.*, *73*, 248(1972).

16. W. H. Pirkle and R. W. Anderson, *J. Org. Chem.*, *39*, 3901(1974).

17. M. Krejci and D. Kourilova, *J. Chromatogr.*, *91*, 151(1974).

18. H. Wiedemann, H. Engelhardt, and I. Halasz, *J. Chromatog.*, *91*, 141(1974).

19. W. H. Pirkle and J. R. Hauske, unpublished work.

Chapter 11

QUALITY CONTROL IN GAS LIQUID CHROMATOGRAPHY (GLC) AND HIGH PERFORMANCE LIQUID CHROMATOGRAPHY (HPLC)

David Sohn, M.D.

Department of Pathology, New York Medical College/
Bird S. Coler Hospital
Roosevelt Island, New York, New York

Julius Simon

Laboratory for Chromatography
Flushing, New York

I. INTRODUCTION

Laboratories employing gas-liquid chromatography (GLC) must take appropriate measures
to institute and maintain adequate quality control. The use of GLC in the analysis
of exogenous substances in biological fluids has passed from the realm of the experimental to a routine procedure for everyday clinical analysis. It is of greatest

value for accurate and rapid quantitative analysis. The discussions which follow
must apply almost entirely to quantitative analysis by GLC. The principles and recommendations included here apply equally to both quantitative and qualitative analyses. Qualitative identification of substances in biological fluids has an extremely
circumscribed and delimited role. Qualitative analysis by GLC may sometimes be used
as an overall initial rapid screening technique. Other independant confirmatory
methodologies would then follow. It may sometimes be employed for confirmation following tentative identification by other means.

In what will follow, a long litany of problems, real and theoretical, are enu-
merated. Despite this very extensive list of possible deficiencies, GLC does work
and, in its areas of competance, can work extremely well.

At this point, we would like to draw an analogy which may put these problems into
proper perspective. Reproductive physiologists often enumerate the multiple, inter-
related factors involving ovulation, fertilization, implantation, pregnancy, and par-
turition which must all operate successfully and in proper sequence to produce a
viable full-term fetus. The remarkable fact is that with all that is involved, pregnancy even occurs. Yet we all recognize that live births are common, frequent, widespread, and "normal" events. With reasonable attention to details, GLC can be a
routine, highly accurate, sensitive, and reproducible technique [1, 2].

II. GAS-LIQUID CHROMATOGRAPHY

The process of GLC can be divided into two major parts:

A. Preparation of the biologic fluid for injection into the chromatograph:
    sampling, storage, extractions, "cleanup" procedures, derivitization, and
    addition of recovery standards and internal standards [3].

B. The injection into the chromatograph, the process of transit of the suitably
    prepared analyte through the chromatograph to the detector, and the recording
    and interpretation of the effluent passing through the detector, which are
    considered the *instrumental* components of the analysis.

A. Sample Preparation, Cleanup, and Extraction

1. Sampling

No result of a determination can be of any meaning unless the material submitted for
analysis is a statistically valid, representative sample of what is to be measured.
This of itself can be a complex problem [3].

If human material is to be used, the choice of fluid, blood or urine, may be of
importance. In some applications, drug analysis, for example, quantitation of amphetamines or of morphine in serum is of relatively little value. Quantitation of

substances in urine has relatively minimal significance. The sample submitted for
quantitation must be a meaningful one. Depending on the distribution of a substance,
whole blood or nonhemolyzed serum may be preferable.

The analyst should be aware of artifacts which may enter the fluid; the analysis
for alcohol of a blood sample after the venipuncture site has been swabbed with alcohol may be challenged. The quantitation of postmortem samples for alcohol days after
the demise of the subject may also have no meaning due to postmortem bacterial action.

2. Storage

Samples may be perishable and subject to decomposition. Enzymatic or bacterial degradation may occur. Stability is extremely important; it may be better to extract
and then store rather than to store, even in the freezer, and extract just before
analysis.

3. Extraction

Homogenation of the sample in a blender or by ultrasonic disintegration may be necessary before extraction. A large variety of solvents or solvent mixtures are available. If a biological substance has metabolites, then the extraction efficiency must
be high for *both* the parent compound and its chief metabolites. The pertinent up-to-
date literature should be consulted for specific substances. Buffering of the material at one or more pH values may be necessary to optimize extraction with organic
solvents or resin columns.

4. Cleanup

Cleanup is the isolation of the desired substance (and its metabolites) from extraneous material so that these substances are in pure enough form to allow detection and
determination without serious interference.

In both extraction and cleanup, organic solvents may be used for liquid-liquid
extraction or for elution of materials adsorbed onto resin columns. Solvents may
vary from nonpolar liquids like chloroform to highly polar solvents. Solvent mixtures
of varying polarity may be used in combination with buffers of different pH values in
serial extractions [4].

5. Recovery

Overall recovery for the whole procedure is important. Generally, this should not
fall below 70% if the method is to be considered reliable. Losses may occur at each
step in the operation. Recovery standards should be added to samples before extraction. It may be desirable to fortify (or "spike") samples with a series of increasing

amounts of the recovery standard to develop a standard recovery curve. Ideally, recoveries should be determined at several concentrations over the *range* in which results
are being reported [3]. These standards can be used to estimate the proportion of
analyte actually extracted from the biologic fluid. They should be different from
both the internal standard and the analyte. They may be added to both the sample and
to known controls run with the analysis.

6. Internal Standards

For accurate quantitative analysis of liquid samples, internal standards are desirable
provided they do not create difficulties in sample preparation. An internal standard
is a marker compound which is added in accurately known concentration to the sample.
It may be added in the last extraction or before injection. Some chromatographers
add the internal standard to the sample directly before cleanup and extraction. They
may or may not add a separate recovery standard. This marker should be a substance,
not expected to be present in the biological fluid examined, which is compatible
with the sample and differs in retention time from the substances to be measured so
that its peak is separate from all the other peaks. It provides a scale value for
quantitation. Ratios of the peaks (either peak heights or peak areas) to that of the
internal standard permit determination of concentration ratios. As with recovery
standards, a standard curve for various concentrations of internal standards may be
prepared in the *range* of the sample concentrations to be determined. Such curves may
compensate for nonlinear detector response [5].

Internal standards can, to a limited extent, be used as controls. In overdose
toxicology, if the *relative retention times* of specific compounds match those of the
unknown, there is a reasonable probability that these were the substances ingested.
Such reasoning may apply where the clinical situation requires rapid analysis and
quantitation.

7. Derivitization

Since quantitation requires good peak separation, it may be necessary to treat the
sample to permit peak separation if several substances in the sample with closely related peaks are present. Derivitization may permit separation due to changes in column adsorption and consequent changes in retention times for the various substances.
The formation of derivitives may permit the measurement of nano- and picogram quantities of substances with the electron capture detector which may not be measurable with
other detectors. Among derivitization techniques are acetylation, methylation, chemical reduction, and bromination. For most quantitative purposes derivitization is
carried out in the final extraction prior to evaporation. *On-column* derivatization
is used primarily for tentative identification by "peak shifts" rather than for
quantitation [6, 7].

8. Calibration Standards

Pure, properly characterized standards at various concentrations should be used frequently to calibrate the instrument. A calibration curve is necessary to permit accurate quantitation as well as to determine the linear portion of the operating curve for the detector and the point where detector overload begins [2, 3].

9. Controls

As in every other chemical analysis, known accurate controls should be run with the samples to verify that each step in the overall procedure is under control. The controls should be in the *same concentration range* as the substances to be measured.

Summary

In summary, meticulous adherence to properly established operational procedures as well as the use of proper controls and internal and recovery standards are necessary if any meaningful approach to quantitative analysis is to be made. In determining whether a substance should ever be examined for quantitation in a particular biologic fluid, special attention must be paid to the following important items:

1. The method of sampling to obtain a statistically meaningful sample
2. Storage of the sample before analysis and the possibilities of bacterial and enzymatic degradation
3. Proper extraction of the substances and of significant metabolites from the biologic matrix
4. Proper sample cleanup procedures to prevent column contamination and rapid deterioration of the column
5. The formation of derivatives to increase sensitivity of detection or to permit separation of peaks so that each peak may be measured
6. The use of recovery standards to monitor the overall efficiency of the various steps in extraction and cleanup
7. The use of internal standards to provide scale values
8. The use, as elsewhere, of proper calibration standards and of known controls

Quality control must start long before the sample is injected into the column. Serious deficiencies which may exist in any of these steps cannot be corrected by even the best controlled column and detector. The column is no better than the raw data in the form of the evaporated extract which is to be placed in the instrument.

B. Instrumental Components and Related Problems

1. Syringe and Injection Techniques

The syringe links the preparative and instrumental parts of the analysis and can greatly influence results. Indiscriminate use of the syringe for many different substances, inadequate cleaning, or poor choice of cleaning solvents may introduce cross-contamination or spurious peaks. In gas analysis (constant volume techniques), precision of injection volume is critical. Methods employing internal standards require their accurate addition to the sample. The use of too long a needle may result in on-column injection and contamination of the glass wool plug.

2. The Gas Chromatograph and Its Components

a. Injector Port

Injections are made through a silicone rubber septum. Air and water vapor may enter through a repeatedly perforated septum to cause accelerated column aging. Additives in the septum may be eluted ("bleed") at high temperatures to produce spurious peaks.

The metal or Vykor-glass inserts which form the lining of the port may become contaminated by fats or waxes in inadequately prepared samples. If not regularly changed, contaminants on the inserts may decompose desired analytes (especially pesticides) or may themselves volatilize to yield spurious peaks and/or contaminate the column by blocking adsorption sites, with decreased column efficiency [2, 3].

If injector temperature is too low, sample components may not all be volatilized; if too high, thermolabile analytes may decompose.

b. The Chromatographic Column

The column is the heart of the gas chromatograph. With all other systems working to perfection, a poorly functioning column will distort the analytical results [1, 2].

(1) Column inlet plug. As with the injector insert, lipids in inadequately cleaned samples may contaminate the glass wool plug and the column resulting in diminished peak heights, tailing of peaks, loss of peak resolution, and breakdown of sensitive substances (especially pesticides). The plugs should be changed regularly (even daily with fatty extracts).

(2) Tubing. Tubing should be of the length specified since overall performance is in direct relation to its length. Metal tubing that has not been silanized may adsorb substances with irregular unpredictable release rendering accurate quantitation impossible.

(3) Column support. The nonreactive granular support for the liquid phase may vary in size. Smaller granules have larger surface area and larger granules have smaller surface area which may be coated by the liquid phase. Hence, the retention time is related to the size and configuration of the granular support.

(4) Liquid phase. A silicone fluid may vary in polarity and boiling point. Supports may have differing "loadings" (concentration of liquid per unit weight of support). Conditioning of the column and contaminants will also influence performance.

Inadequate ("slack") packing of the tubing with coated support will cause a dead volume near the inlet, which will lower column efficiency.

The choice of liquid-phase polarity depends on the substances to be analyzed. Since retention times will tend to increase with the degree of polarity and greater molecular weights of the analytes, reference should be made to the literature for a choice of liquid phases which will optimize retention times and permit adequate separation of peaks. Test runs with the desired analytes should then be made with polar and nonpolar liquid phases to ensure proper separations.

Increasing column temperature in an effort to decrease retention times may result in the elution ("bleeding") of the liquid phase, or its decomposition, or in the degradation of thermolabile analytes. Derivitization or choice of less polar liquid phase is preferable [8, 9].

High column loading (high concentration of liquid phase) may cause column bleed if prior conditioning is inadequate. High loading also increases retention times, and an attempt to shorten retention time by increase of column temperatures will result in bleeding.

Low column loading may result in stripping of liquid phase from the support after continued use with possible interaction between analytes and support.

Samples with fats or lipids may contaminate the injector insert, the column plug, and the first few centimeter of the column packing.

(5) Carrier gases. The inert gases (usually nitrogen or helium) employed to transport the analytes may contain oxygen or water vapor which may oxidize or hydrolyze the liquid phase. Oxidation may cause changes in column polarity and modifications in retention time. Spurious peaks from water vapor will occur with thermal conductivity detectors. These problems are particularly evident when the gas tanks are nearly depleted (gauge pressures below 500 psi).

Flame ionization detectors reflect the presence of carbon atoms. Their presence as contaminants may seriously distort results especially in analysis of hydrocarbon gases. In reference standards, total hydrocarbon content must be known.

Increasing gas flow rate to decrease retention time reduces overall column efficiency.

(6) Oven. Column temperatures are maintained by the oven. Changes in oven temperature may produce changes in baseline retention times or, if transient, spurious responses in the detector.

Decreased temperatures may increase retention time or result in inadequate separation of peaks.

Elevated temperatures cause reduced column efficiency, bleeding of the column or degradation of analytes into fast moving fragments which precede the parent compound and appear as a frontal shoulder of the parent compound or as elevated baseline.

c. The Detector

Detector design and type of detector used are relatively fixed for each assembly.
In quantitative analysis, a serious problem results from detector saturation or
overload where incremental increases for large quantities of sample are less than for
smaller quantities. This results in a response curve which is flat or even concave
downward beyond the linear range. This emphasizes the need for the use of standards
to prepare a linear curve for each substance of interest and to determine its linear
range.

A corollary to detector overload is column overload. These usually occur con-
comittantly. With thermal conductivity detectors, column overload may precede detector overload. This will be seen as a widened peak with longer or shorter retention
time.

Other problems with detectors may occur with *short-term* or *long-term* noise.
Long-term noise may be due to column or septum bleed. The former is due to improper
conditioning of the column. Drafts at the detector may also cause such noise.

Short-term noise is usually electrical: poor grounding, dirty detectors or
contacts.

Electron capture detectors operating at nonoptimum voltages may have an elevated
baseline which cuts off smaller peaks, giving a visually "cleaner" set of curves which
are inaccurate since only part of the total curve is seen and measured.

Quantitation of peaks on noisy, elevated, or fluctuating baselines or on unstable
baselines can give rise to serious errors.

Not only may detectors have a nonlinear range, they may have a restricted range
or a response which decreases with length of use [3].

Gas-liquid chromatography differs from other methods of analysis in common use
in that the system itself interacts with the substances analyzed. In other routine
systems of analysis, optical means are used to pass a beam of light through either a
tube, a flame, or cuvet. These methods of analysis have no memory of the substances
and fluids analyzed. As instrument systems, their useful life equals the normal life-
span of their optical and electronic components.

GLC components interact with the analyte, with remnants of the biological matrix
in which it was contained, and with other elements in the system with the potential:
to contaminate and shorten the longevity of the system; to influence the results of
successive analyses; and to affect the results of each analysis. The factors that
influence such changes have been discussed in terms of the components of the system.

Before commencing routine analyses, 15 to 20 fortified specimens should be run
"blind" over a suitable concentration range to provide data for statistical evaluation of the precision and accuracy of the method with the system used.

III. CORRECTIVE STEPS FOR GOOD QUALITY CONTROL

Some of the many sources of error which may occur in GLC and the types of error which will result have been listed in Sec. II. They all must be considered in any quality control program.

Those potential errors which relate to sampling, specimen storage, extraction, and, where necessary, cleanup can only be alluded to. No specific recommendations can be made other than (1) attention to detail, (2) the regular use of recovery standards, internal standards, and calibration standards, (3) preparation of a calibration curve employing defined standards of known purity and operation within the linear portion of that curve, and (4) the regular use of controls.

It should be emphasized that original samples must be large enough to avoid sampling errors. Subsampling and use of aliquots can reduce the size of the working sample. Too large a working sample will overload the cleanup or detection phase. Too small a sample may cause an unnecessary loss of sensitivity. Appropriate sensitivity can be obtained by adjusting the sample size or aliquot to yield a larger signal-to-noise ratio for enhanced reliability [3].

Specific recommendations and precautions can be given for the various steps involved in the instrumental portion of the analysis.

A. Syringe

Syringes should be of good quality with legible graduations. Preferably, only one syringe should be used for the analysis of a single substance or a group of closely related substances analyzed on the same column. This is particularly true in alcohol analysis, where cleaning solvent residues may introduce spurious peaks.

Syringes can be cleaned by flushing several times with 5% potassium hydroxide in water, rinsing in acetone and then with chloroform, both of reagent grade or better, and drying in an oven (about 85° C). In cleaning, the plunger may be cleaned with tissue paper or towel containing solvent. Only the plunger button should be handled. Touching the plunger may eventually result in freezing of the plunger within the syringe.

For repeatability of injections the *solvent-flush* technique is suggested: the syringe is wetted by repeated filling with solvent. The average needle volume for a microliter syringe is about 0.8 µl. Therefore to obtain 1.0 µl the plunger is dipped into the solvent and pulled back to read 1.2 µl. It is removed from the solvent and further retracted until an air space can be seen at the needle end. The needle is now filled with air.

To obtain sample, immerse the needle in the sample and remove the desired amount of sample by retracting the plunger the corresponding number of scale divisions. Remove the needle and retract the plunger until the sample size can be read in the syringe. The sample, followed by its solvent flush, can now be injected. If air is undesirable during injection, minor changes can decrease the air pockets.

B. Injector

Where possible, all injections of clinical samples should be off-column. The metal or Vykor glass insert will adsorb significant portions of extraneous materials. The glass insert is preferable since the metal insert may enhance oxidation and compound conversion. The glass injection insert should be changed at the end of each working day during which extracts have been injected.

Septa must be of high quality, preferably silicone rubber. They should be checked for bleeding and should be changed at least monthly.

All fittings should be checked for leaks. After initial installation, fittings should be retightened following equilibration and should be checked to ensure that they are gas tight.

C. Column

1. Glass Wool Plugs

Presilanized glass wool plugs measuring 2 cm in length should be added to each end of the column. They should be inserted tightly enough to prevent dislodgment when carrier gas is passed through the column, but they should not impede gas flow.

2. Tubing

Tubing in the precise length should be used. Length should be verified before packing. For most applications, glass tubing will cause fewest problems. If metal tubing is used it should be silanized. Teflon-lined stainless steel might be a good compromise.

3. Column Packing

Columns may be packed with a single material or a mixture of column packing materials. Before packing, the column support is coated with the liquid phase. To best accomplish this, the liquid phase is dissolved in an appropriate solvent in a beaker and the solid support is added. Techniques for preparation include the beaker, rotary vacuum, filtration, and fluidization techniques.

Properly prepared packing is absolutely essential for good results. From the practical standpoint commercially available prepared precoated packing obtained from a reliable source is far superior (and cheaper) to anything that might be accomplished in most laboratories. A chromatogram and a listing of column conditions should be furnished by the supplier for the packing lot supplied to the laboratory.

Each lot of packing should be checked by visual examination to ensure that it is free flowing. If balled or clumped ("roll" test), the packing material should be discarded. The performance of the column should be examined using standard mixtures of analytes.

4. Carrier Gases

Carrier gases must be free from contaminants including oxygen, hydrocarbons, and water vapor. To ensure that the gases received are of sufficient purity, they should be purchased from a reputable manufacturer. If the application requires it, high purity gases with a statement of analysis should be the only gases purchased. A properly purged drying tube should be in line between the gas cylinder and the column. Drying tubes should be purged with the carrier gas according to the manufacturer's instructions before it is inserted in the gas line (usually about 4 hr at a flow rate of 100 ml/min for the average size tube). If the drying tube is of the molecular-sieve type it can be regularly regenerated by heating with a propane torch. The drying tubes must be replaced or regenerated according to the recommendations of the manufacturer. The fittings between the gas cylinder and the column should be checked for leaks. Leaks may introduce air or water vapors into the column.

Water vapor may be present in the bottom of a gas cylinder. Cylinders should, therefore, be changed when they are nearly depleted. A tank valve pressure of 500 psi as the cutoff for replacing a cylinder is generally recommended.

When an instrument has been started in operation, the carrier gas flow should be maintained. If the instrument is turned off for brief periods, the carrier gas flow should nevertheless be maintained. A general rule is that as long as the oven remains hot, gas flow should be maintained. Some users will maintain constant carrier gas flow once proper operating conditions have been achieved. Their rationale is that since it is quite difficult to return to absolutely the same conditions once the gas has been turned off, the relatively small additional volume of gas used is less important than the costs for the time spent in readjusting the carrier gas rates or the sacrifice in accuracy which may result. In any event if the instrument is to be used at relatively frequent intervals during any working period, carrier gas flow should be maintained. Only if the instrument is to be shut down for a long period should the gas flow be reduced.

Carrier gas flow should be adjusted to desired flow rate on the instrument rotameters. The exact flow rate should then be accurately adjusted using a "soap-bubble" tube. At regular intervals, flow rate should be checked with the "soap-bubble" tube.

5. Packing the Column

The column may be packed by hand or mechanical vibrating or by vacuum. Column configuration will dictate the method of choice. To fully pack a coiled column, vacuum is

most effective. Hand vibrating a U-shaped column will produce consistently high
quality columns. Packing can be done by adding about 6 in. of packing through a glass
funnel to each leg of a U-shaped column. The bend of the column is tapped on a pad-
ded surface. This can be repeated, adding 6 in. of packing to each leg, until the
column is filled. A wooden pencil can be used to tap along the length of the tubing
to ensure uniformity.

6. Column Conditioning

The column should be "heat cured." This is best accomplished by connecting the inlet
end in the oven and raising the oven temperature to about $50°$ C higher than the column's normal operating temperature while passing carrier gas through the column. Initial gas flow rate should briefly be at a relatively high rate. The rate should then
be decreased to about 60 ml/min. In general, columns with liquid-phase loading of
less than 5% should be conditioned for at least 24 hr, those above 5%, from 72 to
96 hr. Column conditioning times for certain liquid phases may be shorter. Temperature during column conditioning should not exceed the manufacturer's recommendations.

During conditioning of a column, the column may not be connected to the detector
and the gas may be vented into the oven. Some users employ a spare oven to prevent
contaminating the detector with any column bleed or other volatiles. Others connect
the column to the detector to create "back pressure" and prevent diffusion of air
past the septum or loose fittings into the column. A restrictor attached to the
column exit can serve the dual purpose of creating a back pressure without the fear
of fouling the detector. If the needs of the laboratory require the frequent conditioning of columns, a separate oven for conditioning should be used, otherwise, columns may be conditioned in the chromatograph at night or during the weekend. Of
course, if frequent conditioning is required, this may reflect problems which should
be examined.

When a new column has been installed and is ready for "conditioning," a new gas
cylinder should not be installed. When the column is evaluated for proper operation,
the additional parameter of the gas will not enter. Instead, a reasonably full gas
cylinder, which has been found free of contaminants, should be used.

7. Special Column Treatments

With certain compounds, particularly pesticides such as Endrin, on-column breakdown
may occur. Silylation following conditioning may help prevent this decomposition.
After the oven temperature and carrier gas flow are brought into the approximate
operating range, four consecutive injections, each of 25 µl of Silyl-8 at 30-min
intervals, may be made. One hour should elapse before use with electron capture
detectors to permit elution of traces of silanizing material from the column.

For flame photometric or thermionic detection of organophosphate pesticides
Carbowax treatment to minimize adsorptive problems and increase peak resolution has
been described.

8. Removal of Columns

If a column is to be removed from the chromatograph, the oven should be brought to
room temperature before the column is disconnected. Both ends of the column should
be capped as soon as possible after the ends have been disconnected.

D. DETECTOR

To prevent erratic baselines and noise due to dirty detectors, they should be cleaned
at regular intervals according to the manufacturer's instructions. Flame ionization
detectors can be immersed directly in reagent grade acetone. The ion collector ring
can be cleaned by immersion in hydrogen fluoride. Chlorinated hydrocarbons should
not be used for cleaning. Table 1 lists some common problems and corrective steps.

IV. ESSENTIALS OF GOOD QUALITY CONTROL

If adequate attention has been paid to appropriate sampling techniques, extraction
and cleanup procedures, and good quality syringes and if a gas chromatograph in good
working order with a properly conditioned column containing high quality packing and
high purity carrier gases are used, then the following steps constitute the essen-
tials of a quality control program.

1. The initial status of the column must be evaluated. The column record form
   (Table 2) may be used to list conditioning and operating parameters.
2. A calibration curve for the column should be prepared and operation should be
   within the linear portion of the curve.
3. Regular rigid maintenance includes changing Vykor injector port inserts daily,
   changing column inlet glass wool plugs and septa at least monthly, and maintaining a carrier gas flow rate of about 20 ml/min through the column at nights
   and on weekends when the instrument is not in use. Monitoring of column efficiency requires attention to features of column deterioration, which includes
   depression of peak height response, severe peak tailing, broadening of peaks,
   increased retention time, and on-column breakdown of critical substances injected.
   verified and changes in column characteristics recorded. If column deterioration is noted by comparing records, steps to restore column efficiency
   may be taken. If a record of the nature of specimens and of the number of
   injections made is recorded, it will be possible to determine the longevity
   of a column and adequacy of sample cleanup. Changes in sample cleanup pro-

TABLE 1. Some Problems Associated With GLC Instrumentation and Recommended Corrections

COMPONENT	PROBLEM	CORRECTION
Injector	Septum (a) Repeated use may produce a permanent hole with resulting leaking of air or water vapor into column.	(a) Change septum at regular intervals.
	(b) At high temperature, spurious peaks may result from septum bleeding.	(b) Use teflon-coated septum or purge heads before run.
	(c) Low temperature, low volatilization.	(c) Maintain proper temperature of injection port.
	(d) Elevated temperatures, decomposition of substance: metal sleeve may cause decomposition.	(d) Use silanized glass sleeve.
Column	(a) Glass wool plug: contamination by material with high adipose content or of poorly cleaned sample may result in decreased peak heights, tailing of peaks, and loss of peak resolution [1].	(a) If materials with high lipid content are injected, change plug daily.
	(b) Tubing: adsorption on wall of column with erratic retention or release of substance.	(b) Glass columns should be used. Metal columns and glass wool plugs should be silanized. Porous plugs are available which eliminate the glass wool.
	(c) Support: erratic retention or release of substances, adsorption of substances to be analyzed. Changes in granule size can provide larger or smaller surfaces available for interaction with differences in retention times.	(c) Prepared columns from a reputable manufacturer should be used. If columns are loaded "in-house" support material from one lot should be purchased in bulk from a reputable supplier. If possible mesh size (usually 80-100) should be checked.
	(d) Liquid-phase: polarity. Too polar a liquid phase may result in excessive retention times, or in poor separation. Boiling point and thermostability: liquid phase may be boiled off column (bleeding) or may be destroyed.	(d) Use less polar liquid phase and consult literature for specific application. Maintain column temperature within working range of liquid phase.
	Concentration of liquid phase: Too high a concentration of liquid phase may result in excessive bleeding. Too low a concentration may leave bare support with interaction of analytes with support granules.	Properly condition column maintain record for column and compare retention times of standards and controls as column ages.
Conditioning	Inadequate conditioning may result in bleeding of liquid phase and nonstable condition.	Condition column adequately.

TABLE 1. (Continued)

COMPONENT	PROBLEM	CORRECTION
History	Previously injected materials may contaminate column.	Maintain records on each column, remove contaminants.
	Irreversible adsorption of component on the support may cause distorted tail peak, nonlinear response (concave upward), and increase in retention time with increase in quantity of sample.	Change columns or derivatize to decrease adsorption.
Carrier gas	Tailing of peaks and changes in retention times due to hydrolysis or oxidation of column.	(1) Purchase gas from reputable manufacturer. (2) Do not deplete tank. (3) Have gas dryer or purifier in line (prepurged). (4) Ensure that lines and septum are airtight and free of leaks. (5) Keep gases on when in regular use.
Oven	Changes in oven temperature may cause base line shifts or shifts in retention time.	Check and record oven temperatures periodically with suitable thermosensor. Compare with temperature indicator on instrument sensor.
	Elevated temperatures may cause degrading of compound on column with gradual uneven increase in baseline.	Lower column (oven) temperature, increase flow rate.
Detector	Detector overload with nonlinear response for large quantities of analyte (curve concave downward)	Prepare calibration curves for each substance of interest; operate within linear portion of curve by proper dilution.

cedures may affect the life of a column. A column use form (Table 3) may be
used to record types of biologic fluids injected and the number of injections
to give an indication of column longevity.

V. EVALUATION OF INITIAL STATE OF COLUMN

Following conditioning and connection of the column to the detector, the oven is raised
to the proper operating temperature. The carrier gas flow rate is adjusted with a
soap-film (bubble) device to the recommended rate. The septum is permitted to equili-
brate for at least 1 hr, preferably overnight. At operating temperature and carrier
gas flow a background current (BGC) determination should be made according to the instrument operating manual. This assesses the detector response as affected by the
column being tested. If BGC cannot be brought to levels indicated by the operation
manual, column bleeding may be the cause and additional conditioning may be indicated.

TABLE 2.   Column Record Form

COLUMN #

Date Installed: Instrument:

Tubing: Length: I.D.

Packing   Mesh #   Manufacturer

Liquid Phase   Concentration of Liquid Phase:

Septum: Type   Date Installed:

Conditioning of Column:

Carrier Gas: Flow Rate: Temperature.

 Gas Manufacturer:

Dated Started___________ Time:_________   Date Ended:________________Time:_________

Total Hours of Conditioning:

Usual Operating Conditions:

Injector Temp: Column Temperature (or Program):

Carrier Gas: Flow Rate: Manufacturer:

Detector   Detector Temp.   Detector Gas(s):

           Flow Rate(s)

Oven Temperature (to be checked periodically):

Background Current determinations:

Date   Time   Value

TABLE 3. Column Use Record

DATE	SAMPLE #	TIME	BIOLOGIC FLUID MATRIX	NO. OF INJECTIONS	INT. STANDARD & CONTROL

Contaminated gas may also cause low BGC. If BGC is adequate, a second determination should be done after 2 hr of equilibration. If a further recorder deflection is noted, hourly checks should be made until there is a constant deflection. ,The instrument is now ready for evaluation of efficiency, resolution, component stability, and retention characteristics.

Column efficiency, theoretical plates, should be calculated (the value obtained by measuring, in the same units, the distance from injection to the midpoint of a specific peak, and dividing this value by half the distance, along the baseline, of the projection of the midpeak and the extension of either side of that peak to the baseline). This should give a value of at least 3,000 total plates for a 6-ft column. In high efficient columns, peak width tends to remain tight despite increase in retention time. Separation factor and tailing factor should also be calculated.

With some compounds, particularly pesticides, on-column conversion occurs; for example P,P'-DDT to P,P'-DDD or P,P'-DDE. These changes should be minimal on a new column and the conversion rate increase as a column ages.

Once an initial evaluation has been made, a standard should be injected daily. If, as in pesticide analysis, on-column breakdown is apparent then fouling should be considered.

Injection overloading occurs with fouling of the column by "dirty" samples. The most prominent symptoms are depressed peak-height response, severe tailing of peaks, lowered efficiency and poor peak resolution, increased on-column compound conversion, and changes in retention time.

If known standards and standard mixtures are injected daily and the peak-height and retention time responses recorded and peak tailing and/or deteriorating peak resolution noted, column deterioration can be anticipated. Where on-column breakdown is of importance in the analyses of interest, this must also be evaluated.

In many situations, running the gas chromatograph overnight at operating temperature and carrier gas flow rate may provide significant column regeneration. If cleaning of inlet sleeve, injector, and inlet port do not improve overall response, the column should be replaced. With proper maintenance many thousands of injections may be made with no loss in column performance.

TABLE 4. Column Response Characteristics

DATES	CONTROL(S) and/or MIXTURE(S)	CONCENTRATION OF CONTROL(S)	RETENTION TIME(S)	PEAK HEIGHT(S)	ON COLUMN DECOMP%	REMARKS

Retention characteristics are important since some compounds may be late eluters. If the compounds of interest fall into this category, it may be important to select another column. It is often valuable to use two different columns of differing polarity which can be operated at the same oven temperatures. Table 4, column response characteristics, can be used on a regular basis to record retention times, peak heights, and the presence or absence of on-column decomposition of labile substances. If the same substances are injected at regular intervals, changes such as increased retention times, diminished peak heights, or increased on-column decomposition will be readily apparent.

The changes in these characteristics will enable the operator to detect aging and deterioration of the column so that appropriate corrective measures may be taken. The time intervals for recording characteristics may be daily, weekly, or even less frequently, depending upon the specific application and extraneous materials in the biologic matrix.

In practice, certain operating conditions and characteristics may be more important than others. In some applications, the problem of dealing with samples with high adipose content may never arise and other performance characteristics may be more important. *The analyst should modify his program to take into account the unique problems and needs of his laboratory.*

VI. LABORATORY SAFETY

Though not strictly a factor in quality control, good safety practices should be an intrinsic part of every program. The Occupational Safety and Health Administration (OSHA) states that a gas cylinder containing more than 400 ft^3 must be 25 ft from open flames, ordinary electrical equipment, or sources of ignition [11]. The flame ionization detector would be such a source. Although, one 1A cylinder contains 216 ft^3 and one 1H cylinder, 322 ft^3, two cylinders might be construed as being applicable to these regulations. Gas cylinders must be supported with a base or be chained to the wall so that they cannot fall. Cylinders not connected to a gas line should be capped. Eye baths should be available. Protective goggles must be worn when indicated. Instruments should be grounded. Lighting and ventilation in the work area must be adequate. For maximum safety and efficiency, "washed" (purged with solvent) stainless-steel or copper tubing should be used to connect the gas cylinders to the instruments. The gas lines must be adequately supported. Asbetos gloves to handle hot columns or injection ports, fire extinguishers, fire blankets, and a first aid kit all should be available. The above enumerate some of the safety considerations which go hand-in-hand with any quality control program. Other specific considerations may be of concern in each installation.

VII. SUMMARY OF QUALITY CONTROL PROCEDURES

1. Use appropriate standards (recovery, internal, and calibration) and controls.

2. Use known controls with each series of runs (which may also serve as recovery standards).

3. Prepare calibration curves and operate within the linear portion of these curves.

4. Keep accurate records of each column used to detect aging of column and be able to take corrective action when analysis is "out of control."

5. Adhere to good safety practices.

ACKNOWLEDGMENTS

The authors wish to express particular gratitude to J. F. Thompson for the many insights provided by him which are reflected in this paper, and to the late Sheila Sohn for her unfailing encouragement.

REFERENCES

1. J. F. Thompson, A. C. Walker, and R. F. Moseman, *Ass. Offic. Anal. Chem.*, *52*, 1251(1969).

2. J. F. Thompson (personal communication), 1972.

3. M. S. Schechter and M. E. Getz, *Ass. Offic. Anal. Chem.*, *50*, 1056(1967).

4. D. Sohn, J. Simon, M. A. Hanna, and G. Ghali, *J. Chromatogr. Sci.*, *10*, 294(1972).

5. D. R. Deans, *Chromatographia, 1,* 187(1968).

6. G. Cimbura and J. Kofoed, *J. Chromatogr. Sci.*, *12*, 261(1974).

7. W. W. Fike, *J. Chromatogr. Sci.*, *11*, 25(1973).

8. T. J. Siek, *Analabs Research Notes, 13,* 1(1973).

9. W. R. Supina, in *Theory and Application of Gas Chromatography in Industry and Medicine* (H. S. Kroam and S. R. Bender, eds.), Grune and Stratton, New York, 1968, p. 39.

10. R. E. Johnson, in *Theory and Application of Gas Chromatography in Industry and Medicine* (H. S. Kroam and S. R. Bender, eds.), Grune and Stratton, New York, 1968, p. 18.

11. *Federal Register, 37,* 22164(October 18, 1972).

Chapter 12

AUTOMATION AND QUALITY CONTROL IN HIGH PERFORMANCE LIQUID CHROMATOGRAPHY (HPLC)

Dale R. Baker[*]

E. I. Du Pont De Nemours and Co. (Inc.)
Instrument Products
Scientific and Process Division
Wilmington, Delaware

I. INTRODUCTION

In the past, the liquid chromatographic system did not require a significant amount of the operator's attention since the analysis times were relatively lengthy, thereby allowing the operator to perform other tasks. Recent developments in HPLC hardware and column packings technology have reduced analysis time to the point that almost total operator time is required in sample injections and data reduction. Separation times of less than a minute of multicomponent mixtures have been reported [1], and the trend is toward faster separations. Therefore, automation of HPLC systems is obviously becoming a necessity.

____

* _Presently at_ Analytical Division, Hewlett-Packard, Avondale, Pennsylvania

Future HPLC systems will be directed toward total automation, partly as a result of the recent developments in microprocessor technology which make the instrument more adaptable to automation. Also, automated gas chromatographic (GC) systems have been in use for some time.

An obvious use of an automated HPLC system is in quality control laboratories where the instrument is usually dedicated to a given analysis. However, there are many applications of an automated HPLC system other than quality control. Therefore, this chapter will be divided into three parts: automation, quality control, and equipment for automated HPLC.

II. AUTOMATION IN HPLC

This section will deal with the automation of the total HPLC system with the exception of data handling, which has been discussed in Chap. 13.

An ideal system should be sufficiently automated to require no manual sample handling. Unfortunately, many analyses do require some sample preparation prior to their introduction onto the HPLC column, such as an extraction, protein precipitation, filtering of the sample, etc. Commercially automated systems are available which perform various sample preparation operations, and in the future, instruments will be developed to allow placement of the sample into a sample "pretreater," followed by automatic chromatographic operation. This discussion, however, will be limited to cases where the sample is in suitable form for injection onto the HPLC column [2].

A. Automation of Sampling System

The first requirement of an automated system is that it can accommodate a large number of samples. With the advent of shorter analyses times, many samples can be analyzed in a relatively short period of time; therefore it is desirable for the system to accommodate as many samples as possible, e.g., 100 samples. This need is obvious since the system will often, if not always, be operated unattended overnight or over a weekend.

1. Sample Injection

The mode of sample injection onto an HPLC column is different from that of the GC system: (1) It is desirable to eliminate septa and syringes since septa have a finite lifetime and leakage of mobile phase will occur. Leakage is a more serious problem in HPLC than in GC (especially during unattended operation) since highly flammable solvents (e.g., hexane) are often used as mobile phases. Also, septa material may be deposited onto the column inlet causing plugging of the column. (2) It is desirable to use a system which will provide reproducible sample volume injections. A high-pressure sampling valve with an automatic actuation is available; therefore, reproducible sample volumes can be automatically injected. In order for the sample volume being injected to be reproducible, it is necessary to completely remove the previous sample from the injection valve as well as from all tubing associated with transfer of the sample to prevent any cross-contamination. It is also desirable to have a means of rinsing the tubing and injection valve with a solvent.

The sampling system should also have a capability of replicate analysis without a need of loading the same sample into different vials. The sampling system should recognize that it is not necessary to completely rinse the injection system on the second, and subsequent, injections.

2. Control of Chromatographic Parameters

In addition to controlling sample injection, the system should control gradient elution and data reduction devices. It should also provide automatic shut off of the entire system (including the HPLC pump, the injector, and the data reduction device) after analysis of the last sample is completed.

The automatic sampling system should provide an interface with the data reduction device. This will be dependent upon the sophistication of the device which may be a strip-chart recorder, integrator, or computer.

In the case of the more sophisticated data reduction devices, the system should provide sample-number indexing. In this manner, the operator can quickly identify the sample and its data.

B. Applications

An automated HPLC system is ideal for quality control types of applications, where the instrument is usually "dedicated" to a given analysis. However, there are many applications of such a system to "research"-oriented operation.

1. Kinetic Studies

Automated HPLC can be very useful and time saving for kinetic studies provided that the analysis time is compatible with the rate constant of the reaction to be studied. Henry, et al. [3] have reported the use of HPLC for monitoring the rate of decrease of a mosquito larvicide in simulated saltwater marshes. In this case, the samples could have been loaded into an automatic sampling system and left unattended to perform the analyses.

Automated HPLC is also useful in pharmacokinetic studies. One such area is the measurement of dissolution rates, whereby the amount of dissolved drug (from the dosage form) can be measured automatically at timed intervals.

2. Pharmaceutical Analysis

In the past, pharmaceutical analyses of dosage forms was performed by analyzing an aliquot from a mixture of tablets. However, it has become necessary to demonstrate content uniformity by analyzing individual tablets. The automated analysis of an individual tablet containing aspirin, phenacetin, and caffeine has been described [4] with an analysis performed every seven minutes.

Huettemann and Schroff have reported [5] the rapid analysis of a steroid and its degradation products with analysis time of about 1.5 min (Fig. 1). Figure 2 illustrates the repetitive analysis of several tablets containing the steroid.

Automated HPLC is also applicable to stability monitoring of pharmaceuticals in their dosage form, where a large number of samples must be analyzed after having been stored at varying temperatures, humidity, light exposure, etc., for different lengths of time. These analyses must, obviously, be stability indicating (i.e., be specific for the drug in the presence of excipients and decomposition products of

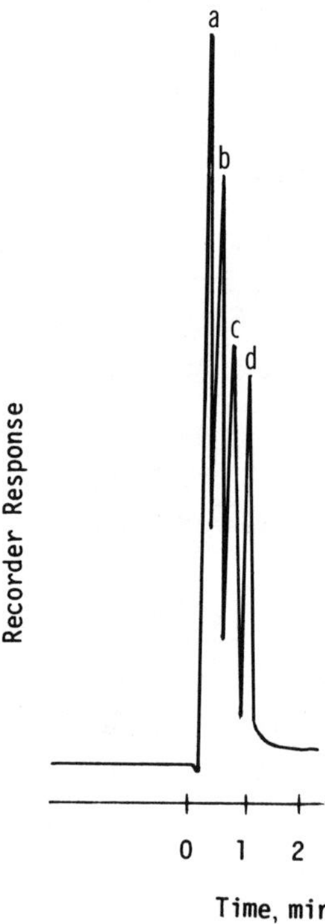

FIG. 1. Liquid chromatographic separation of a steroid and its degradation products: (a) norethindrone; (b) norethindrone oxime; (c) norethindrone acetate; (d) 17α-ethynyl-17β-acetoxy-19-norandrost-4-en-3-one oxime. Column, Permaphase-ODS; mobile phase, methanol:H_2O (1:1); flow rate, 2 ml/min; pressure, 700 psi; temperature, 60° C. Reprinted by permission from the Journal of Chromatographic Science [5].

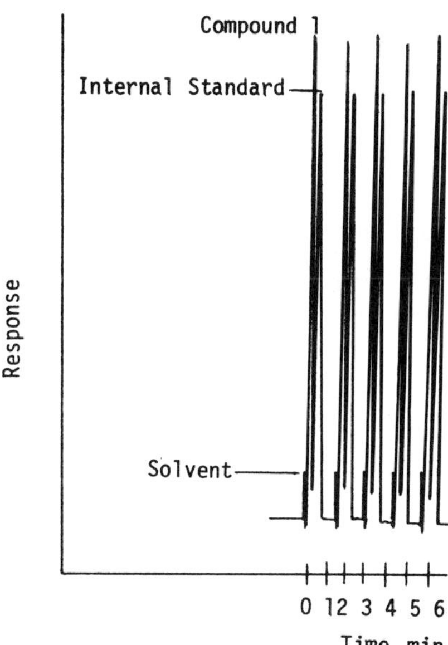

FIG. 2. Replicate analysis of five individual tablets containing a steroid.
Conditions same as Fig. 1. Reprinted by permission from the Journal of Chromatographic Science [5].

the active ingredient). Many such analyses have been reported by HPLC; however, very

little work has been reported using an automated HPLC system.

3. Measurement of Partition Coefficients

Huber et al. [6] have reported the use of liquid-liquid chromatography for the measurement and correlation of partition coefficients. A sample is injected onto a
reversed-phase chromatographic column. Correlation to statically measured (i.e., by
conventional solvent extraction techniques) partition coefficients can then be made
by comparing retention times of compounds with known partition coefficients. In
this manner, one can calculate the partition coefficient on the basis of the retention time of a particular compound under established column and mobile-phase conditions.

4. Methods Development

Automated HPLC can be used for unattended-methods development. A reversed-phase column packing may be used with a mobile-phase gradient from water to methanol, or to
acetonitrile. It is then possible to ascertain whether the compound is retained under
these conditions and to obtain a fair knowledge on the approximate composition of a
mobile phase by which the compound(s) will elute. The only disadvantage of this type
of methods development is that a very nonpolar sample may not be eluted during the
first chromatogram but will elute during a later gradient with a subsequent sample.

5. Qualitative Peak Assignment

One very time consuming task for the chromatographer is peak assignment of complex
mixtures (such as the separation of nucleotides [7] or the PTH derivatives of amino
acids [8]). Usually, the chromatographic parameters are adjusted to obtain the desired number of peaks. If a mixture containing 14 components is separated into 14
peaks, the chromatographer is then faced with the problem of assigning peak identity.
This can be done by "spiking" the mixture with one component, reinjecting the mixture
and observing which peak becomes larger. Alternatively, each component can be injected separately and a correlation between the retention times can be made. However,
strongly retained components in a mixture may act as the stationary phase for early
eluting compounds, and consequently, the same retention time will not be observed for
the compound injected alone as opposed to being in a mixture of strongly retained
components. The "spiking" method does provide less ambiguity in peak assignment;
however, it may be more time consuming. Regardless of which method of peak assignment is used, an automatic sampling system enhances peak assignment as the samples
can be loaded into the vials containing the mixture, each of which is "spiked" with
one of the components, or the vials can be loaded with the individual components.

6. Automated Preparative HPLC

Studies have shown that scaling up from analytical scale (microgram quantities) to
preparative scale (gram quantities) can be accomplished by increasing the column diameter [9, 10]. Quite often sufficient quantities of the component of interest may
not be isolated by one injection onto a preparative scale (i.e., 23-mm or larger i.d.)
column. The larger column still may not provide the capacity, therefore several injections may be needed in order to isolate sufficient quantities of material. This
is especially true when the component of interest is not the major component of the
sample but a minor impurity, decomposition product, metabolite, etc. [11, 12]. Many
injections of the sample through even a large diameter column are necessary to obtain
sufficient quantities of these materials. An automated sampling system coupled with
an automated sample collection device can be quite useful.

Often, sufficient quantities of material can be collected by repetitive injection of the sample onto an analytical scale column. The purification of 5 mg of
progesterone has been reported using an analytical scale column (25 cm x 2.1 mm i.d.)
[9]. Consequently, repetitive injections onto a column of this size could provide a
relatively large amount of purified sample. Also, it was shown that impurities which
have similar molar absorbitivities (at 254 nm) could be detected at levels of 0.001%
(v/v). Therefore, repetitive "semipreparative"-scale separations can be used for
qualitative analysis of product purity prior to production, as well as providing sufficient quantities of purified material for subsequent usage.

7. Size Exclusion Chromatography

The advances in size exclusion chromatography (SEC), also referred to as gel permeation chromatography, have brought such an improvement in column packing materials that very rapid chromatograms are now attainable [13]. SEC is often used for product characterization prior to production; consequently, an automated system is again useful in terms of reducing operator time and attention. Also, with the advent of the newer column packing materials, a small error in retention time measurement can produce a large error in molecular-weight assignment [14]. Therefore, automated SEC with automatic injection marking, or initiation of an integrator or other data system will minimize errors in retention time measurements.

III. QUALITY CONTROL USING AUTOMATED HIGH PERFORMANCE LIQUID CHROMATOGRAPHY

A. Reservoir

The solvent (mobile-phase) reservoir should be large enough to accommodate a large number of analyses. It is desirable to have the capability for stirring the mobile phase in the reservoir during long periods of unattended operation to prevent any demixing of solvents. This is not necessary when a single solvent is used; however, most HPLC separations require at least binary solvent systems. The solvent reservoir should be capable of being sealed to prevent evaporation of volatile solvents as a small change in mobile-phase composition can produce considerable changes in retention time [15]. In adsorption (liquid-solid) chromatography, the mobile phase often consists of a volatile solvent (i.e., methylene chloride or hexane) which contains a small amount of an adsorbent deactivator (i.e., methanol, water, or isopropanol) which is less volatile. Consequently, evaporation of the more volatile solvent can cause a significant change in retention time since the deactivator is usually present at less than 1% (v/v).

Normally, the instrument will be dedicated to a given analysis and it will not be necessary to change columns or mobile phase very often. However, there may be cases where the instrument must be used for an "emergency analysis." Therefore, the instrument should be capable of rapid solvent and column changeover when it must be used in this manner.

B. Injection System

The sample injection system should meet all of the requirements previously described, i.e., accommodate a large number of samples, withstand high operating pressures, provide reproducible sample-volume injections, and be capable of automating the entire liquid chromatographic system. Normally, a fixed-volume-valve injection system is desired as repetitive volumes of the same sample will be injected. Ability to quickly change the sample volume injected (i.e., the injector loop) is desired.

C. Pumping System

A constant flow system is desired to insure constant flow delivery of the mobile phase through the column. Changes in flow rate can cause changes in retention time as well as changes in peak area [16] since most detectors used in HPLC are concentration detectors. Changes in peak area due to changes in flow rate will produce erroneous quantitative data as most quantitative data is obtained on the basis of peak area.

The solvent delivery system should be capable of automatic shutoff after the analysis is completed as well as in the event of solvent depletion or over pressurization of the system due to blockage of the mobile phase flow.

The solvent delivery system should be durable, easy to maintain and repair, and should be compatible with all mobile phases, including concentrated buffers, which are often used in the ion-exchange mode of HPLC. Gradient elution (or solvent pro-programming) is not normally used in quality control HPLC however, in cases where it is used, the gradient system should be capable of providing accurate and reproducible delivery of the mobile-phase composition. This is important because, as mentioned, small changes in mobile phase composition can cause significant changes in retention [15]. If the retention times vary from injection to injection, the data reduction system may not accurately measure the peak as it will be outside of the "window" which has been programmed for that peak. Also, when gradient elution is employed, the system should be capable of initiating the gradient program as well as resetting the mobile-phase composition to initial conditions and allowing sufficient column equilibration prior to injection of the next sample.

D. Column

The column should be loaded with a stable, durable, column packing which will provide extended use with a minimum of cleanup or conditioning. Most columns in use now have chemically bonded stationary phases which provide this stability. A high-efficiency column packing loaded into a relatively short column is desired in order to reduce the analysis time thereby providing the maximum sample throughput.

Depending upon the nature of the sample, it may be necessary to employ a precolumn. The precolumn is usually a very short (3- to 5-cm), narrow-bore (1- to 2-mm-i.d.) column which is installed between the sample injector and the analytical column. The purpose of the precolumn is to serve as a chromatographic filter to retain any materials which would be strongly retained on the analytical column. The precolumn should be loaded with an inexpensive column packing which can be discarded or emptied and reloaded for later use. If the analysis is performed on a small particle column packing, the precolumn may be loaded with a less expensive large-particle packing, with the same stationary-phase functional group. At least two precolumns should be kept so the column can be replaced and the expended column can be emptied,

reloaded, and stored for later use. A disadvantage of the use of a precolumn is that
it will contribute to band spreading and reduce the efficiency of the separation.
Therefore, the chromatographic system must have sufficient resolution of peaks to allow the use of the precolumn.

E. Temperature Control

In most cases, it is desirable to maintain the column at a constant temperature to
prevent any changes in retention time due to changes in temperature.

F. Detector

The detector used for quality control analysis should be durable, stable, easy to
service, trouble free, and, obviously, capable of detecting the components of interest. The most widely used detector in HPLC is the fixed-wavelength (254 nm) ultraviolet (uv) photometer. Other detectors are available (see Chap. 1), including refractive index, fluorescence, multiwavelength filter photometer, and variable-wavelength spectrophotometer [17]. The fixed-wavelength (254 nm) uv photometer is best
suited for meeting the criteria mentioned. It utilizes a low-pressure mercury lamp
which provides a relatively long period of continual operation and the photometer is
relatively simple, sensitive, and trouble free. The optimum detection system will
be dictated by the compounds of interests, for instance, it may be desirable to use
a detector which is more specific (e.g., fluorescence) to minimize sample pretreat-
ment.

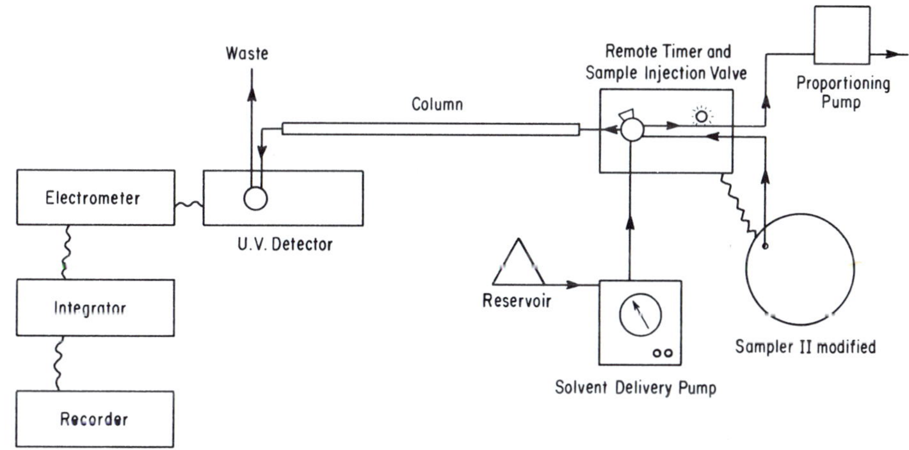

FIG. 3. Schematic diagram of automated liquid chromatograph. Reproduction with
permission of copyright owner [4].

Other criteria for the detection system for quality control analysis are common to detection systems for any HPLC analysis: provide a linear signal output over the concentration region of interest; be insensitive to temperature, flow-rate, and mobile-phase changes; and interface with data reduction devices.

IV. EQUIPMENT

Ascione and Chrekian have described [4] an automatic injector system for the HPLC analysis of aspirin, phenacetin, and caffeine. A schematic of the system is shown in Fig. 3. This system uses an automated, six-port, high-pressure sampling valve. The samples are loaded into vials which are then covered with aluminum foil; a needle pierces the aluminum foil and the sample is pumped to and through the sampling valve. A peristaltic pump (proportioning pump, in Fig. 4) is used to pump the sample. Timing of the sample injection is accomplished through the use of an electric timer and cams. Table 1 shows the good peak-area reproducibility obtained from replicate

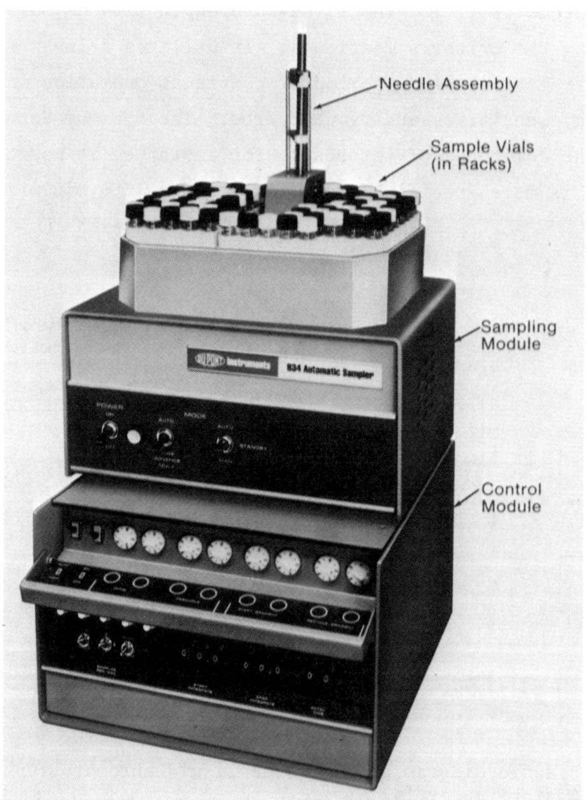

FIG. 4. Du Pont Model 834 automatic sampler.

TABLE 1. Reproducibility of Peak Areas of Aspirin, Phenacetin, and Caffeine Using Automated HPLC[a]

Sample	Peak Area (μV sec)		
	Aspirin (X 10^4)	Phenacetin (X 10^4)	Caffeine (X 10^4)
1	71.21	12.43	42.64
2	71.21	12.45	42.77
3	71.19	12.45	42.44
4	71.15	12.44	42.99
5	71.14	12.45	42.70
6	71.00	12.45	42.44
7	71.10	12.46	42.44
8	71.00	12.45	42.30
9	71.04	12.45	42.66
10	71.08	12.46	42.58
Mean	71.11	12.45	42.58
SD	$\pm$0.081	$\pm$0.0088	$\pm$0.163
RSD	0.11%	0.07%	0.38%

[a]Reproduced with permission of the copyright owner [4].

injections of a standard solution. This system is only an automatic injector and does not provide control over the total liquid chromatographic system.

The Du Pont Model 834 is an automated sampling system which has been designed specifically for HPLC and performs many functions in addition to automatic sample injection. Figure 4 shows the system which consists of three basic components: (a) the sampling valve, (b) the sampling module, and (c) the control module. Figure 5 is a drawing of the sampler plumbing system.

The sampling valve is the familiar six-port valve which is automated through a pneumatic actuator. In Fig. 5, the valve is shown in the inject and load positions.

The sampling module consists of a sample rack which accommodates 95 samples (five each in 19 removable sample racks). Each sample vial has a volume of about 5 ml. Also included in the sampling module is the needle assembly (Fig. 5). A sample vial is moved into position under the motor-driven needle assembly which is lowered, causing the needles to puncture the septum in the screw cap of the vial. The needle assembly consists of two needles: the sample intake needle, which goes below the surface of the liquid almost to the bottom of the vial, and the headspace pressurizing needle, which just penetrates the septum and remains above the level of the liquid (as shown in Fig. 5). The needles are side-port syringe needles, which mini-

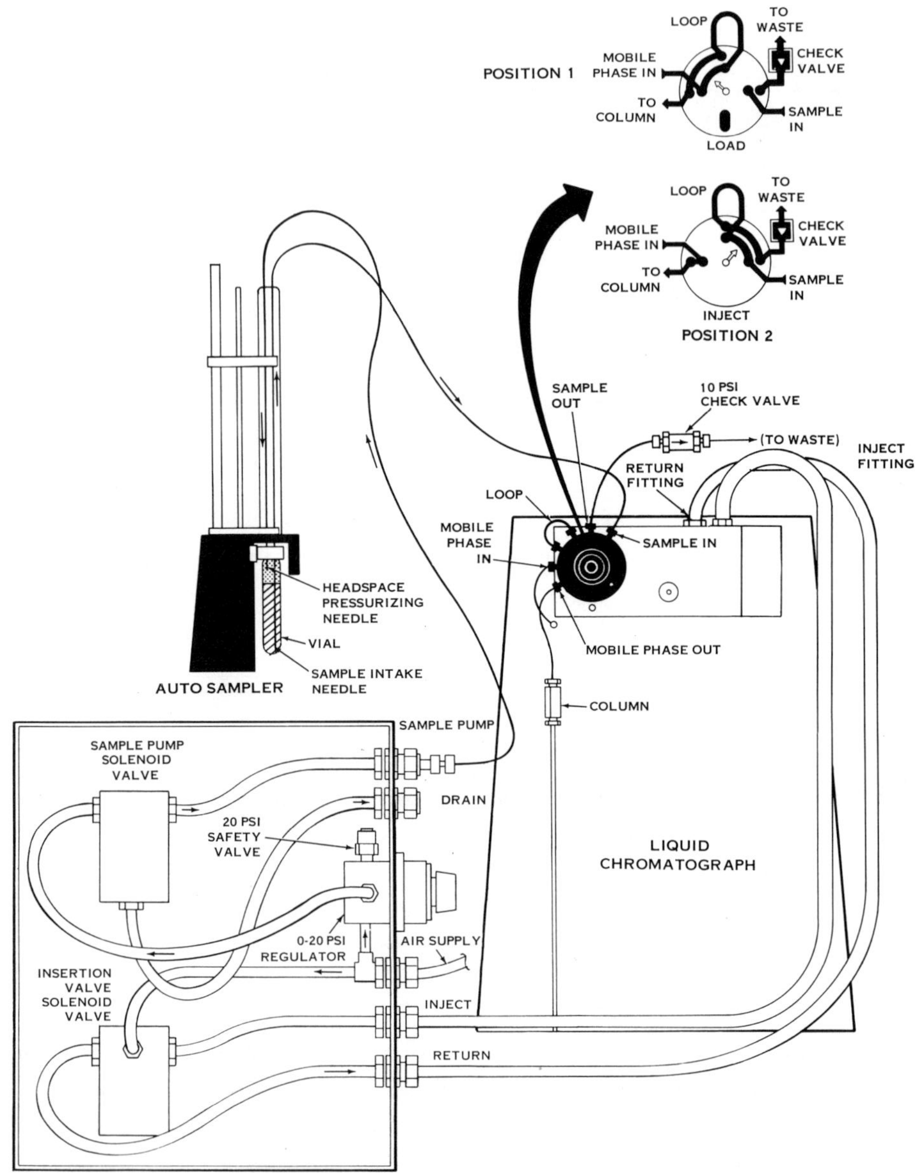

FIG. 5.   Plumbing system of Du Pont Model 834 automatic sampler.

mize needle pluggage from septa material. Air (at 20 psi) is pumped through the headspace pressurizing needle forcing the sample through the sample intake needle to the
sampling valve and out to waste. It is desirable to force the liquid out of the vial
under positive pressure rather than aspirating as aspiration may cause a vacuum in the

results. It is also desirable to keep the vials tightly sealed, especially during long periods of analysis, to prevent evaporation of the sample solvent, which would cause a change in the concentration of the sample. A minimum pumping volume of 1 ml is required in order to completely rinse the tubing and sample loop; however, a minimum sampling volume option is available which requires that only 0.1 ml be loaded into the vial.

It is desirable to load vials which contain only a rinse solvent placed between some or all of the sample vials and that the rinse solvent not be injected onto the column. Vials which have black caps are recognized as sample vials and are injected whereas those with white caps are recognized as rinse vials and are not injected. All of the sample vial racks are white except one, which is black. The sampling module recognizes the black rack as the one containing the last sample, and after the system has completed the programmed sequence on the last vial in this rack, the system (including the liquid chromatograph and, if equipped to accept an external signal, the recorder) is turned off.

The control module is the programmable component of the system. The operator can select either automatic or manual mode of operation. In the automatic mode the instrument will continue sampling and analysis until completion of analysis of the last sample, whereas in the manual mode it will perform only the sequence which has been programmed for one particular vial. The control module allows the operator to select one, two or three injections per sample vial. The length of time the sample is pumped through the tubing and sampling valve prior to injection is also selected by the operator. Samples which are dissolved in high-viscosity solvents (i.e., iso-propanol) require longer pumping times than those dissolved in low-viscosity solvents (i.e., hexane) to insure that the previous sample or solvent, in cases where a rinse vial precedes the sample, is completely rinsed out of the system. In cases where more than one injection per sample is used it is not necessary to completely rinse the entire system for the second or third injection, rather, it is necessary only to pump the sample through the loop since the tubing is already filled with the sample to be injected. The operator can select a re-sample pumping time which applies only to the second and third injections from a given vial and this time can be less than the initial pumping time. The sample and re-sample pumping times are selectable from 0 to 99 sec in 1-sec increments. The samples can be manually moved into position for injection in cases where it is desirable to interrupt the program for a "nonroutine" analysis.

The control module provides initiation and stoppage of a data reduction device as well as BCD output for sample-number indexing. The start and stop integrate times are selectable from 0 to 99.9 min in 0.1-min increments. The total chromatogram time, that is, the time from injection to the next programmed sequence of events, is selectable from 0 to 99 min in 1-min increments. In the event that the analysis

TABLE 2
PEAK AREA REPRODUCIBILITY

A_1	A_2	A_3	A_4
48039	1049	64261	5606
48162	1110	64037	5586
47992	1108	64098	5631
47971	1112	63918	5584
47906	1039	63525	5412
48019	1106	63593	5677
48642	1114	64005	5603
48034	1095	63893	5424
47937	1103	63735	5477
48077	1110	64102	5482
47456	1094	63782	5570
48103	1109	63980	5499
47997	1154	63856	5537
48010	1138	63803	5527
47819	1163	63752	5530
48043	1068	64106	5413
47930	1129	63706	5510
47980	1101	63910	5434
48043	1164	64260	5615
47996	1090	64051	5463
48272	1126	64038	5806
$\sigma = 0.023\%$	$\sigma = 0.43\%$	$\sigma = 0.02\%$	$\sigma = 0.16\%$

TABLE 3
RETENTION TIME REPRODUCIBILITY

t_1	t_2	t_3	t_4
0.40	0.64	1.10	1.54
0.40	0.64	1.10	1.54
0.40	0.64	1.10	1.53
0.40	0.64	1.10	1.53
0.40	0.63	1.10	1.53
0.40	0.64	1.10	1.53
0.40	0.64	1.10	1.54
0.40	0.64	1.10	1.54
0.40	0.64	1.10	1.54
0.40	0.64	1.10	1.54
0.40	0.64	1.10	1.54
0.40	0.64	1.10	1.53
0.40	0.64	1.10	1.54
0.40	0.64	1.10	1.53
0.40	0.64	1.10	1.54
0.40	0.64	1.10	1.54
0.40	0.64	1.10	1.53
0.40	0.64	1.10	1.54
0.40	0.64	1.10	1.53
0.40	0.64	1.10	1.53
0.40	0.64	1.10	1.54
$\sigma = 0\%$	$\sigma = 0.34\%$	$\sigma = 0\%$	$\sigma = 0.32\%$

time is longer than 99 min (which is rare with modern columns and hardware) the operator can place a sample vial (i.e., one with a black cap) which is loaded with mobile
phase between the sample vials. In this manner, the mobile phase would be injected
and the chromatogram would continue.

The control module provides for normal or gradient elution mode of operation.
In the gradient elution mode the operator can select the start gradient time from
0 to 99 min in 1-min increments. This is the time from injection to initiation of
the gradient. In some cases, it is desirable to delay gradient start after injection. Column equilibration between the gradient resetting to initial composition and
injection of the next sample can be accomplished by selecting a total chromatogram
time with a value that is larger than the gradient time. For instance, if the analysis time (i.e., the gradient time) is 20 min and a 10-min column equilibration time
is desired, a total time of 30 min is selected. In this manner, the sample would be
injected, 20 min later the gradient would reset to initial composition, then, 10 min
later a subsequent sample would be injected.

Reproducibility of the system is shown in Tables 2 and 3, which show retention
time and peak area values from replicate injections of a test mixture (Fig. 6).

Baumann et al. [18] have described an instrument for automated sample introduction in chromatography. The Varian Model 8050 LC Autosampler (Fig. 7) was initially
designed for automation of sample introduction onto a gas chromatographic column but
has been modified for use in HPLC.

In the Model 8050, the samples are loaded into screw-cap septum vials which are
then loaded into a four-segment carousel, each segment holds 15 vials, thus the Model
8050 has a 60-sample capacity. Sample injection is made as follows: the sample vial
is moved into position, and a concentric dipper needle penetrates the septum, pres-
surizes the vial and forces the sample through a high-pressure, six-port sampling
valve, which is pneumatically actuated to place the sample onto the column. One, two,
or three injections per vial can be selected. The system can be washed between samples or at the end of a series of analyses by placing wash vials in the carousel,
which are identified by marker pins placed in the racks. The contents of the wash
vials serve only to rinse the tubing and valve and are not injected. The last sample
vial is identified by a marker placed in an empty slot after the last sample and after
analysis of this sample, the Autosampler and the strip-chart recorder are turned off.
The Model 8050 can be operated in the single or multiple modes in which only a single
injection is made or injections of subsequent samples are made until either a stop
marker pin is detected or the sampler is manually turned off, respectively.

Analysis times of 1 to 99 min in 1-min increments are selected by thumbwheel
switches with analysis times of 2 to 198 min in 2-min increments being attainable
after an internal adjustment is made. The sample segment and vial numbers are displayed at the control module and electronic outputs are provided for printing and

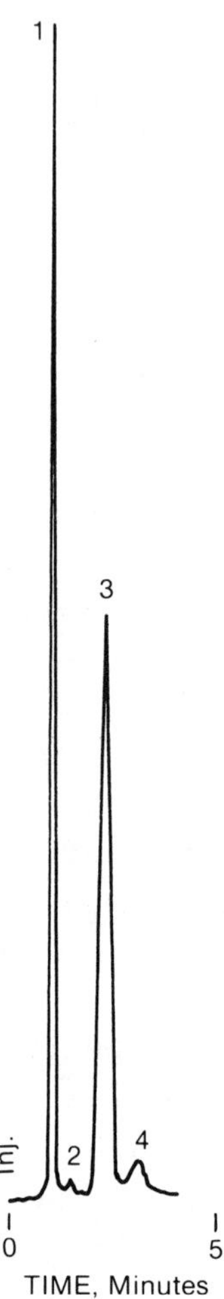

FIG. 6. QC chromatogram of test mixture. Column, Permaphase-ODS; mobile phase, H_2O; flow rate, 1.0 ml/min.

storage of these sample identification numbers by a printer or a computor. An auxiliary timer is present to control external events such as initiation of solvent programming at a time after injection has been made, integration delay, etc.

The Micromeritics Model 725 Automatic injector (Fig. 8) is a sampling system designed specifically for use in HPLC. The Model 725 allows either one, two, or three injections from up to 64 sample vials. The sample volume required is 0.7 ml per vial. In this system, the vial cap is used as a piston to force the sample from the vial to a high-pressure sampling valve.

Analysis times are selected from 1 to 99 min in 1-min increments with rinse or no rinse between injections. The operator selects the location number of the last

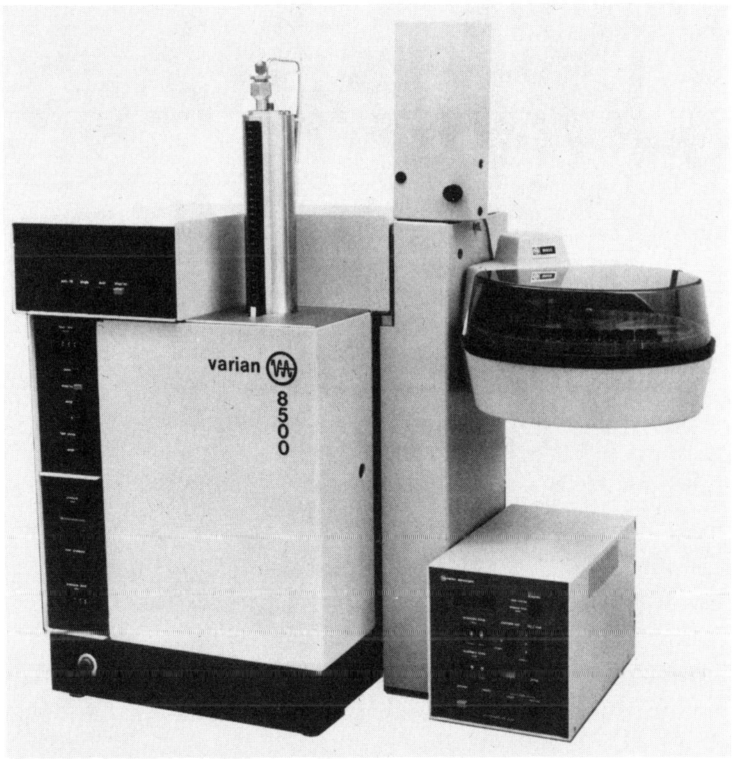

FIG. 7. Varian Model 8050 LC Autosampler attached to a Varian Model 8500 liquid chromatograph. (Photo supplied courtesy of Varian Associates, Palo Alto, California.)

sample and the system automatically turns off after the last sample is completed. The Model 725 can be externally controlled by a computer allowing variation of number of injections from individual sample vials, rinsing of the injection loop on command, and injection of various samples after a search procedure.

Beyer and Gleason have described [19] an automated system for high performance liquid chromatography. This system has a 40-sample capacity as well as last sample stop capability. The sample probe is placed in the sample vial and the sample is pumped to a high-pressure sampling valve by a Milton-Roy minipump. The sampling rate, sampling and rinse time, and the length of the chromatographic time are selected by the operator.

Hewlett-Packard has developed an automatic sampling system for use with Hewlett-Packard 1080 Series Liquid Chromatographs (Fig. 9). In this system the samples are loaded in vials and placed in a vial-transport mechanism which accommodates up to 60 samples. Each link in the vial transport mechanism is numbered and coded to input an identification number to the micro-processor. The operator controls the sampler through the keyboard by entering the number of injections per vial (up to 9) and the number of

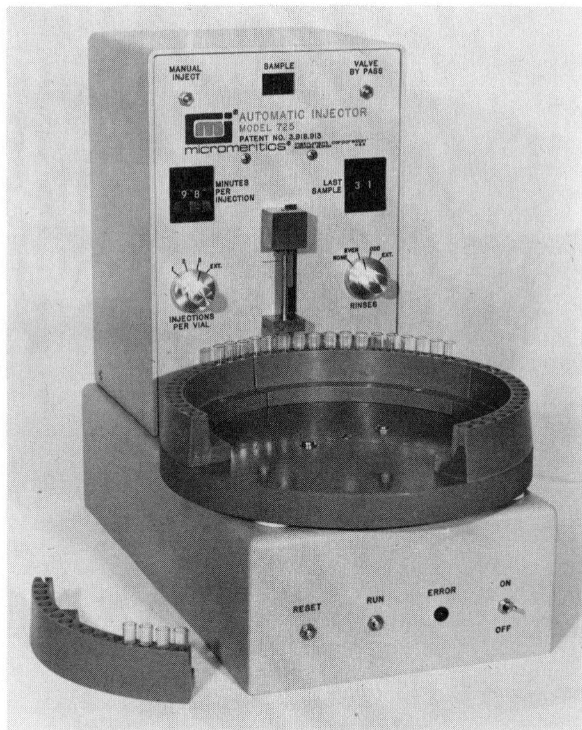

FIG. 8. Micromeritics Model 725 automatic injector. (Photo supplied courtesy of Micromeritics Instrument Corp., Norcross, Georgia.)

the last vial to be sampled. The micro-processor not only controls the liquid chromatograph and the sampler, but also monitors all parameters. Thus, if the system is not ready (e.g., flow not ready, temperature not ready, etc.) the micro-processor will prevent further sample injections until the system is ready. This assures that each sample is injected under reproducible conditions.

Injection is made through the use of a variable volume injector. The operator dials in the volume to be injected (up to 200 µl) and the vial is moved into position, where the injector takes only the amount of sample to be injected and introduces it onto the column at full pressure. The injector injects all of the sample it takes and is continusously purged with mobile phase. A small-volume option is available which leaves 3 µl of sample in the vial. Thus, an injection of 5 µl would require 8 µl of sample.

At the completion of the chromatogram, an analytical report (Fig. 10) is printed, which includes the vial number and the analysis number for that sample.

A recent development from Technicon is the automated HPLC vitamin analysis system shown in Fig. 11.

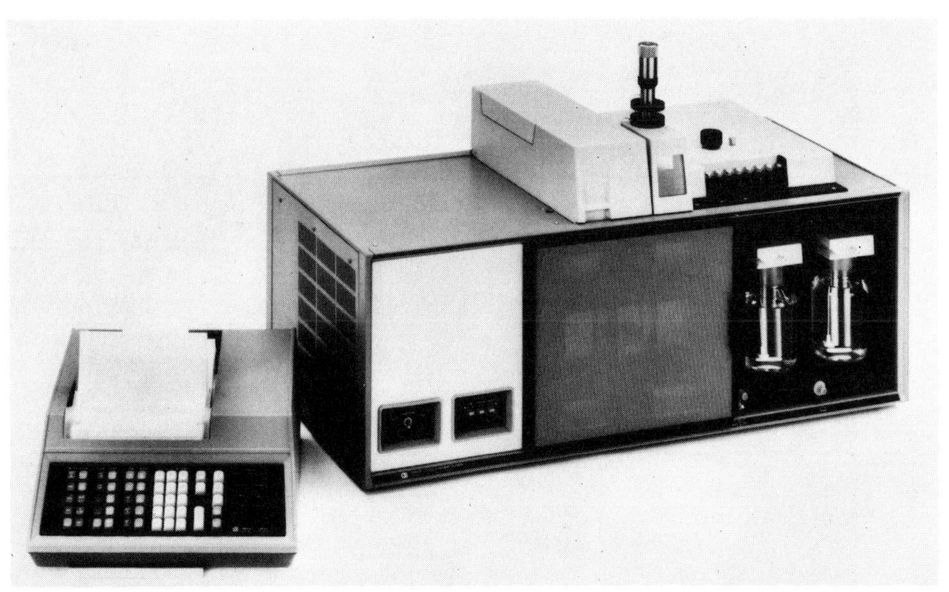

FIG. 9. Hewlett-Packard Model 1084 liquid chromatograph equipped with the HP 79842A automatic sampling system. (Photo supplied courtesy of Hewlett-Packard, Avondale, Pennsylvania.)

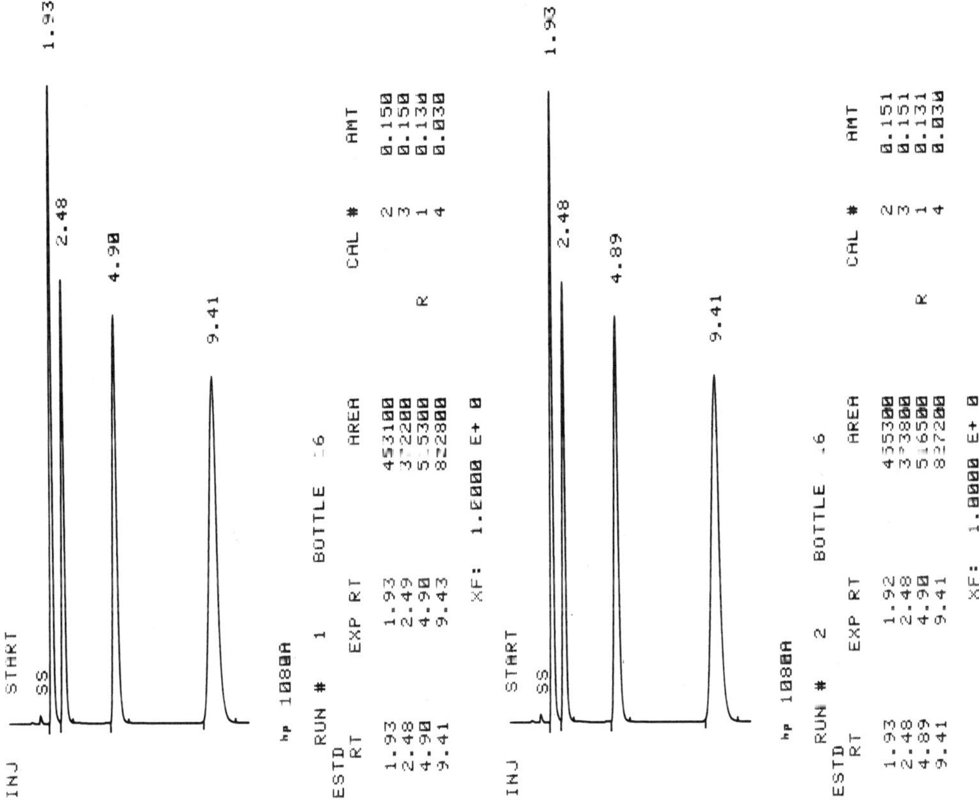

FIG. 10. Analytical report from the Hewlett-Packard Model 1084 liquid chromatograph equipped with the 79842A automatic sampling system.

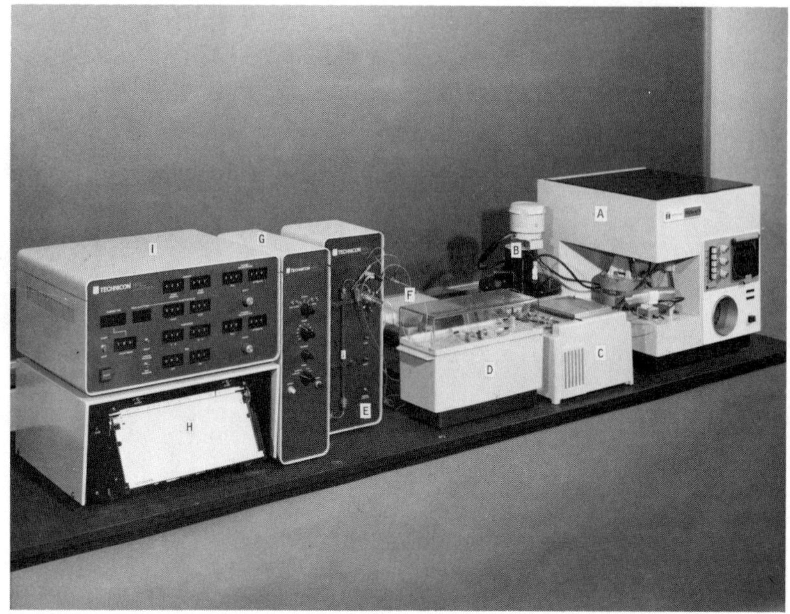

FIG. 11. Technicon automated continuous-flow HPLC system for analysis of fat-soluble vitamins in pharmaceutical preparations. The system consists of nine modules, (A) through (I). Tablets, capsules, or pharmaceuticals in other solid form are disintegrated in the Technicon SOLIDPrep Sampler II (A), with solution heated in the variable temperature bath (B), mixed with reagents and organic solvent delivered by the multi-channel Proportioning Pump III (C), and extracted into solvent in the Analytical Test Cartridge (D).

The stream is then directed to the sample loop of the pneumatically-operated valve on HPLC Cartridge (E), which also contains the liquid chromatography column. HPLC pump (F) drives the mobile phase through the column, and effluent enters the uv detector (G), the signal from which is displayed on the recorder (H). All modules are synchronized by a versatile, solid-state programmer (I).

V. SUMMARY

An automated HPLC should be an inexpensive, basic instrument with capability for more than dedicated analyses. It should have automatic sampling capability with sufficient capacity for overnight or at least one shift of unattended operation. It should be capable of rapid conversion to a "research" instrument, as opposed to an automated or QC-type instrument, for crisis or emergency analyses.

The instrument should be easy enough to operate to be attended by multiple operators during shift work situations and should be basic enough for relatively unskilled personnel to cope with basic maintenance and/or trouble shooting.

Finally, the instrument should be capable of on-line interfacing with data acquisition and reduction devices of all types.

REFERENCES

1. J. J. Kirkland, *J. Chromatogr. Sci.*, *10*, 593 (1972).

2. J. J. Kirkland, *Modern Practice of Liquid Chromatography*, Wiley-Interscience, New York, 1971, p. 173.

3. R. A. Henry, J. A. Schmit, J. F. Dieckman, and F. J. Murphey, *Anal. Chem.*, *43*, 1053 (1971).

4. P. P. Ascione and G. P. Chrekian, *J. Pharm. Sci.*, *64*, 1029 (1975).

5. R. E. Huettemann and A. P. Schroff, *J. Pharm. Sci.*, *64*, 1339 (1975).

6. J. F. K. Huber, E. T. Alderlieste, H. Harren, and H. Poppe, *Anal. Chem.*, *45*, 1337 (1973).

7. R. A. Henry, J. A. Schmit, and R. C. Williams, *J. Chromatogr. Sci.*, *11*, 358 (1973).

8. C. L. Zimmerman, J. J. Pisano, and E. Apella, *Biochem. Biophys. Res. Commun.*, *55*, 1220 (1973).

9. D. R. Baker, R. A. Henry, R. C. Williams, D. R. Hudson, and N. A. Parris, *J. Chromatogr.*, *83*, 233 (1973).

10. J. J. DeStefano and J. J. Kirkland, *Anal. Chem.*, *47*, 1103A (1975); *Anal. Chem.*, *47*, 1193A (1975).

11. J. J. Kirkland, *Analyst*, *99*, 859 (1974).

12. R. C. Williams and J. P. Larmann, *Ind. Res.*, August 1975.

13. J. J. Kirkland, *2nd International Symposium on Column Liquid Chromatography*, Wilmington, May 17-19, 1976.

14. D. D. Bly, H. J. Stoklosa, J. J. Kirkland, and W. W. Yau, *Anal. Chem.*, *47*, 1810 (1975).

15. J. A. Schmit, R. A. Henry, R. C. Williams, and J. F. Dieckman, *J. Chromatogr. Sci.*, *9*, 645 (1971).

16. P. R. Brown, *J. Chromatogr.*, *57*, 383 (1971).

17. D. R. Baker, R. C. Williams, and J. C. Steichen, *J. Chromatogr. Sci.*, *12*, 499 (1974).

18. F. Baumann, R. E. Pescar, B. Wadsworth, and B. Thompson, *Amer. Lab.*, *8* (No. 9), 97 (1976).

19. W. F. Beyer and D. G. Gleason, *J. Pharm. Sci.*, *64*, 1557 (1975).

Chapter 13

COMPUTER INTERFACING AND DATA PROCESSING

John F. Holland

Frank E. Martin

Charles C. Sweeley

Department of Biochemistry
Michigan State University
East Lansing, Michigan

The decision of whether to computerize a gas or liquid chromatograph is a complex one. This chapter will consider the factors involved in making that decision and will discuss the various methods by which computerization can be realized. The style of this chapter will be informal, and the material will be presented in general terms, emphasizing basic concepts and introducing some of the vocabulary in current use. The major objective of this article is to assist the analyst who may be unfamiliar with digital concepts in gaining an appreciation of the processes involved in computerization. Questions regarding the details of interface circuitry or program algorithms should be sought in various publications [1-3].

I. ADVISABILITY OF COMPUTERIZATION

Whether computer assistance is necessary or advisable for the collection and reduction of data from specific chromatographic instruments is often not clearly discernible. In some cases, such assistance is of no value. In other cases, such help may be of marginal benefit and limited scope. In still other situations, the need for assistance in collection and processing of large amounts of data may be real and of considerable urgency. A major problem in all of these cases is that it is not easy to evaluate the actual need and thereby fit the solution to the task in the most effective way.

The major motivations for applying digital devices to chromatographic systems can be divided into two categories, economic and scientific. The economic category can be further divided into two distinct groups, one concerns the cost accountability of the benefits that are to be derived from computer assistance and the second concerns the funding that can be made available for the procurement of the desired devices. It is not uncommon that economic justification for a sophisticated system can readily be documented, but compromise to a lower cost alternative must be made on the basis of existing financial capability. The recommended strategy for implementing the procurement decision is to assemble the economic and scientific justification for the desired options and then weigh these against the funding ability.

A. Economic Justification

The first group of economic motivations can be collectively described simply as money saving. These savings are accrued by

1. The reduction of manpower used in analysis, data handling, and report writing.
2. Automated operation resulting in greater productivity thereby reducing the equipment cost per analysis.
3. The reduction of errors in calculations and interpretations, resulting in fewer costly mistakes.
4. Rapid, concise reporting and superior bookkeeping capabilities, insuring efficient costing and invoicing.
5. The ready capacity for long-term storage which reduces the liability of expensive future challenges to analytical validity. (In many cases, the ability of recall may obviate the necessity for repeating analyses that have already been performed.)

Against these benefits one must assemble the cost factors specifically encumbered by computerization. These factors include

1. Original equipment cost.
2. Installation costs in manpower and money.
3. Lost time for training.
4. Maintenance and cost of operation.
5. Down-time liability (to what extent will a possible computer failure interfere with the laboratory operation once dependence has been cultivated).

Point by point considerations of these economic factors usually lead to a reasonable approximation of the benefit to cost ratio, a high value of which is the best definition of need and a most effective argument for the procurement debate. Indeed, in many quality control and service applications, this is sufficient to certify the need, and attention can be immediately directed toward identifying the most appropriate solution.

B. Scientific Justification

In most research applications, direct cost savings are not sufficient justification for expensive data processing devices. To a large degree, the need must be based upon scientific merit which can be differentiated in the following manner:

1. Extended qualitative capability: Unlike most quality control applications where the nature of the compounds encountered are known, research samples may contain large numbers of unknowns and thereby complicate the identification processes. Automated calculation of retention times and the search of library reference files are beneficial toward this end.
2. Extended quantitative capability: Deconvolutions, area calculations, baseline corrections, and nonlinear response curves can be more precisely accommodated with computer help. In many cases it is only with computer help that they can be accomplished at all.
3. Convenient graphic output capabilities facilitate comprehension of experimental results and thereby catalyze research progress. In addition, rapid processing of analytical results is often mandatory when working with compounds of limited stability.
4. The ability to do sophisticated searches and comparisons on large amounts of data enables the analyst to gain correlations and understandings not otherwise feasible.

Once the level of need for computational assistance has been determined with the aid of economic and scientific factors, specific solutions can be proposed and their benefit to cost ratios can be estimated.

C. Compromise

It is here that the second group of economic concerns are encountered. The case for extensive computer assistance may be impressive and convincing, however, the economic realities may preclude the most efficient solutions. Compromise can be conducted in the usual economic give and take sense; however, two points should always be kept in mind. First, establish a long-range goal for your laboratory and define its component structure as specifically as can be envisioned. Whenever possible, agree upon short-range procurements that relieve some of the current pressing needs and are at the same time desired components of the long-range plan. Secondly, be constantly alert to the forward compatibility of the devices you are obtaining so that they, too, can be readily attached, or interfaced to the final configuration. Never purchase a piece of equipment that will become obsolete as the success of your long-range plan is realized.

Just as there may be several degrees or levels of operational and computational need, there exist several levels of solutions embodying various devices useful in intelligence gathering, storage, and processing. In a sense, once the decision has been made to computerize, a new set of problems surface. These problems are in a dynamic and exciting field, a field which is becoming of increasing importance to the scientist, a field well worth the effort of involvement.

II. METHODS OF COMPUTERIZATION

The computational assistance furnished by small table top or pocket calculators may be adequate for many routine analyses of a limited number of samples of uniform constituency. In cases where the presence of an operator or technician is required for other reasons and the task of running the chromatograph and calculating the results can be readily accommodated within the normal work load, little need for further automation is required.

The first step beyond the hand calculator is often the availability of a teletypewriter or similar device for manual entry of data points after measurement from recorded chromatograms. This procedure is an example of *off-line* data entry. The terminal may be used to punch a paper tape for subsequent processing at a computer center or it may be directly connected to a computer via a telephone line or a direct cable connection. In the latter case, the terminal is referred to as *on-line* to the computer; however, the chromatograph has no direct communicative link to the computer and hence is still off-line. Another example of off-line computer usage is to automatically record the output of the chromatograph on paper cards, paper tape, or magnetic tape for later manual transport to a computer or to an on-line terminal for processing. There always remains a manual or non-automatic linkage somewhere between the instrument and the computer in all cases of off-line data entry.

The obvious next step would be to attach the output of the chromatographic instrument itself to the computer either by direct connection or an on-line terminal. In this mode, the instrument itself is said to be operating *on-line*. Essentially, there are three modes for on-line operation. The computer to which the chromatograph is connected may be shared with several other on-line users all of whom may have active connections at the same time. This arrangement is called *time-sharing* and is made possible by the speed with which the computer can interact. Each user can communicate with the computer in a manner seemingly independent of each others while the computer is rapidly and sequentially servicing each device. Time-sharing is a very efficient procedure and can make very expensive computers available at a relatively low cost per user. The size of the computer may vary from a minicomputer in a laboratory with two to eight instruments to a large computer center system with hundreds of instruments or terminals connected on-line. The nature of the connection between the data generating source and the final processing device may be as simple as a single pair of wires directly linking the two or as complex as an hierarchal network containing multiple computers [4-5]. In the latter systems, the duties of the individual computers may be specialized using small, easily interfaced units for data collection and storage and larger, more powerful units for data processing and outputting. Time-sharing is a very popular mode for chromatographic data systems of all sizes [6-9]. The disadvantages lie in the complexity of the programming that is required and the little control users have over the computer operation. Excessive numbers of users also can degrade the response time of the system thereby reducing its usefulness.

A second mode of on-line operation occurs when a single chromatograph is directly attached to a computer and operated exclusively by that computer. In this mode, both instruments remain discrete and can be operated independently. The computer thus can be made to serve many instruments, but only one at a time [7, 10-14]. The computer is said to be operating in a mutually exclusive mode at any given time as opposed to time-sharing, where the computer is simultaneously shared with other instruments. An obvious advantage of this mode is that programming is much simpler and consequently each new development delay is minimized. This continues to be a very popular mode particularly in situations where the work load on any one instrument fails to justify computerization, but the sum total of several laboratory applications makes precurement feasible.

A third mode of on-line operation occurs whenever the computer is totally dedicated to one instrument and becomes an integral part of that single system (Appendix, 2 and 6). The apparent cost inefficiency of this mode is somewhat overcome by the reduction in the price of the computer, since many features not needed by the specific instrument to which it is attached are not included. In addition, many of these computers are operated in an *in-line* or *closed-loop* mode which may provide significant economy of hardware components. In this mode, the computer completes the communication loop between itself and the instrument to accommodate regulation, control and operational functions. Although the closed loop operation is not limited to dedicated computers, it obviously exerts the greatest cost and component savings in this mode [15, 16]. Microprocessors which are very small computing devices containing only selected components of a computer are very popular in this application; however, even with the decreasing cost of these microcomputers individually computerized instruments still cost more than their standard counterparts and a point of diminishing return is rapidly reached when the alternatives of time-sharing vs totally dedicated systems are considered in situations where numbers of four or more instruments can be anticipated.

There is a large variation in the scope and sophistication of the digital equipment as well as the chromatographic instruments that are used in these systems. These variations will affect any serious deliberations and add another dimension to the benefit to cost evaluation. It should be noted that the order and trends that have been presented in general terms in this discussion may be radically changed when considering unique situations with specialized requirements.

A. System Responsibility

Once the desired level of computer assistance has been determined, the important issue of system responsibility will arise. This is one of the most difficult questions to resolve and is the most sensitive point in the entire computerization process. The difference between success or failure is often determined by this decision which, unfortunately, is often made without due process or considered forethought.

A general rule can be recommended which is easy to remember: If an adequate system is available commercially, buy it. This general rule has few exceptions and one must be certain to have positively identified his position before progressing to other less convenient options.

B. The System

The *turn-key* system refers to an integrated package of hardware and software that will perform a set of designated tasks on selected instruments using a particular computer and peripheral devices. The name turn-key is used because it is a description of the role of the analyst in the fabrication and installation of the system. Several systems of this type are available for use in chromatographic analysis (Appendix 1-10). The major advantage of this approach is that the system has been tested and is of a proven design. In addition, system responsibility rests with a clearly identifiable group outside the laboratory on which considerable pressure can be exerted. Another advantage is the short installation time required, thus allowing an analyst to concentrate on the use and performance of the system rather than details of design, fabrication, and testing. Disadvantages of this approach are that it is relatively expensive (this point can be argued if lost time and realistic internal costs are considered), it is difficult to update or modify, and in several instances, the scientific requirements of a particular application are simply not fulfilled. If the case against the turn-key system is not convincing, it is best to compromise and go with an established system.

If a turn-key system cannot be found that will satisfactorily encompass the needs of the user, an alternate method is to solicit the services of the technical personnel who are made available by the computer manufacturers and other industries for the purpose of designing a custom system. These people will review the situation and submit specific proposals using various combinations of hardware and software packages that are available. In the event that an acceptable configuration contains components available through one vendor, it is likely that they will accept system responsibility. Purchases from more than one vendor greatly diminish the scope of the responsibility that each will assume. Hardware support can usually be realized by means of service contracts directly with the equipment manufacturer. There are many small firms specializing in hardware and software packages, many of whom can be contracted for software design and responsibility [7]. Larger or more complex systems capable of automating an entire laboratory are also available [17]. These are designed for on-line, time-shared operation and provisions for GLC and LC inputs represent only a small part of the total package.

In the event that the commercial packages are inadequate or represent an initial investment that is too large, in-house technical support should be recruited in an effort to compromise the hardware purchase and still accomplish the laboratory objectives. Many pharmaceutical firms have large and skilled internal instrument groups, which in most cases will assist in the planning and installation of the system and accept system responsibility. Programming support is also often available and, when combined with the software, a reliable and efficient package can be assembled. The major disadvantage of this arrangement is that its implementation can be time consuming and inefficient. Familiarity with the equipment varies greatly with the past experiences and other responsibilities of the in-house personnel. Confusion regarding the ultimate component responsibilities may also occur. The advantages of this approach lie mainly in the customization that can be utilized to fulfill the specialized needs of unique laboratory situations. The gradual improvement of in-house capability engendered by this method is a very desirable situation when maintenance or future modifications are to be considered. Future in-house reproductions of a successful system should bring costs down significantly.

The least desirable solution is necessitated when suitable commercial units are not available and in-house support is lacking, while the need to computerize is overwhelming and action must be taken. This approach usually requires the scientist himself to assume the hazardous role of prime contractor repleat with the total system responsibility. This is not an impossible situation, but it will be time consuming and at least initially, very inefficient.

It must be stressed that this is not a viable option for all laboratories. The person assuming the primary responsibility must have a reasonable aptitude for instrumentation, a strong desire to become involved in this type of experience, and plenty of time and energy to devote to the project. Perseverence is a mandatory quality as the initial progress will be slow and for the most part early results will compare unfavorably with commercial units. This alternative is the one with the greatest probability of failure; however, it is not without merit as it is the most challenging and exciting alternative. If successful, this approach is the most psychologically satisfying and has the greatest scientific productivity. At a minimum, the scientist's background will be expanded into a new area and in the extreme, a new career may be germinated. The scientist-computer specialist is fast becoming an acceptable hybrid for which there is a growing demand. The risks are great in charting this course, nevertheless, and with all probabilities carefully weighed, it still remains the least desirable approach to computerization.

III. INFORMATION PATHWAY

In an instrument such as a chromatograph, information can exist in either an analog or digital form and may be transformed from one form to the other using devices called *analog-to-digital converters* (ADC) or *digital-to-analog converters* (DAC) [18]. Systems that intermix analog and digital devices are called *hybrid systems*. This chapter will consider only those analog devices that are a direct part of the information pathway (Fig. 1).

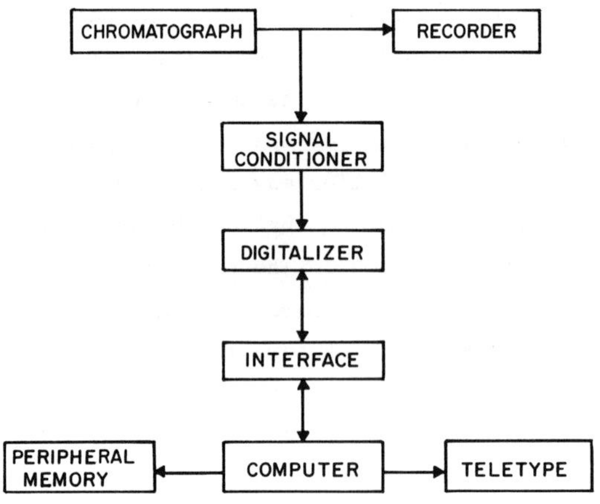

FIG. 1. The information pathway.

The parameters of the measurement processes used in chromatography are reasonably limited in number. In practice, all save one are reduced to a linear time dependence. The elution volume is linearized by constant flow conditions and the temperature parameters are for the most part either isothermal or linearly programmed in time between two different temperatures. Gradients, if employed, are usually either step changes or linear changes created by a programmed action, again based upon a reproducible time scale. Only the output signal from the detector in the typical case is not linearly related to the time axis and this value coupled with an accurate time measurement are the only factors in the two dimensional information scheme of the conventional chromatogram. The data collection process is further simplified because of the relatively slow rate at which sets of data points need to be measured and stored. Data collection rates sufficient so as not to degrade the accuracy of the chromatographic system generally need not be any faster than approximately 10 points/sec and typically lie in the 4 to 5 points/sec range. (Only in recent years has it been considered that 10 points in a second was an extremely "slow" rate). The relatively slow collection rate coupled with the general need for only two values make the chromatograph extremely amenable to on-line, time-shared computerization [19-23]. Even in those configurations where the computer is in-line, the control functions add little additional load to the system.

Another type of intelligence resides in the identification number of the sample, the description of its source and chemical history, and the nature of the analyses that are being performed. This information is usually entered manually and subsequent bookkeeping chores are delegated to the computer, thereby providing one of the major economic benefits of its employment. Record keeping and report generation make a major contribution to the convenience of operation of data systems and their importance is not to be underestimated. This feature is well recognized by the various commercial houses as they direct a good part of their sales literature to this capacity (Appendix 1-10).

The meaning encoded in the time or volume parameter for each data point taken during an elution is based upon its relation to the starting time of the run, which is usually defined as the moment of sample injection. Elution time may be encoded directly in digital form by a device called a *digital clock*. The most common method of monitoring time is to take the data points at regular time intervals. A simple count of the data points can then be transformed into a linear time function. The frequency of the data collection determines the resolution of this parameter, while conditions of computer structure and programming will determine the total range of the time base that can be used. In general, time is easily and accurately measured to within the resolution demanded by chromatographic techniques.

The signal or intelligence from the detector is generally first encoded in analog form and often requires modification or conditioning prior to digitalization.

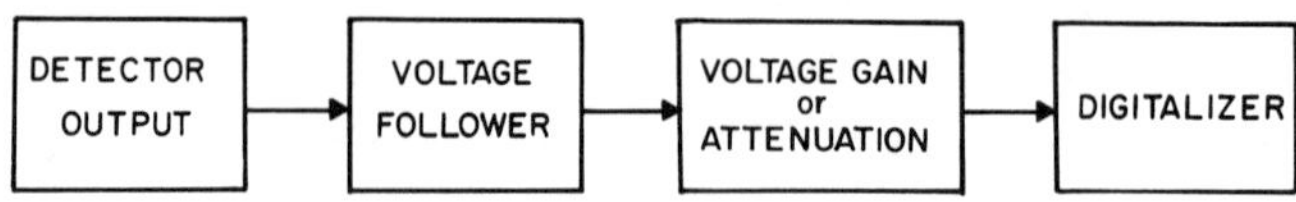

FIG. 2. Analog conditioning scheme.

The change of characteristics of the electrical analog so that it can most effectively
and accurately be digitalized is called *signal conditioning*. A typical analog conditioning circuit is shown in Fig. 2 in block form. The first section, or stage, serves
to isolate the remainder of the interface circuit from the detector and minimize the
effect of these circuits on the performance of the detector itself. Typically, these
stages have a very high input resistance and do not change the size or voltage of the
signal. They are called *voltage followers*. The magnitude of the analog signal must
be compatible with the input range of the digitalizer employed. The second stage sym-
bolizes a device that will change the range of the signal to the necessary levels.
An increase in the voltage of the signal as it passes from the input to the output
of this stage is called *gain*, while a decrease is referred to as attenuation.

A critical restriction on the intelligence derived from the chromatographic detector lies in the dynamic range of the measurement. Dynamic range is the ratio of
the largest signal to the smallest signal that can be measured to within a desired
accuracy. However, in the field of chromatography many authors refer to this parameter as being the ratio of the largest number that can be measured to a number which
represents some fixed relationship to the detector noise or uncertainty. For example,
a common practice is to take this number as twice the noise level [3]. To further
confuse the issue, instrument manufacturers often define dynamic range as the largest
measureable number divided by unity (Appendix 1-10). This is, in reality, the digital
range of the instrument, not the dynamic range. For example, if a device has a digital range of 64,000 and a noise of 10, the dynamic range may be defined as: 64 for
values measured to within 1% accuracy; 6,400 above noise; 3,200 above twice noise;
but may appear as 64,000 in the manufacturer's literature. The dynamic range is an
extremely important consideration in chromatography because of the wide range of signals received from the large and small components within a single sample. Unfortunately, the user must be continuously aware of the existing plurality of terminology
in the definition of this term. In GLC applications in particular the range of signals from a major component to a minor one may be as great as 10,000-fold. If both
of these values are to be measured to within 1%, a minimum discernible scale of
10,000/0.01 or 1,000,000 would be needed. Even with the assumption of a noise-free
signal an accuracy of 1 part in 1 million cannot be attained with existing chromatographic detectors. Fortunately, this restriction can be circumvented. The real problem is to be able to measure on the same system very large peaks and very small peaks

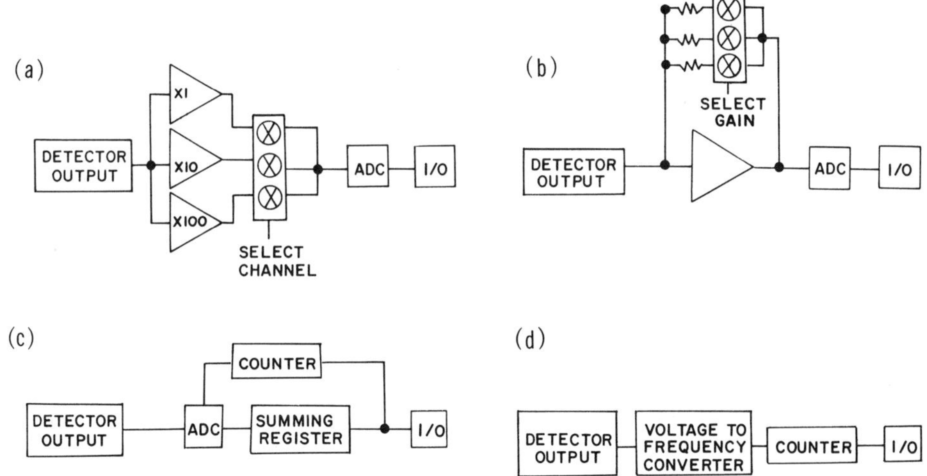

FIG. 3. Dynamic range schemes.

to the same relative accuracy. In normal instrument operation, a large dynamic range
is accomplished by an attenuator switch which serves to change the output of the detector amplifier by preselected factors with the relative accuracy of each range remaining approximately constant.

The dynamic range for computer coupled systems is expanded generally by one of
four methods which are shown in Fig. 3. Circuit (a) shows multiple amplifiers, each
with a fixed gain. The outputs are connected to a switching network, called a *multiplexer*, which enables one channel at a time to be connected to the ADC. The amplifier
that is connected for any given data point depends upon the magnitude of the signal
from the detector and can be selected either by the computer or by a simple digital
circuit [24]. Circuit (b) illustrates a device called a *programmable gain amplifier*.
The voltage gain of this device can be programmed or set again by the computer or by
a digital network [25]. In cases where digital networks are used, information indicating the gain employed must be transmitted to the computer along with each digitalized data value. This configuration is analogous to the manual switching method employed in the normal operation of the chromatograph itself. The third type of range
expander is illustrated in circuit (c). In this circuit, analog to digital conversions are made in rapid, fixed, short intervals (Appendix 10). A preselected number
of these conversions are added together. The sum is treated as a single point in
subsequent data processing. This summation technique produces the wide dynamic range
essential for many forms of chromatographic analysis even though the total range of
the ADC is generally small. Circuit (d) employs a *voltage-to-frequency converter* (VFC)
in the digitalization process [26]. The frequency of the ac output of this device is
dependent upon the dc voltage level at its input. The voltage excursions of the resulting ac signal are independent of the input and the total count then becomes the

number representing the analog dc signal. The total range can be adjusted by varying the length of time used for the frequency count as well as the voltage to frequency conversion factors.

The accuracy of any specific measurement is, in essence, dependent upon the quality of the analog signal and the size of the resulting digital number. The number of binary digits (*bits*) that the digitalizer converts the analog voltage into is the ultimate limit of quality for the digital signal. Ten-bit and 12-bit ADCs are the most common in chromatographic applications, they produce ultimate accuracies to within 0.1% and 0.025%, respectively, both of which are superior to the accuracy of chromatographic systems. It must be emphasized that the digitalization of the analog signal cannot result in a more accurate value than the original signal itself. Electrical filtering to reduce the noise may be applied in the analog sections of the circuit. A technique described as "sample and hold" is used to prevent large signal changes during the critical periods of converter activity.

Digitalization may occur at two sites, either at the chromatograph or at the computer. The advantage of digitalization at the chromatograph rests mainly with the fact that digital information is less likely to be adversely affected by the rigors of transmission over cables of the distances normally encountered. In addition, devices called *optical couplers* can be used with digital circuits. These devices permit the transfer of information via light propagation with no need for direct electrical connections. The use of these couplers minimizes the electrical currents that can pass between different devices that are attached to a common power source. Some of these currents follow the ground paths and are called *ground loops*. Removal of ground loops is an extremely important feature when multiple instruments are to be connected to a single computer, as is often the case. The major drawback of digitalizing at the chromatograph at the present time is in the cost. Each instrument must have its own on-site signal conditioner, ADC, and controller. Alternatively, the analog signals from each instrument may be transmitted to the computer where they are multiplexed into a single ADC for digitalization. This approach is more economical and simpler to install, but it is more subject to transmission noise and grounding problems. The use of special circuits for transmitting and receiving analog signals will minimize these problems, but even under ideal conditions the performance will not approach the noise rejection of digital-signal transmission. The analog signals must be driven in a continuous mode over individual conductors whereas the digital signals can be transmitted either in *parallel* (all bits simultaneously) or *serial* (bits sequentially) modes. The parallel mode is faster and somewhat more reliable, while the serial mode is more economical and requires considerably less cabling (two wires). After signal conditioning, dynamic range expansion, and digitalization, the signals are ready to encounter the principal component of the system, the computer.

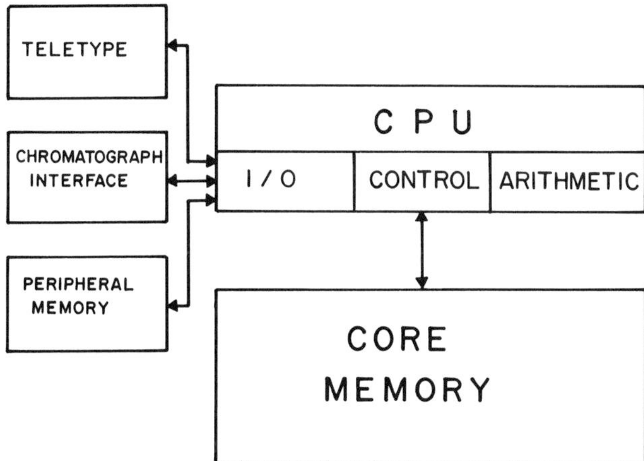

FIG. 4. The digital computer.

IV. DIGITAL COMPUTER

There is little question that the development of solid-state electronics represents
a significant event in the history of mankind. However, rapid change may produce
anxieties as well as facilitate progress. This continually changing market enforces
considerable restraints upon the effective range of future planning.

Figure 4 shows a block diagram of the basic components of a digital computer.
The area for core memory has deliberately been drawn larger than the rest since its
function is the most significant element in the system and its characteristics provide
the digital computer with its major advantages over analog devices. The control unit
is called the CPU or *central processing unit*. It may contain one or more arithmetic
devices which are capable of computational processes. The peripheral devices for
transferring data into and from the CPU and memory are attached to the input/output
sections (I/O). The I/O may be one device, such as a teletype or several devices
such as paper tape readers and punches, magnetic tapes and disks. It is to the I/O
section that the cables from the ADC containing the information from the chromatograph are attached thereby making a communication link to the computer.

Digital computers are actually constructed from only a few types of devices.
They are complicated primarily because of the number and density of the components
used in their fabrication. Digital components are functionally very simple since they
need only to provide for the transfer and storage of numbers. The solid state devices,
that have been designed to perform these duties, are called *gates* and *flip-flops* and,
when added to a few timing devices, are the building blocks of the computers. A gate
is a data transfer device and may be used singly or in combination to perform logic
functions. In its simplest form, the output of a gate may be a transistor which may

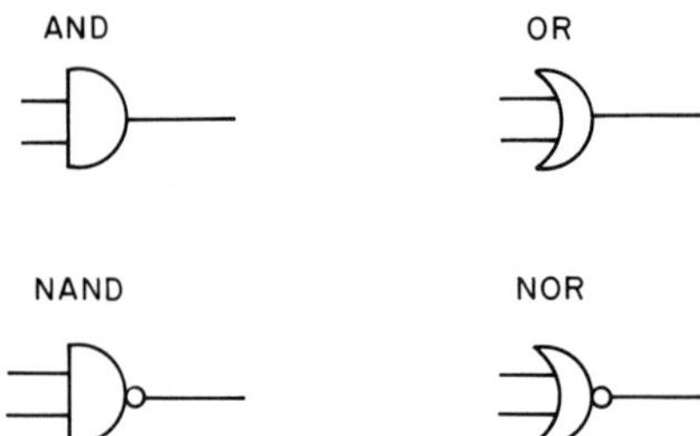

FIG. 5. The basic logical elements.

be conducting (ON) or nonconducting (OFF) giving rise to the two-state or binary nature of digital devices. The actual voltage levels present at the output when the
device is ON are defined as the logic levels. In most recent computers these levels
are 0 to 0.8 v for LOW and +2.6 to +3.5 v for HIGH. In addition to an actual voltage
level, the input and output of a gate are assigned a logical assertion level, 0 or 1.
This duality of assignment leads to confusion unless care is taken to designate voltage levels as H or L and logic levels as 1 or 0. It is important to note that in
some cases an H will be a logical 1, while in other cases an H will be a logical 0.

Figure 5 shows the four basic types of gates. Logical functions are defined by
the straight line on the input side for AND and the curved line for OR. Voltage levels
are designated as a simple intersecting line for H and a circle at the intersection
for a L. The differences between the AND and NAND or the OR and NOR functions lie in
the inversion of the voltage level for the logical assertion from the input to the
output. By changing the assertion assignments, a logical NAND device becomes a logical NOR and vice versa. Since these devices may be interchangable, it is important
that all circuits be drawn to show the logical assertion, not necessarily the nature
of the device used.

The storage elements in computers are divided into two classes, temporary and
permanent. The basic element of temporary storage is called the flip-flop. This is
a device with two stable conditions as is shown by Fig. 6 which represents a NAND-
gate flip-flop. One and only one of these gates can be ON at any time and it holds
the other gate OFF. The flip-flop may be *complemented* or set to the opposite state,

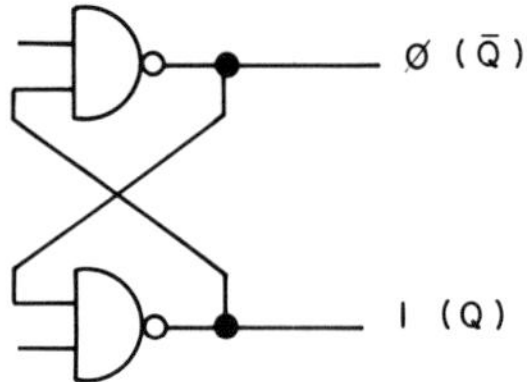

FIG. 6. The NAND gate flip-flop.

an equally stable condition in which the ON and OFF status are reversed. A flip-flop gets its name from this complementing action and generally provides two outputs, one from each gate or leg. To add to the confusion, these outputs are sometimes called 0 and 1. Fortunately, other terminology such as Q and $\bar{Q}$ (Q and not Q) are in increasing usage and obviously preferred. Each flip-flop is capable of storing one bit and its two states may be defined as 0 and 1. Several flip-flops can be aligned to form a register in order to store sequential bits of a larger binary number, the size of which is limited by the number of flip-flops used. The number of bits per word is called the *word length* and is usually 12 or 16 in the types of computers used in chromatographic data systems, but may be much larger in more sophisticated computers. The number stored in a register is called a *word* and can be transferred from one register to another. The 16-bit, and larger, word computers sometimes transfer words by segment or *byte*, a typical size of which is 8 bits. Hence, a 16-bit word can be transferred as two 8-bit bytes.

In addition to the temporary storage function, flip-flop registers may be combined with gates to perform arithmetic processes such addition and subtraction. Multiplication can be accomplished by successive additions and divisions by successive subtraction. These registers are called *accumulators* or *arithmetic registers* and are the essential components in the computational or arithmetic sections of the computer.

Permanent storage devices are found in the memory section. These devices are similar to flip-flops and are aligned into registers that have the same word lengths as the registers in the computer. To date, most of these devices have consisted of little circular magnets which are capable of existing in two stable states dependent upon the direction of the magnetic flux. These states can be defined as 0 and 1 and in this definition each little magnet can store one binary digit. Several coils of electrical conductors are wrapped around these magnets. Electric currents are used to set or WRITE the core magnet to the desired state and other currents are used to sense or READ the status of the core magnet. The name *core memory* originated in reference to these little circular magnets and has become generic to signify that part of memory directly accessible to the CPU and is sometimes used even when referencing newer types of semiconducting memory elements which are not magnetic. One advantage of magnetic core memory is that it is nonvolatile. Once it has been set to a particular state, it will remain in that state until electromagnetic energy is put into the core sufficient to overcome the magnetic hysteresis and change the direction of magnetism. Simple actions such as turning the power off do not destroy the contents of magnetic memory and this feature has been exploited throughout the development of the modern computer.

Core memory can be compared to the old-fashioned country post office. Each mail box or pidgeon hole has a number and each box also has in it some mail or contents. The contents of memory pidgeon holes are numbers or words and each memory register has a number which represents its location in core or its *memory address*. If the

computer can address any location in core and read or write into that location in a
simple direct sequence, called a *memory cycle*, the memory is said to be *random access
memory* (RAM). Other types of memory, which may provide a read capability but not a
write capability, are called *read only memory* (ROM). Still another type of memory,
that provides a read capability within the normal memory cycle and can be written by
using special devices and routines, is called *programmable read only memory* (PROM).
In most cases these latter types of memory are not magnetic; but rather, are comprised
of solid state devices which serve as storage elements. *Solid-state memory* is faster
than magnetic memory but must be replenished intermittently (approximately every
100 μsec) and must be protected against power failures or it will volatilize and lose
its contents. The typical minicomputer has RAM, either solid state or magnetic, freely
accessible to the operator. It also has unaccessible sections of ROM or PROM which
contain small programs called *microprograms* that direct the execution of specific sequential operations which are part of the CPU function of the computer.

These operations which the computer is constructed and/or microprogrammed to
perform are called the *instruction set* and may include from 50 to 150 different operations for the average minicomputer. Each instruction is represented as a word or a
number. The operations range from a simple fetch from a specific memory location, or
test the status of a specific bit, to the performance of a multi-stepped sequence such
as multiplication or division. These instructions can be organized in different sequences, called *programs*, to accomplish a variety of chores. The program is stored in
core memory as instruction words. Data words are also stored in core memory and the
CPU must at all times be able to distinguish between the two lest disaster occur.
From the contents of a memory location, it is impossible for the computer to determine
whether the word is data or instruction. Only by knowing the location in core in which
the program is stored is differentiation possible. This is accomplished by a register
called a *program counter* (PC). The PC is automatically incremented during the execution of each instruction so as to act as a pointer to the memory location of the next
instruction. If a program error, called a *bug*, causes this counter to address incor-
rectly the computer may read a data word or an instruction word out of sequence with
unpredictable results. This is an example of a common type of problem that appears
in the execution of new programs. Identification and correction of these program
errors is called *debugging*.

The program counter is part of the CPU along with numerous gate, register, and
timing circuits, the operation of which are initiated and synchronized by a strobe or
clock. This strobe acts as a heart beat in the life of the computer, all operations
depending upon its repeating pulses. Gates open and close, memory is written and
read, logic decisions made, words transferred, and instructions decoded into actions,
all to the rhythm of the strobe in a continuous procession of internal sequences. The
speed of a computer is usually gauged by its memory cycle time, which is approximately

in the range of 1-4 μsec for magnetic memory and 0.2 to 1 μsec for solid-state memory.
One memory cycle typically requires four beats of the strobe, hence the strobe frequency lies in the range from 0.1 μsec to 1.0 μsec. This signifies for any computer
the least time during which any change of state can occur or any function of the computer be utilized.

The I/O section of the digital computer consists of gating devices for the transmission of data to and from registers in the CPU. This section contains special registers which isolate or *buffer* the internal CPU registers in their transactions with
the outside world. These buffers may act as temporary storage areas for data as it
is transferred back and forth. The outputting of information from the computer is
generally performed under program control, however, several methods may be used for
the input processes. The names may vary from one manufacturer to another, but they
embody analogous features which may be related to one of the following processes:

Program control: A set of instructions are executed by the CPU which effect
data transfer.

Direct memory access by cycle steal: The program being executed is suspended
and the CPU is temporarily placed under the control of a peripheral device
that directs the transfer. When the transfer is complete the CPU is released
to resume the suspended program.

Direct memory access using dual port memory: Data is inserted into memory without passing through CPU registers in a process that is invisible to the program being executed.

Data transfer between the CPU and peripheral devices may be accomplished in
either a serial or a parallel manner. If both devices during the transfer are controlled by the same timing or clock pulses, the transfer is referred to as *synchronous*.
If the timers are separate and transfer is controlled by an exchange of status signals,
such as *ready* and *done*, the transfer is *asynchronous*. The teletype, which is one of
the most popular I/O devices, employs a special 6-, 7-, or 8-bit serial code called
ASCII that has become a standard of the communications industry. To maintain hardware
compatibility with these devices, the computer manufacturers universally furnish
asynchronous serial circuits that will accommodate these coded transfers. In situations where speed is essential, parallel transfers are preferred.

Many of the devices attached to the I/O lines may provide memory capabilities.
Examples of these are paper tape, magnetic tape and magnetic disks. This memory is
not randomly accessible and is contrasted to core memory by the term *peripheral memory*. The fastest read/write responses attainable with this type of memory come from
the magnetic disks which have access times in the milliseconds. These memories are
similar to core memory in that both programs and data may be stored on them. Programs
are not executed from peripheral memory as the time loss would be self defeating.

Prior to execution of a program stored on peripheral memory, it is loaded into some
area of core memory and will be executed from that location. This rapid reading of
a program is an ideal way to increase the scope and size of programs that can be operated in a computer of modest core memory size. Indeed, in many instances a single
program may be too large to fit into memory at one time and must be written in segments or *overlays*. These overlays can be transferred from the peripheral device in
a sequence that will facilitate the functioning of the program as a whole. These
large capacity memory extensions can be powerful additions to a computer system.
Their cost, unfortunately, is often in excess of the cost of the computer itself, a
situation that makes procurement somewhat difficult at the very least.

V. DATA PROCESSING AND OUTPUT

The most critical feature of a good laboratory computer data system lies in the ability to collect and store accurate, meaningful data. Fortunately, the numerous suc-
cesses in both laboratory-designed systems and commercially available installations
have encouraged the development of sophisticated data processing and elegant output
routines. Indeed, the differences between various systems often center on the scope
and capability of their respective *software*. Software is a term which includes all
of the programs available to the operator in normal usage modes. In the turn-key
system, the software as well as the hardware is furnished complete and operational.
For the less fortunate do-it-yourselfer the software represents the greatest obstacle
to total success and is the most time consuming operation.

A. Levels of Language

As previously described, an instruction in its simplest form, is a number. Numbers
that represent .the instruction set form the *machine language*. These numbers are the
only meaningful instructions to the computer and all other program symbols must eventually be translated into them to become useable. A program can be written in machine
language; however, it is very difficult and inefficient, so much so that in practice
only the simplest programs are written in this way.

The next step upward in language sophistication is to use a symbol instead of
a number to represent an instruction. Symbols are easier to remember and to work
with. These symbols, one for each instruction, create a new language and each type
of computer has its own *symbolic or assembly language* representing its instruction
set. Once the program has been written in this language, it is transposed by another
program called an *assembler* into the machine language code ready for computer operation.

It is apparent that programming would be faster if symbols could represent entire
operational sequences consisting of many instructions instead of just one instruction
alone. Many languages of this type have been derived, of which FORTRAN and ALGOL are

examples. They use different sets of symbols to represent the operations encountered in various disciplines. As a result many languages have been devised. Each language has a program called a *compiler* which relates these symbols to the appropriate subroutines and then transposes these into machine language sequences for computer operation. These languages are very easy to use and complex programs can be written in a fraction of the time that the assembly language would require. A shortcoming of these high-level languages is that they are often very inefficient with respect to the amount of core memory they use. In addition they perform I/O tasks at a slower rate than is possible with simpler more direct programming.

Another type of high-level language includes programs called *interpreters*. BASIC, and FOCAL are examples of interpreters. An interpreter combines symbols, some of which may be English or algebraic to represent simple and complex operations just as a compiler does, however, there is a difference in the manner in which these programs are executed. An interpreter program is not compiled into a complete machine language program, but rather is executed as the symbols are being decoded. This is obviously a core and time wasteful procedure, but it is extremely easy to learn and programs can be changed or altered very rapidly without the need for reassembly or recompilition. BASIC is becoming increasingly popular in scientific applications. It is efficient enough to accomplish the mathematical tasks normally encountered and it is simple enough that it can be rapidly mastered and easily reviewed. This is a convenient language, one that a scientist can and will use.

B. Data Collection Programming

The data collection routines for chromatography are usually written in assembly language for a variety of reasons: (1) the maximum efficiency of time and memory is preserved, (2) the actual programs are very simple, involving steps of device selection, digitalization, transfer and storage, and (3) the program contains instructions directly related to the I/O device being serviced and is thus not readily amenable to interpretations by a more sophisticated language. The two most widely used approaches for data acquisition by on-line computer systems are sequential device probing under continuous program control and intermittent program interruptions for the purpose of servicing the external devices. In the first case, the time dependency of the collection is observed by the program itself and the various instruments attached to the system are addressed in a fixed sequence, queried as to their operational status, and, if ready, their output is converted and transferred into the computer and stored. This sequence is repeated in timed cycles and data are stored in core or peripheral memory for each device being operated. In the second mode, the time base is usually furnished by a real time clock. The data collection procedures are initiated by a data ready signal sent to the computer by the external device or instrument. This signal causes

an interruption of the program under execution. Different kinds of computers use
different interrupt procedures, but they all work in a similar fashion. The program
is suspended, the time read from the real time clock, and the attention of the CPU
is turned to a short routine that will handle the transfer processes from the device
creating the interrupt. The manner in which the computer recognizes the interrupting
device varies from automatic program vectoring, in which each device has its own recognizable interrupt to party line servicing where the devices are serially queried
until the interrupting instrument is found. Once identified, it is serviced and the
computer then returns to the program being executed. Since the data transfer routines
are relatively fast, 10-50 µsec, the actual time spent by the computer in servicing
the on-line instruments represents only a small part of the total analysis time. This
situation facilitates the attachment of several chromatographic instruments to a single computer and allows for data processing to be accomplished concurrently with the
data collection routines [8, 19, 21, 23, 27]. A computational program which can be
executed while constantly being interrupted for the servicing of several time-shared,
on-line instruments is called a *background* program. The programs that service the
instruments are called *foreground* programs. In this operation the computers appear
to be running the background/foreground programs simultaneously.

C. Data Processing

The data processing capabilities of a computerized gas or liquid chromatographic data
system may be divided into two classifications, bookkeeping and analytical. The bookkeeping functions of the laboratory can be manually accomplished but generally at a
higher cost. Hence, this function is very important in the economic justification for
computer procurement. Records of sample identification, analysis time, instrument
number, and pertinent experimental parameters combined with the analytical results and
service charges can be conveniently and accurately stored and updated by the computer,
and they can be made available at any desired time and in any desired format. The
range of independent selection of the accounting methods and report forms will vary
from one system to another, but user satisfaction is usually not a severe problem
since custom reprogramming is usually available at a modest cost in time and money,
thus individual preferences can be accommodated.

The greatest scientific benefits accrued by computerizing gas and liquid chromatographic instruments arise from the data processing capabilities. It is in these
capabilities that the greatest differentiation between the various data systems occurs.
This is not to imply that significant hardware variations do not exist, however. In
the correctly installed and properly running system, hardware characteristics are
usually invisible to the user and his judgement most likely will be based upon the
scope of the software analytical capability and the ease of the operation of the
system.

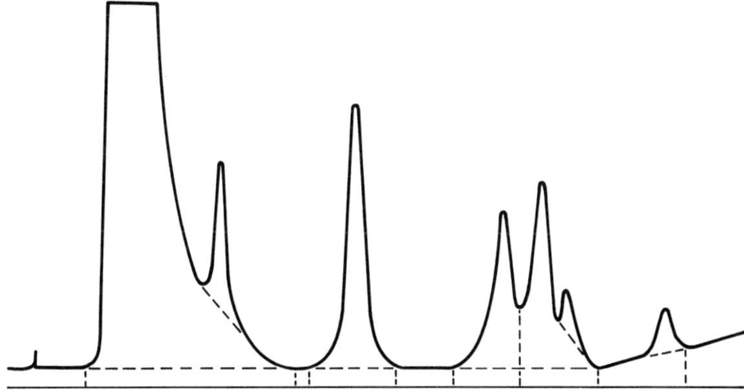

FIG. 7. Composite of chromatograms showing various modes of baseline correction.

In an effort to continue the general nature of this section, the details of data
analysis routines will be omitted. The initial objectives of data processing center
around a limited number of analytical parameters. Retention time, peak height, and
peak area are the fundamental components of analytical intelligence from chromatographic analyses. These three quantities can all be calculated from knowledge of a
single parameter, the detector output. The determination of these three quantities
is the major objective of the greater part of the data processing subroutines.

Figure 7 illustrates the various types of chromatograms normally encountered.
The retention time for any eluting peak is defined as the point where the response
has its maximum value and is generally expressed as the elapsed time from the moment
of injection. There are three steps in the process of detecting a peak. These can
best be illustrated by a series of simple questions:

1. Has a peak started? The appearance of a peak is generally determined in one
of two ways. In one, a threshold value is empirically chosen. When the response
curve rises above this value, a peak is occurring by definition. The threshold can
be an absolute value or a fixed value above an established baseline. In the second
method a value for the slope or derivative of the response curve is chosen. When this
value is exceeded, a peak occurs, again by definition. Since the data are normally
taken at fixed intervals, the slope or derivative can be approximated by calculating
the difference between successive data points. The slope may be an absolute value or
a relative function calculated with respect to the total magnitude of the response.

In both cases a confirmation process should be employed to separate the presence
of a peak from random noise spikes. This usually involves the slope or threshold criteria being exceeded for a given number of successive points.

2. Has the peak maximum occurred? There are three common ways in which this
may be determined. The simplest is to monitor the response curve for its maximum value.

The second is to observe the slope for a change in sign from positive to negative.
The third approach uses sophisticated routines to mathematically determine this point.
Again a verification procedure should be used to discriminate between noise fluctua-
tions and the true peak maximum.

3. Has the peak ended? This is the logical converse of the identification of
the start of the peak. In the simplest case, the threshold or slope criteria is used
in an analogous but opposite sense. A problem may arise if another peak commences
before the conclusion of the first peak. In this case, the response curve does not
return to baseline and the logic scheme must be expanded to recognize this situation.

In the simplest qualitative case, identification may be made by comparing the
retention time of a single component to those of known compounds analyzed under reproducible conditions. In more versatile approaches, retention times are transposed
into a scale relative to one or more internal standards and then compared to a library
of known compounds run under similar conditions [28-30].

Programs to perform retention indexing and compound identification are widely
used and their efficiencies are primarily related to the size of the library file being searched and the diversity of the compounds therein. An obvious limitation lies
in the inability to resolve two compounds with the same retention time. In some cases,
the elution of the same compounds under different experimental conditions can furnish
the necessary resolution. Statistical methods have been applied to time and intensity
data files, in an effort to increase the accuracy and hence, the resolution of retention time determinations. These methods have succeeded in expanding the qualitative
resolution in many instances, but the two-dimensional nature of the data (time, intensity) is a limiting factor in the progress of these programs. These techniques
are becoming increasingly popular in applications of gas chromatography-mass spectrometry where the added dimension of the mass axis greatly expands the scope and resolution of the chromatographic method [31, 32]. In a similar effort to add a third dimension to the data profile, numerous other physical measurements such as uv, ir, nmr,
electrochemical and thermal parameters are being investigated in a concerted effort
to expand the qualitative aspects of chromatography.

D. Quantitative Analysis (Baseline Correction)

Quantitative analysis may be based upon peak height or peak area. Due to the uncertainty of area calculations for very small peaks when the signal to noise ratio is un-
favorable, peak-height quantitation may be preferred. Sharply defined peaks with relatively short elution times are also amenable to peak-height quantitation. This
method of quantitation may be nonlinear over a wide concentration range, but if the
reproducibility is sufficient, the convenience of the technique encourages usage.
However, in the majority of cases, quantitation is based upon area measurements
[33, 34]. The determination of the area of a chromatographic peak involves the
following steps:

1. Identification of the limits of integration: These are normally the points at the start and end of the peak as described above in the discussion on qualitative analyses.

2. Calculation of the total area: Since the data are either taken directly as a function of a linear time axis, or readily transposed into this frame, integrations are performed as simple additions of all the points between the start and the end of the peak.

3. Subtraction of the baseline area: The baseline values at the start and end of the peak must be known and an assumption of the nature of the baseline under the peak must be made. In the simplest case these values are taken as the points immediately preceding and following the start and end points, respectively, as previously determined. The path of the baseline between these two points is usually assumed to be linear resulting in rectangular or trapezoidal baseline contributions to the peak area as indicated by the dotted lines in Fig. 7. The area of the peak is the total area less this baseline contribution. In more stringent situations, the baseline projection may be calculated from the shape of the response curve in the region of the peak.

4. Resolution of fused peaks: There are three popular methods employed to determine the area of each peak in an envelope of two or more unresolved peaks. The most straightforward technique involves the projection of boundaries between the successive peaks by dropping perpendicular dividing lines to the baseline from the point of minimum response between the peaks. Areas are calculated and baseline subtractions are made using these boundaries or limits [35]. This method is accurate only when the peaks are of equal height and when a discernible valley occurs between them. However, it is possible to extend this method to peaks of unequal height by using mathematical corrections that take into account the relative size and separation. Baseline values are defined as the intersections of the perpendicular boundaries with the baseline projection from the beginning to the end of the entire envelope.

A second approach employs a tangential projection of the response curve to separate the contributions of a small peak from an overlapping, large peak. The slope of the response curve at each point on the overlapping peak is compared to the slope of the projection from the point of origin of the small peak to each point of the total response curve. When these two slopes are equal, the small peak is ended. The projection may be a linear extrapolation between the initial and final points of the peak or may be a calculated function derived from curve fitting of the identifiable portion of the large peak. This method works well when the peaks are vastly different in size. The technique is generally called *tangential skim* and the small peak is frequently referred to as a "rider peak." Many computer programs employ both perpendicular drop and tangential skim in the resolution of overlapping peaks as indicated in the three-peak envelope in Fig. 7.

The third general technique for fused peak analysis involves complex deconvolution routines. In this approach, curve-fitting algorithms, which are based upon an assumed response curve shape, are used to mathematically separate the peaks [36-37]. These methods are normally used only as a last resort when experimental conditions cannot be found that will increase the separation so that a recognizable valley exists between the peaks.

E. Data Smoothing

Noisy or scattered data points can adversely affect all of the data processing algorithms. Every device produces a certain noise level below which meaningful interpretation cannot be made. Whenever high sensitivity or high accuracy are necessary, the noise level will often be the limiting factor in analytical accuracy. Electronic filtering of the analog signal may be employed to reduce this level, but the capability to apply mathematical routines to the digital data to remove the effects of random fluctuations is an unquestionable benefit of computerization. Numerical data smoothing may range from simple multiple-point averaging to least-squares curve fitting and other complex statistical routines [38]. The effects of high-frequency random noise are easily corrected but as the noise approaches the frequency of the data sampling rate, benefits of these techniques gradually diminish and they can, in fact, cause considerable distortion of the detector response curve. Care must be taken whenever numerical smoothing is employed, but its use can be of considerable value to subsequent interpretation.

F. Quantitative Calculations

There is no conceptual restriction to the number of mathematical programs that may be written for accomplishing quantitative calculations, dimensional conversions and reporting formats. Nevertheless, there are a few routines that are common to nearly all computer data systems (Appendix 1-10).

Absolute area: output of the raw results of area calculations.

Percent area or relative area: the area of each peak is divided by the total area of all the peaks.

Normalization: the area of each peak is ratioed against the area of a designated peak.

Internal standard: the area of each peak is divided by the area of an internal standard peak and multiplied by a scaling factor.

External standard: the area of each peak is divided by the area of a standard peak from a different chromatographic analysis run under similar instrumental conditions and then multiplied by a scaling factor.

It should be noted that peak number and retention times usually accompany the area outputs. Retention times may likewise be in absolute time or in relative time related to the retention time of a standard compound or compounds.

The complexity of the programs used in the data analysis is often dictated by the peak separation, baseline behavior, signal-to-noise ratio, desired accuracy, and other conditions of the experiment. As a general rule, it is advisable to use a program sufficient to the quantitation with little other capability being included. Excessive sophistication of programming often means wasted time in execution and can introduce confusions where they are neither intended nor significant.

The manner in which the program analyzes the data differs between two extremes. In one case, the incoming data are analyzed in real time. The existence of peaks and valleys are continuously noted by status registers and baselines, slopes and areas are calculated against the time dependent frame of the data collection routines. This approach uses little memory for data storage, saving only the critical information for each peak. On the other extreme, there are programs that store all of the points taken from a chromatographic run and process them after the completion of the entire run. These programs require much larger memory, but facilitate more complicated data smoothing and increase the capacity for resolution of partially resolved peaks. Many data systems lie between these two extremes, intermittantly reducing data files of moderate length during the progress of the chromatographic run.

VI. SUMMARY

Digital computers have brought the scientist three outstanding capabilities: rapid data collection, rapid data processing, and convenient, versatile data storage. The program algorithms employed in the computer analyses of the time dependent detector response are analogous to the thought processes one applies during visual interpretation of a recorded chromatogram. The computer observes the same events as the eye and makes similar decisions. In many cases the logic of the process is directly transposed into computer language and performed as a facsimile of human mental processes. The computer also removes errors due to human bias, except for those introduced by the programmer. Herein lies the greatest contribution of the computerization of chromatographic data. Man's intelligence directs the action of the computer, which, when properly programmed, is capable of objective assessments on large amounts of data in short amounts of time. Results are obtained, comparisons made, and correlations realized that simply would not be done without the aid of the computer.

APPENDIX

1. *Series 3380A Reporting Integrator*, Hewlett Packard, Avondale Division, Route 41, Avondale, Pa.

Hardware: Microprocessor. Memory capacity, 54 peaks, nonexpandable. I/O, keyboard, seven slide switches, built-in printer/plotter. Clock, 333 min maxi-
mum/0.01 min resolution.

Chromatographic capability: Input voltage, 0.01 to 1.0 V. Dynamic range, 10^6
to 1. Attenuator, 1 to 1024 binary and log.

Software: Bookkeeping, retention time printed at apex of each peak; run parameters automatically printed at end of report. Data processing, peak identification by absolute or relative retention time versus reference peak; baseline correction trapezoidal; integration slope sensitivity may be automatically
determined; fused-peak resolution by perpendicular drop or tangent skim. Analytical methods, area %, normalization, internal standard, external standard;
single calibration (maximum of 54 peaks) for all methods. User programming
provisions, none.

2. *5830 A Gas Chromatograph*, Hewlett Packard, Avondale Division, Route 41, Avondale,
Pa.

Hardware: Microprocessor. Memory capacity, 320 peaks, nonexpandable. I/O, keyboard, built-in printer/plotter. Clock, 819 min maximum/0.1 min resolution.

Software: Bookkeeping, key to list all run parameters; retention time printed
at apex of each peak; analytical results printed at end of chromatogram; keyboard entries automatically printed. In-line capabilities (100 function
changes may be specified), detector selection (flame or thermal conductivity),
detector polarity, value switching, rate °C/min chart speed, baseline zero,
attenuation, slope sensitivity, area rejection. Data processing peak identification by absolute or relative retention time vs one reference peak, baseline correction trapezoidal; integration slope sensitivity from 0.01 to 81
mV/min or automatic calculation, automatic increase in isothermal runs; fused
peak resolution by perpendicular drop or tangent skim. Analytical methods,
area %, normalization, internal standard, external standard; one calibration
shared by all methods. User programming provisions, none.

3. *Autolab System 1*, Spectro Physics, 2905 Stenton Way, Santa Clara, Calif.

Hardware: Microprocessor. Memory capacity, 126 peaks, expandable to 297. I/O,
keyboard, built-in printer, teletype optional. Clock, 5 decimal digits
maximum/0.1 sec, 1 sec, 0.01 min, or 0.1 min resolution.

Chromatographic capability: Input voltage, -0.01 V to 1.0 V or -0.1 V to 10 V.
Dynamic range, 10^6 to 1. Attenuator, 1 to 1024 binary and log.

Software: Bookkeeping, report includes all run parameters. In-line capability,
seven time-controlled signals. Data processing, peak identification by absolute retention time; baseline correction horizontal or trapezoidal; integration, 8 decimal digits area; fused-peak resolution by perpendicular drop or

tangent skim. Analytical methods, normalization, internal standard, external standard. User programming provisions, none.

4. *Autolab System IVB*, Spectra-Physics, 2905 Stenton Way, Santa Clara, Calif.

   Hardware: Microprocessor. Memory capacity, 32 peaks, expandable to 160 peaks. I/O, keyboard and built-in printer, teletype, optional. Clock, 99.999 sec maximum.

   Chromatographic capability: Four chromatographs may be attached. Digitalization integrating ADC. Input voltage, -0.01 to +1.0 V or -0.1 to 10.0 V. Dynamic range, 10^6 to 1. Attenuator, 1 to 512 binary and log.

   Software: Bookkeeping, Report contains channel number, sample ID, autosampler output, retention times, peak areas, baseline correction codes. Data processing, peak identification by absolute or relative retention time; baseline correction, trapezoidal; integration slope sensitivity approximate initial peak width calculated automatically; fused-peak resolution, perpendicular drop, tangent skim; Maximum area, 8 decimal digits; normalized area, 5 decimal digits. Ananytical methods. User programming provisions, none.

5. *Minigrator*, Spectra-Physics, 2905 Stenton Way, Santa Clara, Calif.

   Hardware: Microprocessor. Memory size, 0.5 K RAM, nonexpandable. I/O, keyboard, built-in printer, teletype, optional. Clock, 4 decimal digits maximum/1 sec or 0.1 min resolution.

   Chromatographic capability: Digitalization, voltage to frequency converter and counter. Input voltage, 0 to 1V or 0 to 10V. Dynamic range, 10^6 to 1. Attenuator, 1 to 1024 binary and log.

   Software: Bookkeeping, event markers on chromatogram start and end of run, start and stop of peak integration; report includes run number, sample ID, run parameters, retention times, peak areas, baseline correction codes. In-line capabilities, four time-controlled signals. Data processing, peak identification by absolute retention time; peak height or area; baseline correction, trapezoidal; integration slope sensitivity automatically computed; small peak rejection by area; fused peak resolution by perpendicular drop or tangent skim; maximum of four fused peaks. Analytical methods, normalization, internal standard, external standard. User programming provisions, none.

6. *Series 1089 High Performance Liquid Chromatographs*, Hewlett Packard, Avondale Division, Route 41, Avondale, Pa.

   Hardware: Microprocessor. I/O, keyboard, built-in printer/plotter.
   Software: Bookkeeping, key to list all run parameters; retention time printed at apex of each peak; analytical results printed at end of chromatogram; keyboard entries automatically printed. Data processing, peak identification by

absolute or relative retention time; fused-peak resolution by perpendicular
drop or tangent skim. Analytical methods, area %, normalization, internal
standard, external standard. User programming provisions, none.

7. *The Varion Aerograph Chromatography Data System 220/240*, Varion Aerograph, No. 25
Route 22, Springfield, N. J.

> Hardware: Data 620 computer, 16 bit words. Memory size, 12 K expandable to
> 32 K; Peripheral memory, 60 to 1000 K. I/O, teletype, seven maximum.
> Chromatographic capabilities: Multiplexer, 40 channel. Dual-slope ADC. Auto-
> ranging amplifier, 3 x 10^7 to 1 dynamic range.
> Software: Bookkeeping, report includes sample ID, date, instrument number,
> method ID, peak number, name, type, composition by % wt., absolute and rela-
> tive retention time. Data processing, peak identification by absolute or
> relative retention time vs any number of reference peaks; digital smoothing
> combination of 1 to 512 point average and a 5-point moving average and 11-
> point convoluted least squares differentiation; baseline correction trape-
> zoidal; integration-peak rejection by minimum width and area; fused-peak res-
> olution by perpendicular drop, tangent skim, or by gaussian fit. Analytical
> methods, any number of methods may be prepared with a conversational mode
> language. User programming provisions, multiuser BASIC.

8. *EAI Lab PACE* (Models LP-25 and LP-50), Electronic Associates, Inc., 185 Monmouth
Parkway, West Long Branch, N. J.

> Hardware: Data PACER DP50, 16-bit words. Memory size, 16 K (LP-25), 24 K
> (LP-50). I/O, teletype.
> Chromatographic capabilities: Multiplexer, 8 channels (LP-25), 16 channels
> (LP-50); scan rate 3.75, 7.5, or 15.0 points per second per channel. Suc-
> cessive approximation ADC. Auto-ranging amplifier, 1 to 1024 binary, dynamic
> range, 2 x 10^6 to 1.
> Software: Bookkeeping, report includes date, channel number, run parameters,
> sample ID, and comments. In-line capabilities: LP-50 contact closures for
> column switching, temperature stepping, etc. Data analysis, peak identifica-
> tion by relative retention time; baseline correction, trapezoidal; integra-
> tion slope sensitivity automatically increased during run, digital filtering
> applied during integration; fused-peak resolution by perpendicular drop or
> tangent skim. Analytical methods, area %, internal standard, external stand-
> ard. User programming provisions, none.

9. *EAI LabPACE* LP-1000, Electronic Associates, Inc., 185 Monmouth Parkway, West Long
Branch, N. J.

Hardware: Data PACER 100, 16-bit words. Memory Size, 64 K. Peripheral memory, 2.2 M expandable to 8.8 M. I/O, up to 12 teletypes, line printer optional.

Chromatographic capabilities: Multiplexer, 64 channels; scan rates 3.75, 7.5, 15.0, or 30.0 points per second; separate analog filter for each channel. ADC, 16-bit bipolar, autoranging; 1000 conversions per second; dynamic range, 2×10^6 to 1.

Software: Bookkeeping, report includes run title, analytical method title, channel number, date and time, peak name, concentration %, absolute and relative retention times, peak area, response factor, and fusion code. Data processing, peak identification by absolute or relative retention times; baseline correction, trapezoidal; integration slope sensitivity increased automatically during run, peak threshold to reject small peaks; fused-peak resolution by perpendicular drop, tangent skim, or gaussian fit, inflection points monitored to detect unresolved peaks. Analytical methods, unnormalized area, normalized area, internal standard, external standard. User programming provision, foreground/background operation, symbolic assembler, PACE Operation Interpreter, FORTRAN IV, text editor.

10. *3352 B Laboratory Data System*, Hewlett Packard, Avondale Division, Route 41, Avondale, Pa.

Hardware: 2100 A Computer, 16-bit words. Memory size, 12 K expandable to 32 K. I/O, 8 teleprinters, line printer, paper tape punch, and high speed paper tape reader.

Chromatographic capability: Transmission loop with up to 15 instruments. Continuously integrating, 23-bit ADC for each chromatograph, 1024 conversions per second, dynamic range 8×10^6 to 1.

Software: Bookkeeping, report includes report number, title, channel number, comments, run parameters, retention times, peak name, wt. %, area. Data processing, peak identification by absolute or relative retention times; baseline correction, trapezoidal; integration, all parameters may be set automatically; fused-peak resolution, perpendicular drop, tangent skim. Analytical methods, area %, normalization, internal standard, external standard; up to 248 methods may be stored. User programming provisions, LAB BASIC.

ACKNOWLEDGMENTS

The authors wish to acknowledge support for this undertaking from NIH Research Grant RR-480 and a Post Doctoral Fellowship from the Smith, Kline, and French Corporation to Dr. Martin.

REFERENCES

1. J. W. Frazer and F. W. Kunz, *Computerized Laboratory Systems*, ASTM, Philadelphia, 1975.

2. G. A. Korn, *Minicomputers for Engineers and Scientists*, McGraw-Hill, New York, 1973.

3. S. P. Perone and D. O. Jones, *Digital Computers in Scientific Instrumentation*, McGraw-Hill, New York, 1973.

4. I. Wehling, *Chromatographia*, *5*, 197 (1972).

5. C. E. Klopfenstein, *J. Chromatogr. Sci.*, *10*, 22 (1972).

6. J. W. Frazer, *Anal. Chem.*, *40*, 26A (1968).

7. S. P. Perone, *Anal. Chem.*, *43*, 1288 (1971).

8. A. W. Westerberg, *Anal. Chem.*, *41*, 1595 (1969).

9. P. Vestergaard, L. Hemmingsen, and P. W. Hansen, *J. Chromatogr.*, *40*, 16 (1969).

10. G. Lauer and R. A. Osteryoung, *Anal. Chem.*, *40*, 30A (1968).

11. L. Ramaley and G. S. Wilson, *Anal. Chem.*, *42*, 606 (1970).

12. W. Simon, W. P. Costelli, and D. D. Rutstein, *J. Gas Chromatogr.*, *5*, 578 (1967).

13. J. Baudish, *Chromatographia*, *12*, 443 (1968).

14. W. J. MacLean, *J. Chromatogr.*, *99*, 425 (1974).

15. M. F. Burke and R. G. Thurman, *J. Chromatogr. Sci.*, *8*, 39 (1970).

16. R. G. Thurman, K. A. Mueller, and M. F. Burke, *J. Chromatogr. Sci.*, *9*, 77 (1971).

17. R. S. Juvet, Jr. and S. P. Crom, *Anal. Chem.*, *46*, 101R (1974).

18. H. V. Malmstadt, C. G. Enke, and S. R. Crouch, *Instrumentation for Scientists Series, Module 3*, W. A. Benjamin, Inc., Reading, 1973.

19. F. Beumann, A. C. Brown, and M. B. Mitchell, *J. Chromatogr. Sci.*, *8*, 20 (1970).

20. J. W. Frazer, *Chem. Instrum.*, *2*, 271 (1970).

21. J. Hettinger and J. Hubbard, *Amer. Lab.*, *6*, 99 (1974).

22. S. P. Perone, *J. Chromatogr. Sci.*, *7*, 714 (1969).

23. A. W. Westerberg, *Anal. Chem.*, *41*, 1595 (1969).

24. C. C. Sweeley, B. D. Ray, W. I. Wood, J. F. Holland, and M. I. Kirchevsky, *Anal. Chem.*, *42*, 1505 (1970).

25. G. S. Wilson and L. Ramaley, *Anal. Chem.*, *42*, 611 (1970).

26. S. P. Perone, K. Ernst, H. R. Brand, and J. W. Frazer, *Computerized Laboratory Systems*, ASTM STP 578, American Society for Testing and Materials, 1975.

27. J. D. Hettinger, J. R. Hubbard, J. M. Gill, and L. A. Miller, *J. Chromatogr. Sci.*, *9*, 710 (1971).

28. E. Kovats, *Helv. Chim. Acta*, *41*, 1915 (1958).

29. R. L. Jolley and C. D. Scott, *J. Chromatogr.*, *47*, 272 (1970).

30. P. A. Klein and B. A. Kunze-Kalknen, *Anal. Chem.*, *37*, 1245 (1965).

31. H. Nau and K. Biemann, *Anal. Chem.*, *46*, 426 (1974).

32. C. C. Sweeley, N. D. Young, J. F. Holland, and S. C. Gates, *J. Chromatogr.*, *99*, 507 (1974).

33. D. D. Chilcote and J. Mrochek, *Clin. Chem.*, *18*, 744 (1972).

34. C. D. Scott, D. D. Chilcote, and W. W. Pitt, Jr., *Clin. Chem.*, *16*, 637 (1970).

35. K. Kishimato, H. Miyauchi, and S. Musha, *J. Chromatogr. Sci.*, *10*, 220 (1972).

36. A. W. Westerberg, *Anal. Chem.*, *41*, 1770 (1969).

37. H. A. Hancock, Jr., L. A. Dahm, and J. E. Muldoon, *J. Chromatogr. Sci.*, *8*, 57 (1970).

38. A. Savitzky and M. J. E. Golay, *Anal. Chem.*, *36*, 1627 (1969).